Problems in Advanced Organic Chemistry

Problems in Advanced Organic Chemistry

Contributors

William Kofie, Stephen Caddick et al.

www.aurisreference.com

Problems in Advanced Organic Chemistry

Contributors: William Kofie, Stephen Caddick et al.

Published by Auris Reference Limited

www.aurisreference.com

United Kingdom

Problems in Advanced Organic Chemistry

ISBN: 978-1-78154-883-7

British Library Cataloguing in Publication Data
A CIP record for this book is available from the British Library

Printed in the United Kingdom

Exclusively distributed by CBS Publishers & Distributors Pvt. Ltd.

Sales & Distribution Rights only for India, Pakistan, Bangladesh, Sri Lanka, Nepal and Bhutan. This book is not to be sold outside these territories.

Contents

List of Abbreviations

NHC	N-heterocyclic carbenes
TLC	layer chromatography
CAN	ceric ammonium nitrate
TLC	thin-layer chromatography
ESI	electrospray ionization
CDKs	Cyclin-Dependent Kinases
TCS	Tetrachlorosilane
NMR	Nuclear Magnetic Resonance
DEE	diethyl ether
MOE	Molecular Operating Environment
PTSA	p-toluenesulfonic acid
DMF	dichloroacetic acid
AM	amylose
AP	amylopectin
DSC	Differential Scanning Calorimetry
NFE	nitrogen-free extract

List of Contributors

William Kofie
Centre for Drug Design and Development, Department of Pharmaceutical Chemistry, Faculty of Pharmacy and Pharmaceutical Sciences, Kwame Nkrumah University of Science and Technology, Kumasi, Ghana
Centre for Biomolecular Design and Drug Development, The Chemistry Laboratory, CPES, University of Sussex, Brighton, UK

Stephen Caddick
Centre for Biomolecular Design and Drug Development, The Chemistry Laboratory, CPES, University of Sussex, Brighton, UK
Department of Chemistry, University College London, London, UK

Prasanna K. Vuram
Laboratory of Bioorganic Chemistry, Department of Biotechnology, Indian Institute of Technology Madras, Chennai, India

C. Kabilan
Laboratory of Bioorganic Chemistry, Department of Biotechnology, Indian Institute of Technology Madras, Chennai, India

Anju Chadha
Laboratory of Bioorganic Chemistry, Department of Biotechnology, Indian Institute of Technology Madras, Chennai, India
National Centre for Catalysis Research, Indian Institute of Technology Madras, Chennai, India

Md. Razzak
Department of Chemistry, Tennessee State University, Nashville, USA

Mohammad R. Karim
Department of Chemistry, Tennessee State University, Nashville, USA

Md. Rejaul Hoq
Department of Chemistry, Tennessee State University, Nashville, USA

Aminul H. Mirza
Department of Chemistry, Faculty of Science, Universiti Brunei Darussalam, Bandar Seri Begawan, Brunei Darussalam

Hamdi M. Hassaneen
Department of Chemistry, Faculty of Science, Cairo University, Giza, Egypt

Zakaria Ahmed Gomaa
Department of Chemistry, Faculty of Science, Cairo University, Giza, Egypt

Mohammed Saleh Tawfik Makki
Department of Chemistry, Faculty of Science, King Abdulaziz University, Jeddah, Saudi Arabia

Reda Mohammdy Abdel-Rahman
Department of Chemistry, Faculty of Science, King Abdulaziz University, Jeddah, Saudi Arabia

Faisal Mohammed Aqlan
Department of Chemistry, Faculty of Science, King Abdulaziz University, Jeddah, Saudi Arabia

Samy B. Said
Chemistry Department, Faculty of Science, Damietta University, New Damietta, Egypt

Mohammad M. A. Mashaly
Chemistry Department, Faculty of Science, Damietta University, New Damietta, Egypt

Ahmed M. Sheta
Chemistry Department, Faculty of Science, Damietta University, New Damietta, Egypt

Saad S. Elmorsy
Chemistry Department, Faculty of Science, Mansoura University, Mansoura, Egypt

Mahmoud Sayed Bashandy
Chemistry Department, Faculty of Science (Boys), Al-Azhar University, Nasr City, Cairo, Egypt

Mohammed Saleh Tawfek Makki
Chemistry Department, Faculty of Science, King Abdul Aziz University, Jeddah, Kingdom of Saudi Arabia

Reda Mohammady Abdel Rahman
Chemistry Department, Faculty of Science, King Abdul Aziz University, Jeddah, Kingdom of Saudi Arabia

Ola Ahmad Abu Ali
Chemistry Department, Faculty of Science, King Abdul Aziz University, Jeddah, Kingdom of Saudi Arabia

Asmaa S. Salman
Department of Chemistry, Faculty of Science, Al-Azhar University, Girls' Branch, Nasr City, Cairo, Egypt

Naema A. Mahmoud
Department of Chemistry, Faculty of Science, Al-Azhar University, Girls' Branch, Nasr City, Cairo, Egypt

Anhar Abdel-Aziem
Department of Chemistry, Faculty of Science, Al-Azhar University, Girls' Branch, Nasr City, Cairo, Egypt

Mona A. Mohamed
Biochemistry Division, Faculty of Science, Al-Azhar University, Girls' Branch, Nasr City, Cairo, Egypt

Doaa M. Elsisi
Department of Chemistry, Faculty of Science, Al-Azhar University, Girls' Branch, Nasr City, Cairo, Egypt

Marco A. Obregón-Mendoza
Instituto de Química, Universidad Nacional Autónoma de México, Ciudad Universitaria, México City, México

María Miriam Estévez-Carmona
Instituto de Química, Universidad Nacional Autónoma de México, Ciudad Universitaria, México City, México

Carolina Escobedo-Martínez
[2]Departamento de Farmacia, División de Ciencias Naturales y Exactas del Campus Guanajuato, Universidad de Guanajuato, México

Manuel Soriano-García
Instituto de Química, Universidad Nacional Autónoma de México, Ciudad Universitaria, México City, México

Raúl G. Enríquez
Instituto de Química, Universidad Nacional Autónoma de México, Ciudad Universitaria, México City, México

Louis Korbla Doamekpor
Department of Chemistry, University of Ghana, Accra, Ghana

Raphael Kwaku Klake
Department of Chemistry, University of Ghana, Accra, Ghana

Vincent Kodzo Nartey
Department of Chemistry, University of Ghana, Accra, Ghana

Takehiko Yamato
Department of Chemistry, Saga University, Saga, Japan

Oti Gyamfi
Ghana Atomic Energy Commission, Accra, Ghana

Dennis Adotey
Ghana Atomic Energy Commission, Accra, Ghana

Bidyut Kumar Senapati
Department of Chemistry, Prabhat Kumar College, Contai, India

Dipakranjan Mal
Department of Chemistry, Indian Institute of Technology, Kharagpur, India

Fatemeh Darvish
Department of Chemistry, K.N.Toosi University of Technology, P. O. Box 15875-4416, Tehran, Iran

Shima Khazraee
Department of Chemistry, K.N.Toosi University of Technology, P. O. Box 15875-4416, Tehran, Iran

Reda M. Abdel-Rahman
Department of Chemistry, Faculty of Science, King Abdul Aziz University, Jeddah, KSA

Mohammed S. T. Makki
Department of Chemistry, Faculty of Science, King Abdul Aziz University, Jeddah, KSA

Abeer N. Al-Romaizan
Department of Chemistry, Faculty of Science, King Abdul Aziz University, Jeddah, KSA

Maira Rubi Segura-Campos
Facultad de Ingeniería Química, Universidad Autónoma de Yucatán, Mérida, México

Sonia Marina López-Sánchez
Facultad de Ingeniería Química, Universidad Autónoma de Yucatán, Mérida, México

Arturo Castellanos-Ruelas
Facultad de Ingeniería Química, Universidad Autónoma de Yucatán, Mérida, México

David Betancur-Ancona
Facultad de Ingeniería Química, Universidad Autónoma de Yucatán, Mérida, México

Luis Chel-Guerrero
Facultad de Ingeniería Química, Universidad Autónoma de Yucatán, Mérida, México

Abeer N. Al-Romaizan
Department of Chemistry, Faculty of Science, King Abdul Aziz University, Jeddah, KSA

Mohammed S. T. Makki
Department of Chemistry, Faculty of Science, King Abdul Aziz University, Jeddah, KSA

Reda M. Abdel-Rahman
Department of Chemistry, Faculty of Science, King Abdul Aziz University, Jeddah, KSA

Preface

Organic chemistry is a chemistry sub discipline involving the scientific study of the structure, properties, and reactions of organic compounds and organic materials, i.e., matter in its various forms that contain carbon atoms. The text *Problems in Advanced Organic Chemistry* focuses on various types of problems in advanced organic chemistry. First chapter focuses on palladium/imadazolium salt mediated cyclisations for the synthesis of heterocyclic compounds. A simple and green alternative protocol for the synthesis 3-indolyl-3-hydroxy oxindoles in moderate to excellent yields has been reported in second chapter. Eight novel Schiff bases of 6,6'-diformyl-2,2'-Bipyridyl with O, S, N and F containing amines have been synthesized in third chapter. Utility of styrylpyrazoloformimidate in the synthesis of fused heterocyclic compounds has been discussed in fourth chapter. In fifth chapter, we evaluate the protein kinase inhibiting and cytotoxic activity of fluorinated heterobicyclic systems containing 1,2,4-triazine moiety. New method for preparation of 1-amidoalkyl-2-naphthols via multicomponent condensation reaction utilizing tetrachlorosilane under solvent free conditions has been presented in sixth chapter. A recent work on the synthesis and chemistry of bioactive sulfur bearing 1,2,4-triazinone moiety has been reported in seventh chapter. Eighth chapter deals with the synthesis of novel fluorine substituted isolated and fused heterobicyclic nitrogen systems bearing 6-(2'-phosphorylanilido)- 1,2,4-triazin-5-one moiety as potential inhibitor towards HIV-1 activity. The objective of ninth chapter is to evaluate the physicochemical and functional properties of velvet bean depigmented starch. In tenth chapter, we report catalytic activity of $FeCl_3$ for the preparation of 1-substituted tetrazoles from a wide variety of primary amines with trimethylsilylazide and trimethylorthoformate under solvent-free conditions. Synthetic studies of Naphtho [2, 3-B] furan moiety which present in diverse bioactive natural products have been proposed in eleventh chapter. Twelfth chapter reports new and convenient methods for the synthesis of heterocyclic ring systems that are required to medicinal chemistry utilizing 1-(4-(pyrrolidin-1-ylsulfonyl)phenyl)ethanone as a starting material. Thirteenth chapter describes a facile and simple nucleophilic attack of amino group bearing fluorine substituted 1,3,4-thiadiazolo[2,3-c][1,2,4]triazine(1) to active electrophilic reagents via addition, fluorinated acylation/ aroylation, phosphorylation and/or alkylation reactions. Last chapter discusses on synthesis, reactions and antimicrobial activity of some new 3-substituted indole derivatives.

Chapter 1

PALLADIUM/IMADAZOLIUM SALT MEDIATED CYCLISATIONS FOR THE SYNTHESIS OF HETEROCYCLIC COMPOUNDS

William Kofie[1,2], Stephen Caddick[2,3]

[1]Centre for Drug Design and Development, Department of Pharmaceutical Chemistry, Faculty of Pharmacy and Pharmaceutical Sciences, Kwame Nkrumah University of Science and Technology, Kumasi, Ghana

[2]Centre for Biomolecular Design and Drug Development, The Chemistry Laboratory, CPES, University of Sussex, Brighton, UK

[3]Department of Chemistry, University College London, London, UK

ABSTRACT

The use of intramolecular reactions involving palladium/imidazolium salts to synthesize heterocyclic compounds is described. Reactivity of phenyl, ethyl and methyl substituents leading to isolation of various isomeric products is also illustrated. Rearrangement of phenyl intermediates to furnish benzoxazoles is also mentioned.

INTRODUCTION

Palladium mediated reactions play significant role in organic synthesis. Intramolecular Heck reactions in particular have become useful for the synthesis of carbocylic and heterocyclic rings [1] [2]. Conventional coupling methods require the use palladium phosphine complexes as catalyst for Heck reactions. The mechanism for these palladium catalyzed coupling reactions involves four key stages namely, i) oxidative addition of aryl halide to Pd(0); ii) intramolecular carbometallation; iii) β-hydride elimination and lastly iv) reductive elimination to regenerate the Pd(0) catalyst for the continuation of the catalytic cycle [3] .

The choice of ligand is very important in these palladium catalyzed reactions. In recent times, the development of novel ligands for transition metal catalysis and the study of their mechanistic pathway have received greater attention [4] [5]. Sterically hindered phosphine ligands for palladium-catalysed reactions in particular, have improved greatly the applications of many metal mediated transformations [6].

Despite their wider application in palladium mediated synthesis, phosphine ligands do suffer limitations. This has necessitated the search for practical alternatives to phosphines as ligands in metal mediated cyclisation reactions [7].

N-heterocyclic carbenes (NHC) are now potential candidates as alternative ligands in transition metal catalyzed reactions due to their advantages over their conventional phosphine counterparts. They are known to be excellent donors and tend to form strong bonds with metals, giving them very good catalytic activity [7] [8].

NHCs have also found use in palladium catalysis. In complexes where NHC is coordinated to palladium (II), catalytic activity of these species have been examined and confirmed. In particular, catalytic properties of palladium-carbene complexes have been proven in amination, Heck and Suzuki reactions [9].

Hartwig and Nolan [10] [11], have reported that palladium-NHC complexes may be produced as intermediates when imidazolium salts are employed as additives in palladium catalyzed transformations. Although a full mechanistic rational is yet to be established, it is assumed that, the palladium and imidazolium salts generate either neutral or cationic palladium-carbene complexes in situ during the course of the metal catalytic cycle [12]. Nevertheless it can be appreciated that palladium-NHC complexes are proving valuable in synthetic organic chemistry [13].

We previously reported the successful use of Pd-carbene complexes in intramolecular Heck reactions involving aromatic chlorides. This indicates the effectiveness of these Pd-NHC complexes in metal catalyzed reactions, particularly involving the less reactive aryl chlorides, [14]. Improved yields were achieved with addition of tetrabutyl ammonium salts. These reactions require generation of the carbene in situ from imidazolium salts and subsequent completion of the catalytic cycle leading to the generation of heterocyclic compounds.

We describe herein our findings on palladium/imidazolium salt mediated protocols for intramolecular cyclisation reactions on to aryl iodides. Our studies have revealed that palladium/imidazolium salt catalysis can be used to synthesize a range of heterocyclic compounds from cheaper and readily available starting materials and precursors. Yields of these reactions are particularly good, and the products are free from contamination by phosphine related by-products, making them easily isolable.

EXPERIMENTAL

Chemicals and Instruments

Melting points were carried out on Gallenkemp melting point apparatus and are uncorrected. IR spectra were recorded on a Perkin Elmer transform instrument as thin film neat of a solution dissolved in nujol. All nmr spectra were recorded on Bruker Fourier transform instruments with frequencies quoted in mega Hertz recorded in all experimental data. All coupling constants (J values) were quoted in Hertz. Chemical shifts are reported upfields in parts per million. Mass spectra were recorded on Kratos or Fisons double focusing spectrometers. X-ray crystallography experiments were carried out using Enraf Nonius diffractometer. Thin layer chromatography (tlc) was performed using Merck Kieselgel precoated silica gel plates and visualized with UV light, vaporized iodine or potassium permanganate solution. Flash chromatography was carried out using Merck Kieselgel 230-400 mesh. Organic solvents were purified by distilling over drying agents. Throughout the experimental, imidazolium salt refers to the ligand 1,3-bis-(2,6-diisopropylpheynyl)imidazolium chloride.

General Method for the Alkylation of Benzyl Alcohol

A suspension of 2-iodobenzylalcohol (1.0 eq) and NaH (2.0 eq) in THF was stirred at room temperature for 30 min. Alkenyl halide (2.0 eq) was added and the reaction mixture stirred for a further 16 h. The reaction mixture was then filtered through celite, washed with water and the aqueous layer extracted with Et_2O. The combined organic fractions were dried ($MgSO_4$), filtered and solvent removed in vacuo to give the product as an oil.

General Method for the Synthesis of Phenylacrylamide

To a solution of acyl chloride (1.0 eq) in Et_2O at 0°C was added slowly 2-iodoaniline (2.0 eq) and the reaction mixture left to warm to room temperature for 2 h. This was then quenched with HCl (0.5 M), washed with $NaHCO_3$ solution followed by water and the aqueous layer extracted with Et_2O. The combined organic fractions were dried ($MgSO_4$), filtered and solvent removed in vacuo to give a crude sample which was purified by flash chromatography (Silica gel) [eluent-Pet:Et_2O, 20:1 - 5:1] to give the title compound as a solid.

General Method for the Synthesis of Isochromene and Isochroman

Iodobenzyl ether (1.0 eq), Cs_2CO_3 (1.5 eq), $Pd_2(dba)_3$ (1 mol%) and imidazolium salt (1 mol%) were placed in a reaction vessel and purged with nitrogen under vacuum. N,N-dimethylacetamide was then added and the mixture heated at 140°C for 15 h. The reaction mixture was allowed to cool to room temperature, dissolved in diethyl ether and filtered through celite. The filtrate was washed with water and aqueous layer extracted with diethyl ether. Combined organic fractions were dried (MgSO$_4$), filtered and solvent removed to give crude sample which was purified by flash chromatography (silica gel) [eluent-Pet:Et$_2$O, 20:1 - 10:1] to give the desired compound.

General Method for the Synthesis of Benzoxazole

Iodo-benzamide (1.0 eq), Cs$_2$CO$_3$ (1.5 eq), Pd$_2$(dba)$_3$ (1 mol%), and imidazolium salt (1 mol%), were place in a reaction vessel and purged with nitrogen under vacuum. N,N-dimethylacetamide was then added and the mixture heated at 140°C for 10 h. This was allowed to cool to room temperature, dissolved in diethyl ether and filtered through celite. The filtrate was washed with water and the aqueous layer extracted with ether. Combined organic fractions were dried (MgSO$_4$), filtered and solvent removed to give crude sample which was purified by flash chromatography (silica gel) [eluent-Pet:Et$_2$O, 20:1 - 5:1] to give the desired compound.

SPECTRAL AND ANALYTICAL DATA

1-Allyloxymethyl-2-iodobenzene (3.1)

R$_f$ Pet:EtOAc, 6:1 (0.8).

ν_{max}cm^{-1} 3406, 3048, 2935, 1695, 1451, 1345, 1090, 1004, 915, 736.

δ_H (300 MHz, CDCl$_3$) 7.72 (1H d, J = 7.7, ArH), 7.40 (1H, d 7.7, ArH), 7.22 (1H, dd, J = 7.7, 7.3 ArH) 6.87 (1H, app. t, J = 7.7, ArH), 5.90 (1H, ddt J = 17.1, 10.3, 6.7, -CH$_2$OCH$_2$CHCH$_2$), 5.28 (1H, d J = 10.3, -CH$_2$OCH$_2$CHCH$_2$), 5.12, (1H, J = 17.1, -CH$_2$OCH$_2$CHCH$_2$), 4.4 (2H, s, -CH$_2$OCH$_2$CHCH$_2$) 4.02 (2H,d J = 6.7, -CH$_2$OCH$_2$CHCH$_2$).

δ_C (75 MHz, CDCl$_3$) 141.04 (C), 139.53 (CH), 134.13 (CH), 130.66 (CH), 129.87 (CH), 128.84 (CH), 117.76 (CH$_2$), 98.15 (C), 72.07 (CH$_2$), 69.55 (CH$_2$).

m/z (EI) 274 (M$^+$, 25%), 231 (M$^+$-C$_3$H$_5$, 63%) 217 (100%), 90 (75%).

1-But-2-enyloxymethyl-2-iodobenzene (3.2)

R$_f$ Pet:Et$_2$O, 5:1 (0.88).

ν_{max} cm^{-1} 3021, 2920, 1562, 1438, 1350, 1092, 1090, 1005, 959, 743.

δ_H (300 MHz, CDCl$_3$) 7.70 (1H d, J = 7.9, ArH), 7.33 (1H, d 8.2, ArH), 7.22 (1H, dd, J = 7.7, 7.3 ArH) 6.83 (1H, dd, J = 7.7, 7.3 ArH), 5.69 (1H, dq J = 15.2, 6.0, -CH$_2$OCH$_2$CHCHCH$_3$), 5.58 (1H, dt J = 15.2, 6.0 -CH$_2$OCH$_2$CHCHCH$_3$),

4.36, (2H, s, -CH$_2$OCH$_2$CHCH$_3$), 3.39 (2H, d, J = 5.8, (-CH$_2$OCH$_2$CHCHCH$_3$), 1.63 (3H, d J = 6.0, -CH$_2$OCH$_2$CHCHCH$_3$).

δ_C (75 MHz, CDCl$_3$) 141.15 (C), 139.51 (CH), 130.25 (CH), 129.16 (CH), 128.72 (CH), 127.81N(CH), 127.08 (CH), 98.26 (C), 71.84 (C), 66.43 (CH$_2$), 18.33 (CH$_3$).

m/z (EI) 288 (M$^+$, 20%), 244 (M$^+$, 40%) 217 (M+-OC$_4$H$_7$, 100%), 36 (60%).

1-Cynnamyloxymethyl-2-iodobenzene (3.3)

R$_f$ Pet:Et$_2$O, 9:1 (0.55).

v_{max}cm^{-1} 3033, 2848, 1557, 1443, 1354, 1107, 1090, 960, 746, 687.

δ_H (300 MHz, CDCl$_3$) 7.92 (1H d, J = 7.5, ArH), 7.60 (1H, d J = 7.9, ArH), 7.42 (2H, dd, J = 7.5, 7.4 ArH), 7.32-7.10 (4H, m, ArH), 7.07 (1H, apt. t J = 7.9 ArH), 6.57 (1H, d, J = 16.0, -CH$_2$OCH$_2$CHCHPh), 4.65 (2H, s, -CH$_2$OCH$_2$CHCHPh), 4.37 (2H, d J = 6.0, -CH$_2$OCH$_2$CHCHPh).

δ_C (75 MHz, CDCl$_3$) 141.108 (C), 139.68 (CH), 137.18 (C), 133.13 (CH), 129.71 (CH), 129.34 (CH), 129.10 (CH), 128.76 (CH), 128.25 (CH), 127.07 (CH), 126.35 (CH), 98.47 (C), 76.43 (CH$_2$), 71.79 (CH$_2$).

m/z (ES) 386 (MNH$_4^+$, 100%), 305 (100%).

N-(2-Iodophenyl)-3-phenylacrylamide (3.6)

R$_f$ Pet:EtOAc 6:1 (0.41).

Mp. 160°C - 162°C.

v_{max}cm^{-1} 3189, 1655, 1620, 1451, 1374, 731.

δ_H (300 MHz, CDCl$_3$) 8.3 (1H, appt. d J = 7.5, NH), 7.74-7.72 (1H, m, ArH), 7.68 (1H, d J = 15.8, -COCHCH-), 7.56-7.49 (3H, m, ArH), 7.36-7.18 (4H, m, ArH), 6.79 (1H, appt. t J = 7.9, ArH), 6.52 (1H, d J = 15.5 -COCHCH-).

δ_C (75 MHz, CDCl$_3$) 164.30 (C=O), 143.41 (C), 139.24 (C), 138.70 (CH), 134.82 (CH), 130.60 (CH), 129.75 (CH), 129.32 (CH), 128.50 (CH), 126.44 (CH), 122.46 (CH), 120.98 (CH), 90.05 (C).

m/z (EI) 349 (M$^+$, 30%), 222 (M$^+$-I, 80%), 131 (M$^+$-C$_6$H$_5$NI, 100%).

N-(2-Iodophenyl)-benzamide (3.7)

R$_f$ Pet:Et$_2$O 6:1 (0.39).

Mp. 140°C - 142°C.

v_{max}cm^{-1} 3214, 1639, 1506, 1456, 1373, 1287, 739, 704.

δ_H (300 MHz, CDCl$_3$) 8.40 (1H, d J = 8.2, ArH), 8.22 (1H, s, NH), 7.90 (2H, dd, J = 8.2, 7.9, ArH), 7.74 (1H, d J = 7.9, ArH), 7.56-7.40 (3H, m, ArH), 7.34 (1H, dd, J = 8.2, 7.3, ArH), 6.81 (1H, dd, J = 7.5, 7.7, ArH).

δ_C (75 MHz, CDCl$_3$) 165.70 (C=O), 139.21 (C), 138.66 (C), 134.92 (CH), 132.60 (CH), 130.57 (CH), 129.83 (CH), 129.36 (CH), 127.57 (CH), 126.44 (CH), 122.14 (CH), 90.59 (C).

m/z (EI) 323 (M$^+$, 10%), 196 (M$^+$-I, 62%), 122 (87%), 105 (M$^+$-C$_6$H$_5$NI, 100%).

4-Methyl-1H-isochromene (4.1)

R$_f$ Pet:Et$_2$O, 9:1 (0.87).

ν_{max}cm^{-1} 3048, 2844, 1640, 1484, 1354, 1448, 1140, 1107, 1007, 936, 72.

δ_H (300 MHz, CDCl$_3$) 7.18, (1H d, J = 7.5, ArH), 7.07 (1H dd J = 7.5, 7.4 ArH), 7.01 (1H, d, J = 7.5 ArH), 6.93 (1H, dd J = 7.5, 7.3, ArH), 6.39 (1H, s, -CH$_2$OCHC-Me), 4.91 (2H, s, -CH$_2$OCH-), 1.84 (3H, s, CH$_3$).

δ_C (75 MHz, CDCl$_3$) 141.28 (CH), 131.34 (CH), 127.68 (C), 127.02 (CH), 125.61 (CH), 122.68 (CH), 119.26 (CH), 110.36 (C), 67.20 (CH$_2$), 12.07 (CH$_3$).

m/z (EI) 146 (M$^+$, 85%), 117 (100%), 91 (15%).

4-Ethyl-1H-isochromene (4.2)

R$_f$ Pet:Et$_2$O, 10:1 (0.88).

ν_{max}cm^{-1} 3062, 2960, 2834, 1628, 1483, 1448, 1139, 1107, 935, 747.

δ_H (300 MHz, CDCl$_3$) 7.18, (1H d, J = 7.0, ArH), 7.09 (2H dd J = 7.7, 7.3 ArH), 6.95 (1H, d, J = 7.9 ArH), 6.40 (1H, s, -CH$_2$OCHCCH$_2$CH$_3$), 4.89 (2H, s, -CH$_2$OCHCCH$_2$CH$_3$), 2.31 (2H, q J = 7.5, -CH$_2$OCHCCH$_2$CH$_3$) 1.09 (3H, t J = 7.5, -CH$_2$OCHCCH$_2$CH$_3$).

δ_C (75 MHz, CDCl$_3$) 142.26 (CH), 131.92 (2XC), 129.41 (CH), 128.39 (CH), 126.87 (CH), 124.36 (CH), 120.58 (CH), 118.10 (C), 68.65 (CH$_2$), 21.14 (CH$_2$), 13.92 (CH$_3$).

m/z (EI) 160 (M$^+$, 85%), 117 (100%), 91 (55%), 49 (80%).

4-Benzyl-1H-isochromene (4.3)

R$_f$ Pet:Et$_2$O 9:1 (0.8).

Mp. 53°C - 55°C.

ν_{max}cm^{-1} 29684, 1623, 1454, 1448, 1141, 956.

δ_H (300 MHz, CDCl$_3$) 7.27, (1H d, J = 7.0, ArH), 7.22-7.19 (3H, m, ArH), 7.08-7.05 (3H, m, ArH), 6.95 (2H, appt. t, J = 7.7 ArH), 6.40 (1H, s, -CH$_2$OCHC-), 4.96 (2H, s, -CH$_2$OCHCC-), 3.61 (2H, s-CH$_2$Ph).

δ_C (75 MHz, CDCl$_3$) 144.79 (CH), 140.02 (C), 131.74 (2XC), 129.16 (CH), 128.81 (2 × CH), 128.78 (CH), 128.42 (CH), 127.07 (CH), 126.58 (CH), 124.29 (CH), 121.31 (CH), 115.10 (C), 68.81 (CH$_2$), 34.36 (CH$_2$).

m/z (EI) 222 (M$^+$, 55%), 57 (57%), 43 (100%), 28 (60%).

4-Benzylidene-isochroman (4.4)

R_f Pet:Et$_2$O 9:1 (0.5).

Mp. 78°C - 80°C.

Elemental Analysis: Expected; C 86.45%, H 6.35%, O 7.20%.

Found; C 86.62%, H 6.24%, O 7.14.

v_{max}cm^{-1} 3379, 2983, 1712, 1596, 1492, 1447, 1268, 1100, 764.

δ_H (300 MHz, CDCl$_3$) 7.66 (1H, d J = 7.3, ArH), 7.27, (2H appt t, J = 7.5, ArH), 7.19 - 7.12 (5H, m, ArH), 7.08 (1H, s, -CH$_2$OCH$_2$CCH-Ph), ArH), 6.95 (1H, d, J = 7.3 ArH), 4.68 (2H, s, -CH$_2$OCH$_2$CCH-Ph), 4.66 (2H, d J = 1.5 -CH$_2$OCH$_2$CCH-Ph).

δ_C (75 MHz, CDCl$_3$) 137.09 (C), 135.27 (C), 132.68 (C), 132.57 (C), 129.72 (CH), 128.78 (CH), 128.04 (CH), 127.60 (2 × CH), 127.59 (2 × CH), 126.58 (CH), 125.15 (CH), 123.81 (CH), 123.66 (CH), 69.01 (CH$_2$), 67.28 (CH$_2$).

m/z (EI) 222 (M$^+$, 100%), 178 (60%), 115 (95%).

2-Styryl-benzooxazole (4.5)

R_f Pet:EtOAc 3:1 (0.7).

Mp. 90°C - 92°C.

v_{max}cm^{-1} 2984, 1637, 1530, 1453, 1372.

δ_H (300 MHz, CDCl$_3$) 7.81 (1H, d J = 16.3, -CHCHPh), 7.74 (1H, d J = 7.5 ArH) 7.73 (2H, dd J = 7.6, 7.4, ArH), 7.54 (1H, d J = 7.5 ArH), 7.43 (2H, dd J = 7.6, 7.6 ArH), 7.36-7.34 (3H, m ArH), 7.10 (1H, d J = 16.3 - CHCHPh).

δ_C (75 MHz, CDCl$_3$) 161.76 (-NC), 149.37 (C), 141.15 (C), 138.42 (CH), 134.1 (C), 128.74 (CH), 127.94 (CH), 127.33 (CH), 126.52 (CH), 126.20 (CH), 124.18 (CH), 123.47 (CH), 118.84 (CH), 112.91 (CH), 101.29 (CH).

m/z (EI) 221 (M$^+$, 80%), 220 (M$^+$ - H 100%), 191 (45%).

2-Phenyl-benzooxazole (4.6)

R_f Pet:EtOAc 6:1 (0.81).

Mp. 110°C - 112°C.

v_{max}cm^{-1} 2986, 1611, 1550, 1458, 1373, 1233, 1045, 731.

δ_H (300 MHz, CDCl$_3$) 8.22-8.17, (2H, m, ArH), 7.72-7.69 (1H, m, ArH), 7.47-7.44 (4H, m, ArH), 7.30-7.27 (2H, m, ArH).

δ_C (75 MHz, CDCl$_3$) 149.73 (C), 141.07 (C), 130.49 (C), 127.89 (2 × CH), 126.59 (2 × CH), 126.13(C), 126.20 (CH), 124.08 (CH), 123.55 (CH), 118.99 (CH), 109.57 (CH).

m/z (EI) 195 (M$^+$, 100%), 167 (50%), 77 (40%), 63 (50%).

RESULTS AND DISCUSSION

Starting materials were obtained by standard synthetic transformations, where 2-iodobenzylalcohol was deprotonated and the anion subsequently quenched with the corresponding alkenylbromide [15] [16] , to furnish the desired alkenyloxymethyl-2-iodobenzene as the cyclisation precursors, 3.1-3.3 (Scheme 1). To obtain oxyindoles as final heterocyclic products, acrylamide precursors were synthesized by direct acylation of iodoaniline with enoyl chlorides to give desired products 3.4-3.6 (Scheme 2), [17] .

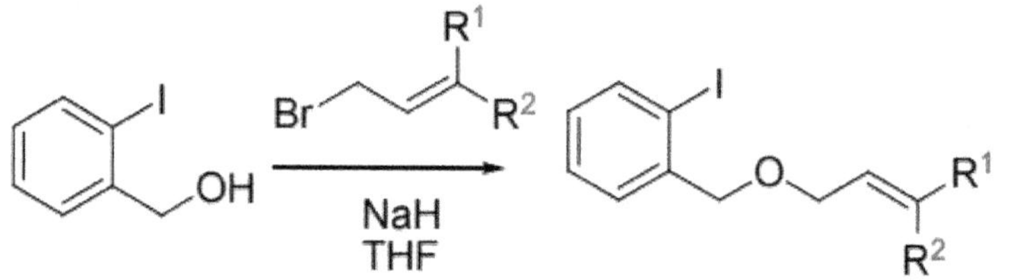

3.1 R^1=H, R^2=H 99%
3.2 R^1=H, R^2=Me 99%
3.3 R^1=H, R^2=Ph 87%

Scheme 1: Synthesis of benzyloxy precursors used in the synthesis of bicyclic rings.

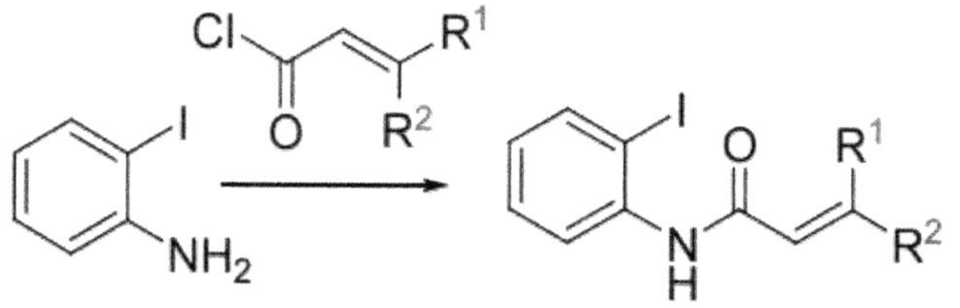

3.4 R^1=H, R^2=H 91%
3.5 R^1=H, R^2=Me 76%
3.6 R^1=H, R^2=Ph 89%

Scheme 2: Synthesis of amide precursors which gave oxazoles upon cyclisations.

When alkenyloxy precursors 3.1 and 3.2 were subjected to reaction conditions using the palladium/imidazo- lium salt protocol, the desired bicyclic isochromene products 4.1 and 4.2 were isolated in very good yields following purification by column chromatography. The cyclisations proceeded possibly via 6-exo fashion followed by rapid isomerization to give the products obtained, (Scheme 3).

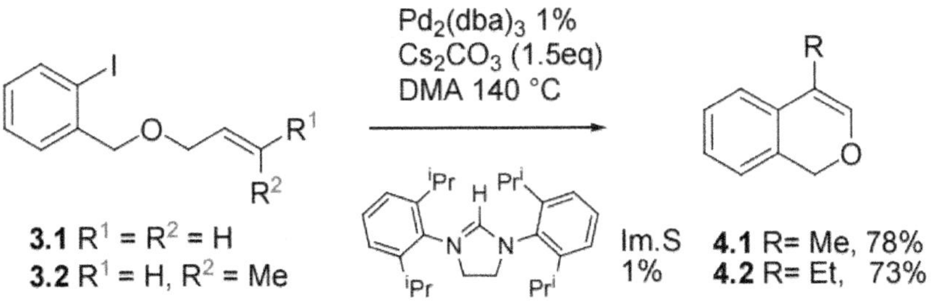

Scheme 3: Pd/ Im.S mediated cyclisations to produce bicyclic ring systems (Im.S = imidazolium salt).

When the phenyl substituted precursor 3.3 was subjected to the same reaction conditions, however, two isomeric products were isolated [18] . Benzylisochromene 4.3 [19] , and benzylideneisochroman 4.4 [20] [21] , nuclei were obtained in yields of 22% and 63% respectively, (Scheme 4).

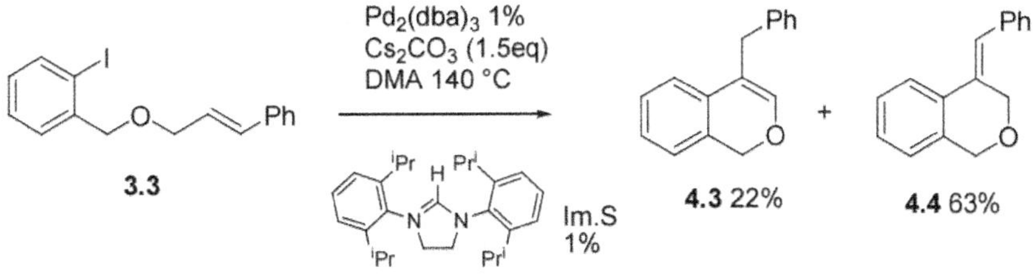

Scheme 4: Phenyl substituted precursor gave isomers 4.3 and 4.4 when subjected to Pd/ Im.S conditions.

Isolation of benzylideneisochroman 4.4 as the major product in this case means the reaction probably proceeded via 6-exo cyclisation. Unlike the previous examples where isomerization is presumed to have taken place following cyclisation to give the kinetically stable products, benzylideneisochroman 4.4 (Figure 1) happen to be the major isolated product from cyclisation of 1-cynnamyloxymethyl-2-iodobenzene, 3.3. Although there is a possible 1,3-hydrogen shift to generate isochromene 4.3, the isolation of isochroman 4.4 as the major product may be due to the conjugation system that extends from the alkene bond onto the phenyl ring to give a more stable isomer.

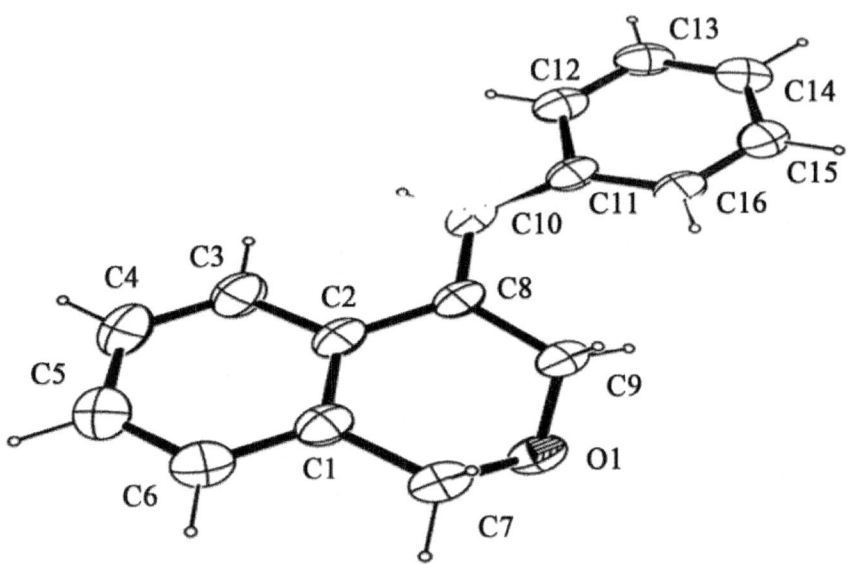

Figure 1: Crystal structure of benzylideneisochroman 4.4, major isomer from phenyl substituted precursor.

It was disappointing to observe that, treating but-2-enoic phenyl amide and phenyl acrylamide substrates with palladium/imidazolium salt failed to give the desired oxyindoles. Surprisingly, what seemed to be loss of the al- kenoyl moiety leading to isolation of 2-iodoaniline was observed, (Scheme 5).

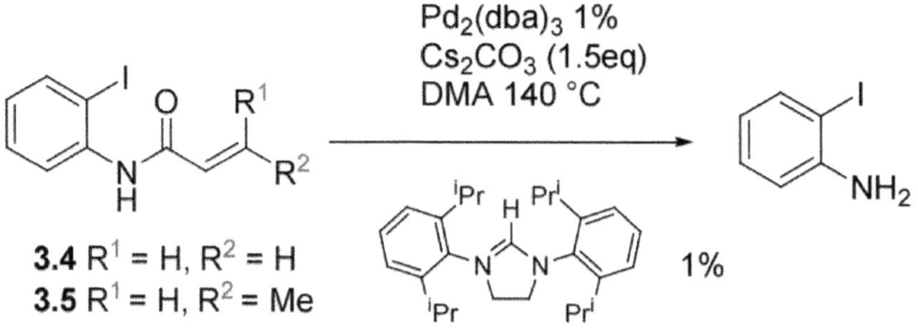

Scheme 5: Reaction of alkenoyl moiety upon treatment with Pd/Im.S.

When N-(2-iodophenyl)-3-phenylacylamide **3.6** was subjected to palladium/imidazolium salt protocols, there was conversion of starting material to product after heating for 3 h. It was revealed by spectral analysis that the desired oxyindole had not been formed, instead cyclisation had occur to give 2-styryl-benzooxazole 4.5, [22] [23] , (Scheme 6), (Figure 2) [24] .

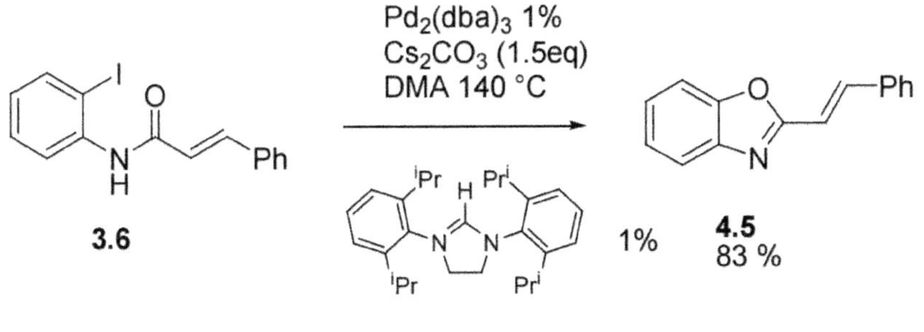

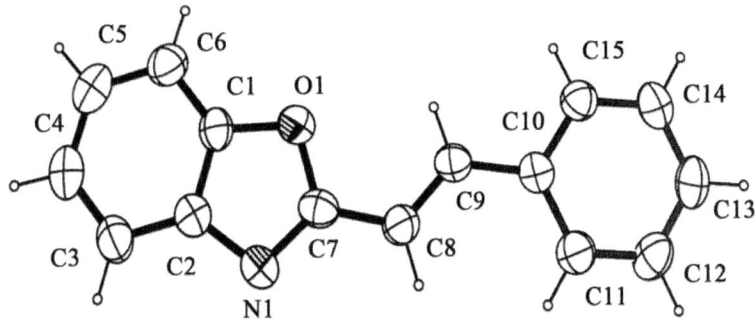

Scheme 6: Cyclisation to give 2-styryl-benzooxazole using Pd/Im.S.

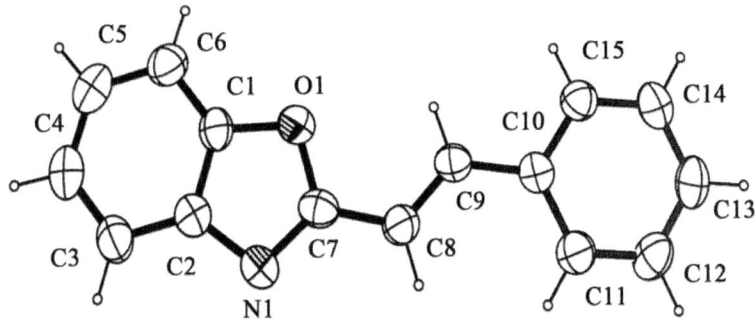

Figure 2: Crystal structure of 2-styryl-benzooxazole, 4.5.

To further extend the scope of these findings, N-(2-iodophenyl)benzamide 3.7 and N-(2-iodophenyl) aceta- mide were subjected to the same reaction conditions. While the benzamide reacted possibly via o-arylation to produce 2-phenylbenzooxazole 4.6 in good yield, no product was isolated when the acetamide derivative was subjected to the same reaction conditions, (Scheme 7).

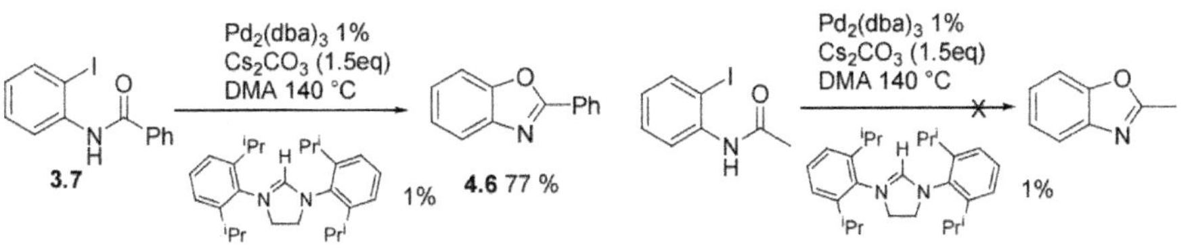

Scheme 7: Under Pd/Im.S condition, iodophenylbenzamide gave Phenyl benzooxazole, 4.6 whiles the acetamide intermediate gave no desired product.

We believe the ability of the phenyl substrates to efficiently undergo cyclisation may be due to the phenyl group providing strong steric and electronic effects leading to a stable rotamer with a configuration that allows the carbonyl oxygen to be within close proximity to the iodine, and thus favouring an attack on to the ring, (Scheme 8).

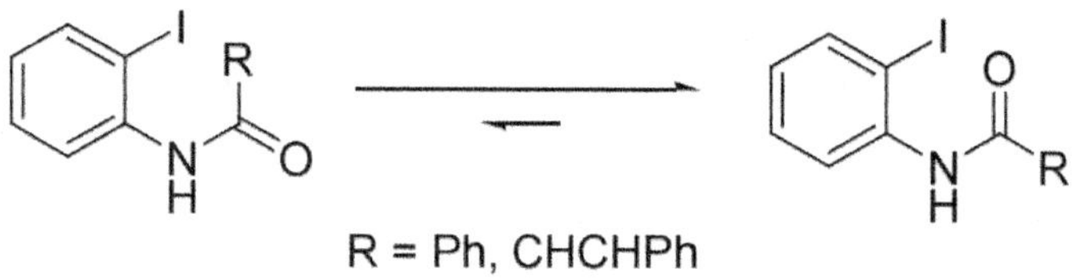

R = Ph, CHCHPh

Scheme 8: Rotation around the carbonyl showing a shift in equilibrium to the more favourable rotamer.

CONCLUSION

In conclusion, we have shown that palladium imidazolium salt protocols can be used to promote intramolecular cyclisation reactions, enabling the synthesis of fused bicyclic heteroaromatics with great efficiency. Phenyl substituents have strong influence on reaction outcomes as a result of steric and electronic factors. Iodophenylbenzamide substrates undergo cyclisation to give substituted benzooxazoles. The ease with which phenyl containing amides undergo cyclisation to produce substituted benzooxazoles is noteworthy.

ACKNOWLEDGEMENTS

We gratefully acknowledge Dr. John Hayler at GSK Tonbridge for his encouragement, support and helpful comments throughout this project. We also gratefully acknowledge support of this project from GSK and EPSRC. We are grateful to EPSRC Mass Spectrometry Service at Swansea and Drs Abdul-Sada, Avent and Hitchcock of The University of Sussex for their contributions to this work. We gratefully acknowledge the support by members of staff of department of pharmaceutical chemistry KNUST, Kumasi Ghana. We are also grateful to Lagray Chemicals, Ghana Limited for their regular support.

CITE THIS PAPER

WilliamKofie, StephenCaddick,11, (2015) Palladium/Imadazolium Salt Mediated Cyclisations for the Synthesis of Heterocyclic Compounds. *International Journal of Organic Chemistry*,**05**,223-231. doi:10.4236/ijoc.2015.54022

REFERENCES

1. Belestkaya, I.P. and Cheprakov, A.V. (2000) The Heck Reaction as a Sharpening Stone of Palladium Catalysis. Chemical Reviews, 100, 3009-3066. http://dx.doi.org/10.1021/cr9903048

2. 2. Oestreich, M., Dennison, P.R., Kodanko, J.J. and Overman, L.E. (2001) Thwarting β-Hydride Elimination: Capture of the Alkylpalladium Intermediate of an Asymmetric Intramolecular Heck Reaction. Angewandte Chemie International Edition, 40, 1439-1442. http://dx.doi.org/10.1002/1521-3773(20010417)40:8<1439::AID-ANIE1439>3.0.CO;2-F

3. 3. Tsuji, J. (1997) Palladium Reagents and Catalyst; Innovations in Organic in Organic Synthesis. Wiley & Sons.

4. 4. Alcazar-Roman, L.M. and Hartwig, J.F. (2001) Mechanism of Aryl Chloride Amination:? Base-Induced Oxidative Addition. Journal of the American Chemical Society, 123, 12905-12906. http://dx.doi.org/10.1021/ja016491k

5. 5. Alcazar-Roman, L.M. and Hartwig, J.F. (2002) Mechanistic Studies on Oxidative Addition of Aryl Halides and Triflates to Pd(BINAP)2 and Structural Characterization of the Product from Aryl Triflate Addition in the Presence of Amine. Organometallics, 21, 491-502. http://dx.doi.org/10.1021/om0108088

6. 6. Muci, A.R. and Buchwald, S.L. (2002) Practical Palladium Catalysts for C-N and C-O Bond Formation. Topics in Current Chemistry, 219, 131-201. http://dx.doi.org/10.1007/3-540-45313-X_5

7. 7. Northeton, M.R. and Fu, G.C. (2001) Air-Stable Trialkylphosphonium Salts: Simple, Practical, and Versatile Replacements for Air-Sensitive Trialkylphosphines. Applications in Stoichiometric and Catalytic Processes. Organic Letters, 3, 4295-4298. http://dx.doi.org/10.1021/ol016971g

8. 8. Herrmann, W.A., Weskamp, T. and Bohm, V.P.W. (2002) Metal Complexes of Stable Carbenes. Advances in Organometallic Chemistry, 48, 1-69. http://dx.doi.org/10.1016/S0065-3055(01)48001-4

9. 9. Bohm, V.P.W., Gstottmayr, C.W.K., Westkamp, T. and Herman, W.A. (2000) N-Heterocyclic Carbenes: Part 26. N-Heterocyclic Carbene Complexes of Palladium(0): Synthesis and Application in the Suzuki Cross-Coupling Reaction. Journal of Organometallic Chemistry, 596, 186-190. http://dx.doi.org/10.1016/S0022-328X(99)00590-2

10.   10. Staueeer, S.R., Lee, S., Sambuli, J.P., Hauck, S.I. and Hartwig, J.F. (2000) High Turnover Number and Rapid, Room-Temperature Amination of Chloroarenes Using Saturated Carbene Ligands. Organic Letters, 2, 1423-1426. http://dx.doi.org/10.1021/ol005751k

11.   11. Huang, J., Grasa, G. and Nolan, S.P. (1999) General and Efficient Catalytic Amination of Aryl Chlorides Using a Palladium/Bulky Nucleophilic Carbene System. Organic Letters, 1, 1370-1309. http://dx.doi.org/10.1021/ol990987d

12.   12. Grundeman, S., Albrecht, M., Kovacevic, A., Faller, J.W. and Crabtree, R.H. (2002) Bis-Carbene Complexes from Oxidative Addition of Imidazolium C-H Bonds to Pal-ladium(0). Journal of the Chemical Society, Dalton Transactions, 2163-2167. http://dx.doi.org/10.1039/b110964b

13.   13. Igau, A., Grutzmacher, H., Baceiredo, A. and Bertrand, G. (1988) Analogous α,α'-Bis-Carbenoid, Triply Bonded Species: Synthesis of a Stable λ3-Phosphinocarbene-λ5-Phosphaacetylene. Journal of the American Chemical Society, 110, 6463-6466. http://dx.doi.org/10.1021/ja00227a028

14.   14. Caddick, S. and Kofie, W. (2002) Observations on the Intramolecular Heck Reactions of Aromatic Chlorides Using Palladium/Imidazolium Salts. Tet-rahedron Letters, 43, 9347-9350. http://dx.doi.org/10.1016/S0040-4039(02)02340-7

15.   15. Menicagli, R., Malanga, C., Dell'Innocenti, M. and Lardicci, L. (1987) Triisobutylaluminum-Assisted Reductive Rearrangement of Alkyl 1-Alkenyl Acetals: An Easy Synthesis of β-Alkoxy Alcohols. The Journal of Organic Chemistry, 52, 5700-5704. http://dx.doi.org/10.1021/jo00235a010

16.   16. Kaiwar, V., Rees, C.B., Gray, E.J. and Neidle, S. (1995) Synthesis of 9-[cis-3-(hydroxymethyl)cyclobutyl]-adenine and -guanine. Journal of the Chemical Society, Perkin Transactions, 2281-2288. http://dx.doi.org/10.1039/p19950002281

17.   17. Eremeev, A.V., Polyak, F.D., Vosekalna, I.A., Chervin, I.I., Nasibov, S.S. and Kostyanovskii, R.G. (1984) Absolute Configuration of Diastereomeric Derivatives of N-Substituted Aziridine-2-Carboxylic Acids. Chemistry of Heterocyclic Compounds, 20, 1102-1107. http://dx.doi.org/10.1007/BF00503598

18.   18. Bankston, D., Fang, F., Huie, E. and Xie, S. (1999) Palladium(II) Acetate-Tris(triphenylphosphine)rhodium(I) Chloride: A Novel Catalytic

Couple for the Intramolecular Heck Reaction. The Journal of Organic Chemistry, 64, 3461- 3466. http://dx.doi.org/10.1021/jo982058q

19. 19. Marimoto, K., Hirano, K., Sato, T. and Miura, M. (2011) Synthesis of Isochromene and Related Derivatives by Rhodium-Catalyzed Oxidative Coupling of Benzyl and Allyl Alcohols with Alkynes. The Journal of Organic Chemistry, 76, 9548-9551. http://dx.doi.org/10.1021/jo201923d

20. 20. Giles, R.G.F., Rickards, R.W. and Senanayake, S.B. (1998) Synthesis of Isochroman-3-Ylacetates and Isochromane- γ-Lactones through Rearrangement of Aryldioxolanylacetates. Journal of the Chemical Society, Perkin Transactions, 3949-3956. http://dx.doi.org/10.1039/a807005i

21. 21. Young, D.W. (2010) Synthetic Chemistry: An Upfront Investment. Nature Chemical Biology, 6, 174-175. http://dx.doi.org/10.1038/nchembio.325

22. 22. Terashima, M. and Ishii, M. (1982) A Facile Synthesis of 2-Substituted Benzoxazoles. Synthesis, 1982, 484-485. http://dx.doi.org/10.1055/s-1982-29847

23. 23. Minami, T., Isonaka, T., Okada, Y. and Ichikawa, J. (1993) Copper(I) Salt-Mediated Arylation of Phosphinyl-Stabi- lized Carbanions and Synthetic Application to Heterocyclic Compounds. The Journal of Organic Chemistry, 58, 7009-7015. http://dx.doi.org/10.1021/jo00077a018

24. 24. Park, H.J., Park, M.S., Lee, T.H. and Park, K.H. (2013) Synthesis of 2-Styrylbenzoxazole Derivatives by the Reaction of Styrylphenolic Schiff Bases with Thianthrene Cation Radical. Journal of Heterocyclic Chemistry, 50, 663-667. http://dx.doi.org/10.1002/jhet.1707

Chapter 2

CATALYST AND SOLVENT-FREE MICROWAVE ASSISTED EXPEDITIOUS SYNTHESIS OF 3-INDOLYL-3-HYDROXY OXINDOLES AND UNSYMMETRICAL 3,3-DI(INDOLYL) INDOLIN-2-ONES

Prasanna K. Vuram[1], C. Kabilan[1], Anju Chadha[1,2]

[1]Laboratory of Bioorganic Chemistry, Department of Biotechnology, Indian Institute of Technology Madras, Chennai, India

[2]National Centre for Catalysis Research, Indian Institute of Technology Madras, Chennai, India

ABSTRACT

A simple and efficient method for the synthesis of 3-indolyl-3-hydroxy oxindoles and unsymmetrical 3,3-di(indolyl)indolin-2-ones using microwave irradiation without catalyst and solvent is described. A series of 3-indolyl-3-hydroxy oxindoles and unsymmetrical 3,3-di(indolyl) indolin-2- ones have been synthesized in very short reaction times of 5 and 10 minutes and in yields ranging from 31% to 98% and from 53% to 78% respectively. This method offers a significant advantage over the conventional methods in terms of simplicity and shorter reaction time. To the best of our knowledge compounds N-allyl-3-hydroxy-3-(1-methyl-indol-3-yl)indolin-2-one (6c), N-allyl-3- hydroxy-3-(5-methoxy-indol-3-yl) indolin-2-one (8c), N-benzyl-3-hydroxy-3-(1-methyl-indol-3-yl) indolin-2-one (10c), N-propargyl-3-hydroxy-3-(1-methyl-indol-3-yl)indolin-2-one (13c), N-propargyl- 3-hydroxy-3-(5-methoxy-indol-3-yl)indolin-2-one (14c), 3-(5-methoxy-1H-indol-3-yl)-3-(1H-indol- 3-yl)indolin-2-one (1e), 3-1-methyl(5-methoxy-1H-indol-3-yl)-3-(1H-indol-3-yl)indolin-2-one (2e), 3-1-allyl(5-methoxy-1H-indol-3-yl)-3-(1H-indol-3-yl)indolin-2-one (3e), 3-1-benzyl(5-methoxy-1H-in- dol-3-yl)-3-(1H-indol-3-yl)indolin-2-one (4e) and 3-1-(prop-2-ynyl)(5-methoxy-1H-indol-3-yl)-3(1H- indol-3-yl)indolin-2-one (5e) are reported here for the first time. All the compounds are characterized by IR, ^{1}H, ^{13}C NMR and HRMS.

INTRODUCTION

Indoles and their derivative oxindoles are a privileged class of molecules in synthetic as well as biological chemistry. The hybrid molecules of indole and oxindole are also important due to their prevalent biological activities. For instance, the spermicidal activity of 3,3-bis(5-methoxy-1H-indol-3-yl)indolin-2-one and 3,3-bis(3-car- boxymethyl-1H-indol-2-yl)indolin-2-one is higher than that of the standard spermicide "Nonoxynol-9" (N-9) [1] ; di(indolyl) indolin-2-one derivatives showed strong and selective cytotoxicity against cancer cells [2] . These molecules are also known to possess antibacterial, anti-inflammatory and antiprotozoal activities [3] . In addition to these manifold biological applications, very recently 3-indolyl-3-hydroxy oxindoles (hybrid molecule of isatin and indole) were used as precursors in the synthesis of natural products (+)-gliocladin [4] and (+)-folicanthine [5] and in the synthesis of the heterocyclic analogue of BINAP [6] . The synthesis of 3-indolyl-3-hydroxy oxindoles involves a Friedel-Crafts type of electrophilic substitution between the electron-rich third position of the indole and electron deficient carbonyl group of the isatin [7] . This reaction usually results in the formation of symmetrical 3,3'-diindolyl oxindoles in a single step. Controlling the reaction at monosubstituted 3-indolyl-3-hydroxy oxindole stage is quite challenging.

The synthesis of symmetrical 3,3-di(indolyl)indolin-2-ones has been reported [8] - [22] , while in the case of unsymmetrical 3,3-di(indolyl)indolin-2-ones, so far only three reports are available in the literature to the best of our knowledge. The synthesis involves a stepwise process. Wang and Ji reported an ultrasound irradiation method in the presence of ceric ammonium nitrate (CAN) for 1 to 5 h [23] ; Moghadam and co-workers reported the reaction in the ionic-liquid N,N,N,N-tetramethylguanidinium trifluoroacetate (TMGT) for 1 h [24] and Nikpassand and co-workers reported it with a montmorillonite in 30 - 35 minutes [25] .

In the case of 3-indolyl-3-hydroxy oxindoles, Kumar and co-workers reported the synthesis of various 3-in- dolyl-3-hydroxy oxindoles by supramolecular catalysis (β-Cyclodextrin) in 45 - 190 minutes [7] ; Shanthi and co-workers reported a time of 60 - 120 minutes using K_2CO_3 as a catalyst [26] ; Meshram and co-workers reported a time of 15 minutes in the presence of Triton B [27] ; Hosseini and Tavakolian reported a reaction time of 2.5 h using ZnO nanorods in aqueous medium [28] . EtOH and water medium (60:40) in the presence of Lewis acid ($FeCl_3·6H_2O$) and ultrasonic irradiation in 5 - 25 minutes was reported by Khorshidi and Tabatabaeian [29] while Makarem and co-workers reported an electrochemical method using EtOH/Propanol (60 - 240 minutes) [30] . Srihari and Murthy reported a heterogeneous catalyst (Kaolin/KOH) in the presence of MeOH as a solvent in 138 - 470 minutes

[31] . Using various ionic-liquids, Moghadam and co-workers reported a reaction time of 10 - 20 minutes [24] . Jing Deng and coworkers reported an enantioselective version by using cupreine [32] . Nadine and coworkers also reported the reaction using chiral scandium(III) and indium(III) pybox complexes [33] . Recently, Pravathaneni Sai Prathima and co-workers reported the synthesis in aqueous medium using a base catalyst-diethanolamine [34] . The focus of our lab is to develop "green" methods for organic synthesis. In addition to using biocatalysts for organic transformations, we have now used microwave irradiation for organic synthesis as reported in the present study. Use of solvent-free and catalyst-free reaction conditions is an attractive proposition as seen in numerous microwave-assisted reactions. To the best of our knowledge, this is the first report for catalyst and solvent-free synthesis of 3-indolyl-3-hydroxy oxindoles and unsymmetrical 3,3- di(indolyl)indolin-2-ones under microwave irradiation in 5 and 10 minutes respectively.

RESULTS AND DISCUSSION

Optimization of reaction conditions for the synthesis of 3-indolyl-3-hydroxy oxindoles, was carried out using a isatin (1a) [1 equivalent] and an indole (1b) [1.2 equivalents]. Using a microwave oven temperature at 100°C and irradiation for 5 minutes, gave only the symmetrical 3,3-di(indolyl)indolin-2-ones as product (Scheme 1). Decreasing the time of irradiation to 3 minutes and then to 1 minute also resulted in the formation of the symmetrical 3,3-di(indolyl) indolin-2-one product and in 1 minute, 3-hydroxy-3-(1H-indol-3-yl)indolin-2-one (1c) and unreacted starting materials isatin (1a) and indole (1b), were also detected. Then we envisioned that 100°C was not a suitable temperature for controlling the reaction at the mono substituted hydroxyl stage. The reaction was therefore carried out at reduced temperatures: at 90°C and 80°C for 5 minutes, a mixture of 3-hydroxy-3-

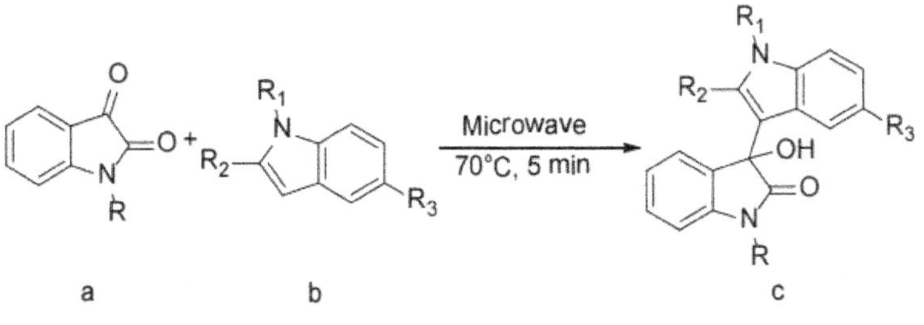

Scheme 1: Synthesis of various 3-indolyl-3-hydroxy oxindoles.

(1H-indol-3-yl)indolin-2-one (1c), symmetrical 3,3-di(indolyl)indolin-2-one in addition to unreacted starting material were observed. At 70°C for five minutes, only 3-hydroxy-3-(1H-indol-3-yl)indolin-2-one (1c) and unreacted starting materials were detected. But an increase in the reaction time from the sixth minute onwards, resulted in the formation of symmetrical 3,3-di(indolyl) indolin-2-one, even at 70°C. Based on this observation, 70°C and 5 minutes were optimized as the reaction conditions for synthesizing 3-hydroxy-3-(1H-indol-3-yl) indolin-2-one (1c), but the isolated yield was only 31%. Further increasing the amount of one of the starting materials i.e. three and five equivalents of indole (1b), showed no improvement in the yield of 3-hydroxy-3-(1H-indol-3-yl)indolin-2-one (1c). Also with N-allyl, and N-benzyl isatins, the yields of N-allyl-3-hydroxy-3- (1H-indol-3-yl)indolin-2-one (5c) and N-benzyl-3-hydroxy-3-(1H-indol-3-yl)indolin-2-one (9c) were only 34% and 40% respectively. Based on the nucleophilicity index of the indole ring [35] , N-methyl indole, 2-methyl indole and 5-methoxy indole were selected [these are electron-rich at the third position]. In addition, 5-nitro indole which is electron deficient at the third position was also selected. The reaction between 5-methoxy indole and simple isatin, gave a yield of 64% for 5-Methoxy-3-hydroxy-3-(1H-indol-3-yl)indolin-2-one (3c). Interestingly excellent yields were obtained for products synthesized from the reactions between N-allyl, N-benzyl, N-pro- pargyl and N-methyl isatins and 5-methoxy indole which is electron rich at the third position. Thus, N-al- lyl-3-hydroxy-3-(5-methoxy-indol-3-yl)indolin-2-one (8c), N-benzyl-3-hydroxy-3-(5-methoxy-indol-3-yl) indolin-2-one (12c), N-propargyl-3-hydroxy-3-(5-methoxy-indol-3-yl) indolin-2-one (14c) and N-methyl-3-hydroxy- 3-(5-methoxy-indol-3-yl) indolin-2-one (15c) gave yields of 97%, 98%, 89% and 96% respectively. 76% yield was obtained in the reaction of 2-methyl indole with simple isatin to give 3-hydroxy-3-(2-methyl-1H-indole- 3-yl)indolin-2-one (2c), while N-allyl isatin and N-benzyl isatin gave N-allyl-3-hydroxy-3-(2-methyl-indol-3- yl) indolin-2-one (7c) and N-benzyl-3-hydroxy-3-(2-methyl-indol-3-yl)indolin-2-one (11c) with 90% and 92% yields respectively. Moderate yields were obtained in the case of N-methyl indole with various isatins. The yields were 56% for N-allyl-3-hydroxy-3-(1-methyl-indol-3-yl)indolin-2-one (6c) 56% for N-benzyl-3-hydroxy- 3-(1-methyl-indol-3-yl)indolin-2-one (10c) and 58% for N-propargyl-3-hydroxy-3-(1-methyl-indol-3-yl) indolin-2-one (13c). No product was obtained in the case of electron withdrawing nitro group present at fifth position of the indole ring with simple isatin even after a prolonged reaction time of 10 minutes and at a temperature of 100°C (Table 1).

The proposed reaction mechanism for the synthesis of 3-hydroxy-3-(1H-indol-3-yl)indolin-2-one (1c) is shown in Scheme 2.

Entry 8 (Table 1) was selected as a representative example for comparing reaction rates in various solvents such as acetonitrile (ACN), 1,4-dioxane, water (H_2O), ethanol (EtOH) and 1,2-dichloroethane (1,2-DCE). Moderate yield was observed only in the case of ACN (53%) in five minutes which improved to 96% in a reaction time of 15 minutes. Comparatively less yields were observed in the case of H_2O (46%), EtOH (46%), 1,2-DCE (36%) and 1,4-dioxane (15%).

Synthesis of Unsymmetrical 3,3-Di(indolyl)indolin-2-ones

Formation of symmetrical diindolyl was observed as a major product at 100°C in the synthesis of 3-hydroxy-3- (1H-indol-3-yl)indolin-2-one (1c). The same conditions were adopted for synthesizing unsymmetrical 3,3-di (indolyl) indolin-2-ones. For optimization, 5-methoxy-3-hydroxy-3-(1H-indol-3-yl) indolin-2-one (1d) [1 equivalent] and indole [1.2 equivalents] were selected. At a microwave oven temperature of 150°C and irradiation for 5 minutes, 53% product 3-(5-methoxy-1H-indol-3-yl)-3-(1H-indol-3-yl)indolin-2-one (1e) was formed. At a temperature higher than 150°C and a reaction time of 10 minutes, no significant increase in the yield was observed. When the same conditions were applied to the reaction [i.e. 150°C for 10 minutes] using various N- substituted isatins, good yields were obtained in all the cases (Table 2, Scheme 3).

All the compounds were characterized by IR, ^{1}H, ^{13}C NMR and HRMS. To the best of our knowledge com-

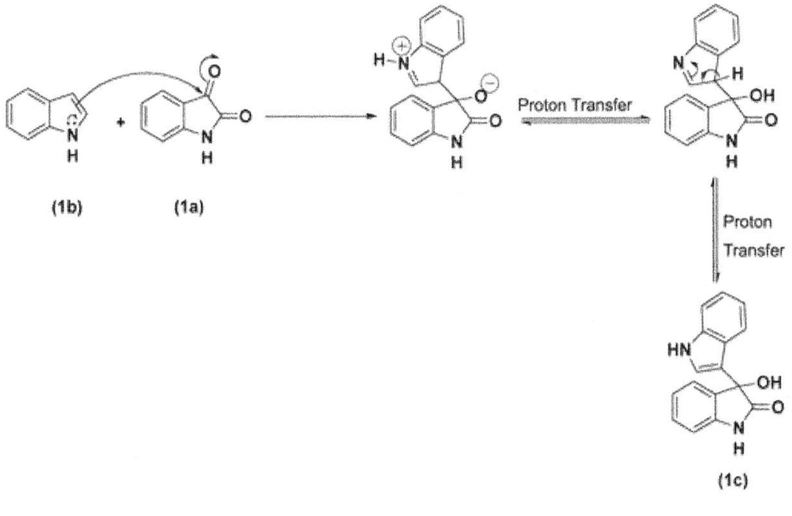

Scheme 2. Proposed mechanism for the formation of 3-indolyl-3-hydroxy oxindole.

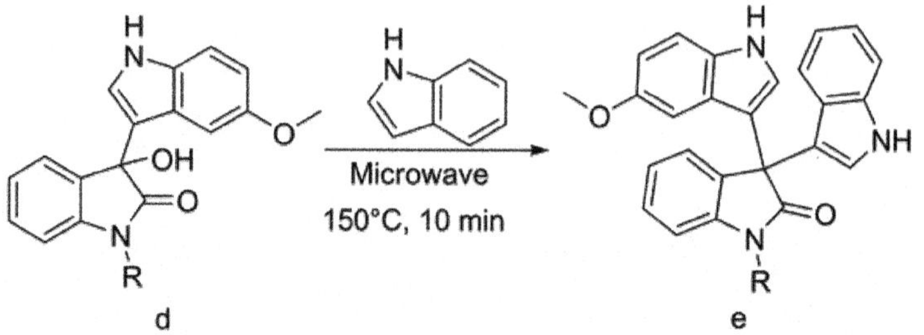

Scheme 3. Synthesis of various unsymmetrical 3,3-di(indolyl)indolin-2-ones.

Table 1: Synthesis of various 3-indolyl-3-hydroxy oxindoles

Sl. No.	Isatin a	Indole b	Product c	Yield %[a]
1	R = H	R_1 = H, R_2 = H, R_3= H	1c	31
2	R = H	R_1 = H, R_2 = Me, R_3 = H	2c	76
3	R = H	R_1 = H, R_2 = H, R_3 = OMe	3c	64
4	R = H	R_1 = H, R_2 = H, R_3 = NO$_2$	4c	00
5	R = allyl	R_1 = H, R_2 = H, R_3= H	5c	34
6	R = allyl	R_1 = Me, R_2 = H, R_3= H	6c	56
7	R = allyl	R_1 =H, R_2 = Me, R_3 =H	7c	90
8	R = allyl	R_1 = H, R_2 = H, R_3 = OMe	8c	97
9	R = benzyl	R_1 = H, R_2 = H, R_3= H	9c	40
10	R = benzyl	R_1 = H, R_2 = H, R_3= H	10c	56
11	R = benzyl	R_1 =H, R_2 = Me, R_3 =H	11c	92
12	R = benzyl	R_1 = H, R_2 = H, R_3 = OMe	12c	98
13	R = propargyl	R_1 = Me, R_2 = H, R_3= H	13c	58
14	R = propargyl	R_1 = H, R_2 = H, R_3 = OMe	14c	89
15	R = propargyl	R_1 = H, R_2 = H, R_3 = OMe	15c	96

[a] Isolated yield.

Table 2: Synthesis of various unsymmetrical 3,3-di(indolyl)indolin-2-ones.

Sl.No.	Substrate (R)d	Product e	Yield %[a]
1	H	1e	53
2	methyl	2e	72
3	allyl	3e	78
4	benzyl	4e	77
5	propargyl	5e	77

[a] Isolated yield

pounds   N-Allyl-3-hydroxy-3-(1-methyl-indol-3-yl)indolin-2-one
(6c),   N-Allyl-3-hydroxy-3-(5-methoxy-indol-   3-yl)indolin-2-one   (8c),
N-Benzyl-3-hydroxy-3-(1-methyl-indol-3-yl)indolin-2-one   (10c),
N-Propargyl-3-hy-   droxy-3-(1-methyl-indol-3-yl)indolin-2-one   (13c),
N-propargyl-3-hydroxy-3-(5-methoxy-indol-3-yl)indolin-2-   one   (14c);
3-(5-methoxy-1H-indol-3-yl)-3-(1H-indol-3-yl)indolin-2-one(1e),
3-1-methyl(5-methoxy-1H-in-dol-3- yl)-3-(1H-indol-3-yl)indolin-2-one (2e),
3-1-allyl(5-methoxy-1H-indol-3-yl)-3-(1H-indol-3-yl)indolin-2-one   (3e),
3-1-benzyl(5-methoxy-1H-indol-3-yl)-3-(1H-indol-3-yl)indolin-2-one   (4e)
and 3-1-(prop-2-ynyl)(5-methoxy-1H- indol-3-yl)-3-(1H-indol-3-yl)indolin-
2-one (5e) are reported here for the first time.

MATERIAL AND METHODS

Chemicals were obtained from Spectrochem, and used without further
purification. All known products were identified by comparison of their
physical and spectral data with those of authentic samples. ¹H and ¹³C NMR
spectra were recorded on a Bruker AV-400 spectrometer operating at 400 and
100 MHz, respectively. The spectra were calibrated on the solvent residual peak
(DMSO-d6: δ = 2.50 ppm for ¹H and δ = 39.52 ppm for ¹³C; CDCl₃: δ = 7.26
ppm for ¹H and δ = 77.0 ppm for ¹³C). Analytical thin layer chromatography
(TLC) was performed on Kieselgel 60 F254 aluminum sheets (Merck 1.05554).
Column chromatography was performed on silica gel (100 - 200 mesh). FT-IR
spectra were recorded on a Nicolet-6700. High-resolution mass spectra were
recorded with Thermo Scientific-Orbitrap Elite (Electro spray Ionization).
Anton Paar microwave synthesizer was used at 600 rpm and temperature was
kept constant at 70°C and 150°C.

General Procedure for the Synthesis of 3-Indolyl-3-hydroxy
Oxindoles (1c to 15c)

A mixture of the isatin (0.6797 mmol, 100 mg) and indole (0.8156 mmol,
96 mg) was added to a microwave-oven reaction vial then irradiated for 5
minutes at 70°C. After completion of the reaction (as indicated by TLC) the
residue was washed with 2 mL of DCM and mixed with silica gel and then
evaporated the solvent under reduced pressure, and the mixture was purified
by column chromatography using EtOAc in hexanes. All the known products
have spectral and physical data consistent with those reported in literatures.

3-Hydroxy-3-(1H-indol-3-yl)indolin-2-one (1c)

White solid; mp: 294 - 296 (lit[26]: 294°C - 296°C). ^{1}H NMR (400 MHz, DMSO-d$_6$): δ 10.98 (s, 1H), 10.34 (s, 1H), 7.35 (t, J = 9.6 Hz, 2 H), 7.27 - 7.24 (m, 2H), 7.09 (d, J = 2.4 Hz, 1H), 7.03 - 7.02 (m, 1H), 6.98 - 6.86 (m, 4H), 6.36 (s, 1H). ^{13}C NMR (100 MHz, DMSO-d$_6$): δ 178.5, 141.7, 136.8, 133.5, 129.1, 124.9, 124.8, 123.5, 121.7, 121.1, 120.3, 118.5, 115.5, 111.5, 109.6, 74.9. IR (KBr): υ 3264, 1711, 1617, 1549, 1467, 1424, 1335, 1226, 1184, 1107, 1071, 938, 910, 745, 686, 655 cm^{-1}. HRMS: Calcd for C$_{16}$H$_{12}$O$_2$N$_2$ [M + Na]$^+$ 287.0899, found 287.0795.

3-Hydroxy-3-(2-methyl-1H-indol-3-yl)indolin-2-one (2c)

White solid; mp: 178°C - 180°C (lit[26]: 176°C - 178°C). ^{1}H NMR (400 MHz, DMSO-d$_6$): δ 10.86 (s, 1H), 10.33 (s, 1H), 7.25 - 7.17 (m, 3H), 6.96 - 6.88 (m, 4H), 6.74 - 6.70 (m, 1H), 6.26 (s, 1H), 2.34 (s, 3H). ^{13}C NMR (100 MHz, DMSO-d$_6$): δ 178.7, 141.6, 134.8, 134.1, 133.4, 129.0, 126.6, 124.9, 121.7, 119.8, 119.2, 118.2, 110.2, 109.6, 109.4, 75.8, 13.3. IR (KBr): υ 3348, 3229, 3034, 1693, 1611, 1487, 1454, 1431, 1374, 1345, 1291, 1239, 1211, 1163, 1024, 996, 954, 927, 893, 853, 832, 737, 696 cm^{-1}. HRMS: Calcd for C$_{17}$H$_{14}$O$_2$N$_2$ [M + Na]$^+$ 301.0948; found 301.0953.

5-Methoxy-3-hydroxy-3-(1H-Indol-3-yl)indolin-2-one (3c)

White solid; mp: 196°C - 198°C (lit[26]: 196°C - 198°C). ^{1}H NMR (400 MHz, DMSO-d$_6$): δ 10.84 (s, 1H), 10.33 (s, 1H), 7.28 - 7.23 (m, 3H), 7.02 (d, J = 2.8 Hz, 1H), 6.98 (t, J = 7.6 Hz, 1H), 6.92 (d, J = 7.26 Hz, 1H), 6.86 (d, J = 2 Hz, 6.71 (dd, J = 2.4 Hz, J = 8.8 Hz, 1 H), 6.34 (s, 1H), 5.76 (s, 1H), 3.62 (s, 3H). ^{13}C NMR (100 MHz, DMSO-d$_6$): δ 178.5, 152.7, 141.7, 133.4, 132.0, 129.0, 125.4, 124.8, 124.2, 121.7, 115.0, 112.0, 110.9, 109.6, 102.7, 75.0, 55.2. IR (KBr): υ 3318, 3166, 1701, 1614, 1581, 1465, 1433, 1344, 1304, 1235, 1207, 1173, 1039, 927, 837, 779, 752, 650 cm^{-1}. HRMS: Calcd for C$_{17}$H$_{14}$O$_3$N$_2$ [M + Na]$^+$ 317.0897; found 317.0902.

N-allyl-3-hydroxy-3-(1H-indol-3-yl)indolin-2-one (5c)

White solid; mp: 148°C - 150°C. ^{1}H NMR (400 MHz, DMSO-d$_6$): δ 11.01 (s, 1H), 7.35 - 7.33 (m, 4H), 7.06 - 7.04 (m, 4H), 6.89 - 6.86 (m, 1H), 6.49 (s, 1H), 5.89 - 5.82 (m, 1H), 5.24 - 5.16 (m, 2H), 4.38-4.27 (m, 2H). ^{13}C NMR (100 MHz, DMSO-d$_6$): δ 176.4, 142.2, 136.8, 132.7, 132.0, 129.0, 124.9, 124.5, 123.6, 122.3, 121.1, 120.4, 118.5, 117.0, 115.2, 111.5, 109.1, 74.6, 41.4. IR (KBr): υ 3219, 3056, 1688, 1603, 1485, 1459, 1435, 1365, 1332, 1294, 1241, 1223, 1182, 1104, 981, 921, 905, 826, 735, 682 cm^{-1}. HRMS: Calcd for

$C_{19}H_{16}O_2N_2$ [M + Na]$^+$ 327.1109; found 327.1110.

N-allyl-3-hydroxy-3-(1-methyl-indol-3-yl)indolin-2-one (6c)

White solid; mp: 136°C - 138°C. ^{1}H NMR (400 MHz, CDCl$_3$): δ 7.67 (d, J = 8 Hz, 1H), 7.52 (d, J = 7.2 Hz, 1H), 7.33 - 7.19 (m, 3H), 7.09 - 7.05 (m, 2H), 6.97 (s, 1H), 6.90 (d, J = 7.6 Hz, 1H), 5.91 - 5.81 (m, 1H), 5.30 - 5.22 (m, 2H), 4.44 (dd, J = 4.8 Hz, J = 16 Hz, 1H), 4.28 (dd, J = 4.8 Hz, 16.4 Hz, 1H), 3.69 (s, 3H), 3.42 (m, 1H). ^{13}C NMR (100 MHz, CDCl$_3$): δ 176.9, 142.4, 137.7, 131.2, 129.6, 127.7, 125.4, 124.9, 123.1, 122.1, 120.8, 119.7, 117.8, 113.8, 109.5, 109.4, 75.5, 42.5, 32.8. IR (KBr): υ 3349, 3046, 1699, 1606, 1462, 1412, 1365, 1329, 1208, 1178, 1137, 1113, 1072, 984, 934, 888, 743, 669 cm^{-1}. HRMS: Calcd for $C_{20}H_{18}O_2N_2$ [M + Na]$^+$341.1260; found 341.1261.

N-allyl-3-hydroxy-3-(2-methyl-indol-3-yl) indolin-2-one (7c)

White solid; mp: 164°C - 166°C (lit[26]: 164°C - 166°C). ^{1}H NMR (400 MHz, DMSO-d$_6$): δ 10.91 (s, 1H), 7.32 (t, J = 7.6 Hz, 1H), 7.26 (d, J = 7.2 Hz, 1H), 7.19 (d, J = 8 Hz, 1H), 7.01 (t, J = 7.6 Hz, 2H), 6.92 - 6.85 (m, 2H), 6.71 (t, J = 7.2 Hz, 1H), 6.40 (s, 1H), 5.91 - 5.81 (m, 1H), 5.27 - 5.16 (m, 2H), 4.39 - 4.26 (m, 2H), 2.38 (s, 3H). ^{13}C NMR (100 MHz, DMSO-d$_6$): δ 176.6, 142.2, 134.8, 133.7, 133.3, 131.9, 129.0, 126.6, 124.7, 122.4, 119.8, 119.2, 118.2, 117.1, 110.3, 109.1, 75.5, 41.5, 13.3. IR (KBr): υ 3401, 1703, 1607, 1523, 1488, 1461, 1431, 1370, 1331, 1304, 1222, 1181, 982, 931, 895, 759, 673 cm^{-1}. HRMS: Calcd for $C_{20}H_{18}O_2N_2$ [M + Na]$^+$ 341.1261; found 341.1266.

N-allyl-3-hydroxy-3-(5-methoxy-indol-3-yl)indolin-2-one (8c)

White solid; mp: 102°C - 104°C. ^{1}H NMR (400 MHz, DMSO-d$_6$): δ 10.88 (s, 1H), 7.34 (brs, 2H), 7.24 (d, J = 8.8 Hz, 1H), 7.08 - 7.06 (m, 3H), 6.78 (s, 1H), 6.70 (d, J = 8.4 Hz, 1H), 6.48 (s, 1H), 5.89 - 5.82 (m, 1H), 5.25 (s, 1H), 5.18 (t, J = 10.8 Hz, 1H), 4.32 (dd, J = 14 Hz, J = 35.6 Hz, 2 H), 3.60 (s, 3 H). ^{13}C NMR (100 MHz, DMSO-d$_6$): δ 176.4, 152.8, 142.2, 132.6, 132.0, 131.9, 129.1, 125.3, 124.6, 124.3, 122.4, 117.0, 114.7, 112.1, 111.2, 109.0, 102.2, 74.6, 55.1, 41.5. IR (KBr): υ 3324, 1702, 1611, 1485, 1463, 1435, 1361, 1210, 1177, 1109, 1064, 990, 9224, 899, 840, 805, 752, 704, 674, 631 cm^{-1}. HRMS: Calcd for $C_{20}H_{18}O_3N_2$ [M + Na]$^+$ 357.1215; found 357.1202.

N-benzyl-3-hydroxy-3-(1H-indol-3-yl)indolin-2-one (9c)

White solid; mp: 120°C - 124°C (lit[26]: 120°C - 124°C). ^{1}H NMR (400 MHz, DMSO-d$_6$): δ 11.03 (s, 1H), 7.37 - 7.24 (m, 9H), 7.10 (s, 1H), 7.05 - 6.96 (m, 3H), 6.82 (t, J = 7.6 Hz, 1H), 6.57 (s, 1H), 4.92 (s, 2H). ^{13}C NMR (100 MHz,

DMSO-d$_6$): δ 176.8, 142.1, 136.8, 136.4, 132.7, 129.0, 128.5, 127.4, 124.8, 124.5, 124.5, 123.6, 122.5, 121.1, 120.4, 118.4, 115.1, 111.5, 109.1, 74.7, 42.7. IR (KBr): υ 3296, 3029, 1697, 1607, 1488, 1458, 1428, 1342, 1238, 1212, 1166, 1067, 991, 901, 738, 694 cm^{-1}. HRMS: Calcd for C$_{23}$H$_{18}$O$_2$N$_2$ [M + Na]$^+$ 377.1266; found 377.1251.

N-benzyl-3-hydroxy-3-(1-methyl-indol-3-yl)indolin-2-one (10c)

White solid; mp: 126°C - 128°C. ^{1}H NMR (400 MHz, DMSO-d$_6$): δ 7.38 - 7.25 (m, 9H), 7.12 - 7.10 (m, 2H), 7.02 (t, J = 7.6 Hz, 1H), 6.97 (d, J = 8 Hz, 1H), 6.87 (t, J = 7.6 Hz, 1H), 6.54 (s, 1H), 4.91 (s, 2H), 3.73 (s, 3H). ^{13}C NMR (100 MHz, DMSO-d$_6$): δ 176.6, 142.1, 137.2, 136.3, 132.6, 129.0, 128.4, 127.9, 127.3, 125.1, 124.5, 122.4, 121.1, 120.6, 118.5, 114.3, 109.6, 109.1, 74.5, 42.7, 32.3. IR (KBr): υ 3302, 1693, 1610, 1549, 1486, 1464, 1373, 1342, 1211, 1164, 1112, 1077, 990, 928, 893, 855, 747, 702, 669, 628 cm^{-1}. HRMS: Calcd for C$_{24}$H$_{20}$O$_2$N$_2$ [M + Na]$^+$ 391.1422; found 391.1420.

N-benzyl-3-hydroxy-3-(2-methyl-indol-3-yl)indolin-2-one (11c)

White solid; mp: 96°C - 98°C. ^{1}H NMR (400 MHz, DMSO-d$_6$): δ 10.91 (s, 1H), 7.36 - 7.18 (m, 7H), 7.19 (d, J = 8 Hz, 1H), 6.99 (t, J = 7.2 Hz, 2H), 6.92 - 6.88 (m, 1H), 6.80 (d, J = 8 Hz, 1H), 6.68 - 6.64 (m, 1H), 6.47 (s, 1H), 4.92 (ABq, J = 16 Hz, J = 20 Hz, 2H), 2.37 (s, 3H). ^{13}C NMR (100 MHz, DMSO-d$_6$): δ 177.0, 142.1, 136.3, 134.8, 133.7, 133.4, 129.0, 128.5, 127.5, 126.5, 124.8, 122.5, 119.8, 119.2, 118.1, 110.3, 109.1, 109.2, 109.1, 75.6, 42.7, 13.3. IR (KBr): υ 3314, 3288, 3092, 1700, 1610, 1551, 1530, 1486, 1461, 1430, 1350, 1302, 1237, 1173, 1074, 993, 917, 745, 698, 634 cm^{-1}. HRMS: Calcd for C$_{24}$H$_{20}$O$_2$N$_2$ [M + Na]$^+$ 391.1417; found 391.1418.

N-benzyl-3-hydroxy-3-(5-methoxy-indol-3-yl)indolin-2-one (12c)

White solid; mp: 204°C - 206°C. ^{1}H NMR (400 MHz, DMSO-d$_6$): δ 10.89 (s, 1H), 7.30 - 7.23 (m, 8H), 7.05 - 7.02 (m, 2H), 6.97 (d, J = 7.6 Hz, 1H), 6.70 - 6.67 (m, 2H), 6.55 (s, 1H), 4.91 (dd, J = 15.6 Hz, J = 22 Hz, 2H), 3.49 (s, 3 H). ^{13}C NMR (100 MHz, DMSO-d$_6$): δ 176.8, 152.7, 142.2, 136.4, 132.7, 131.9, 129.1, 128.5, 127.3, 125.2, 124.7, 124.3, 122.5, 114.7, 112.1, 111.2, 109.1, 102.2, 74.7, 54.9, 42.6. IR (KBr): υ 3354, 3283, 3047, 1702, 1611, 1582, 1527, 1486, 1461, 1350, 1306, 1211, 1174, 1065, 995, 931, 902, 854, 751, 698, 673 cm^{-1}. HRMS: Calcd for C$_{24}$H$_{20}$O$_3$N$_2$ [M + Na]$^+$ 407.1372; found 407.1357.

N-propargyl-3-hydroxy-3-(1-methyl-indol-3-yl)indolin-2-one (13c)

White solid; mp: 158°C - 162°C. ^{1}H NMR (400 MHz, DMSO-d$_6$): δ 7.41 - 7.33 (m, 4H), 7.19 (d, J = 8 Hz, 1H), 7.13 - 7.08 (m, 3H), 6.91 (t, J = 7.2 Hz, 1H), 6.60 (s, 1H), 4.56 (ABq, J = 16.4 Hz, J = 33.2 Hz, 2 H), 3.72 (s, 3 H), 3.31 (s, 1H). ^{13}C NMR (100 MHz, DMSO-d$_6$): δ 175.7, 141.2, 137.2, 132.5, 129.2, 127.9, 125.1, 124.6, 122.8, 121.3, 120.6, 118.7, 114.2, 109.7, 109.3, 78.0, 74.5, 32.4, 28.8. IR (KBr): υ 3322, 3254, 1711, 1609, 1535, 1463, 1426, 1369, 1335, 1244, 1213, 1170, 1110, 1071, 997, 934, 892, 744, 671, 628. HRMS: Calcd for C$_{20}$H$_{16}$O$_2$N$_2$ [M + Na]$^+$ 339.1109; found 339.1097.

N-propargyl-3-hydroxy-3-(5-methoxy-indol-3-yl)indolin-2-one (14c)

White solid; mp: 181°C - 184°C. ^{1}H NMR (400 MHz, DMSO-d$_6$): δ 10.9 (s, 1H), 7.41 (t, J = 7.6 Hz, 1H), 7.33 (d, J = 7.6 Hz, 1H), 7.25 - 7.18 (m, 2H), 7.12 - 7.07 (m, 2H), 6.75 (s, 1H), 6.70 (d, J = 8.4 Hz, 1H), 6.55 (s, 1H), 4.57 (ABq, J = 18 Hz, J = 45.2 Hz, 2H). ^{13}C NMR (100 MHz, DMSO-d$_6$): δ175.8, 152.9, 141.3, 132.5, 131.9, 129.1, 125.1, 124.6, 124.3, 122.8, 114.5, 112.1, 111.3, 109.2, 102.0, 78.1, 74.7, 74.5, 55.2 28.8. IR (KBr): υ 3411, 3325, 3276, 3089, 3052, 1705, 1611, 1583, 1530, 1486, 1463, 1438, 1360, 1291, 1242, 1209, 1177, 1111, 1059, 991, 924, 894, 835, 752, 702, 667 cm^{-1}. HRMS: Calc. for C$_{20}$H$_{16}$O$_3$N$_2$ [M + Na]$^+$ Calcd 355.1059; found 355.1047.

N-methyl-3-hydroxy-3-(5-methoxy-indol-3-yl)indolin-2-one (15c)

White solid; mp: 102°C - 104°C. ^{1}H NMR (400 MHz, DMSO-d$_6$): δ 10.85 (s, 1H), 7.39 - 7.31 (m, 2H), 7.23 (d, J = 8.8 Hz, 1H), 7.09 - 7.06 (m, 2H), 7.01 (s, 1H), 6.81 (s, 1H), 6.70 (d, J = 8.8Hz, 1H), 6.40 (s, 1H), 3.62 (s, 3H), 3.16 (s, 3H). ^{13}C NMR (100 MHz, DMSO-d$_6$): δ176.6, 152.7, 143.1, 132.7, 131.9, 129.2, 125.3, 124.4, 124.2, 122.4, 114.7, 112.1, 111.0, 108.4, 102.3, 74.7, 55.1, 25.9. IR (KBr): υ 3292, 3060, 1700, 1611, 1466, 1346, 1301, 1212, 1172, 1066, 997, 934, 903, 848, 752, 695, 668, 635 cm^{-1}. HRMS: Calcd for C$_{18}$H$_{16}$O$_2$N$_2$ [M + Na]$^+$ 331.1053; found 315.1064.

General Procedure for the Synthesis of Unsymmetrical 3,3Di(indolyl)indolin-2-ones (1e to 5e)

A mixture of 3-indolyl-3-hydroxy oxindoles (0.3400 mmol, 100 mg) and indole (0.4100 mmol, 48 mg) was added to a microwave-oven reaction vial, and then irradiated for 10 minutes at 150°C. Then the residue as washed with 2 mL of EtOAc and mixed with silicagel and then evaporated the solvent under reduced pressure, and the mixture was purified by column chromatography using EtOAc in hexanes.

3-(5-Methoxy-1H-indol-3-yl)-3-(1H-indol-3-yl)indolin-2-one (1e)

White solid; mp: 280°C - 282°C. ^{1}H NMR (400 MHz, DMSO-d$_6$): δ 10.95 (s, 1H), 10.79 (s, 1H), 10.59 (s, 1H), 7.35 (d, J = 8 Hz, 1H), 7.25 - 7.22 (m, 4H), 7.03 - 6.98 (m, 2H), 6.93 (t, J = 7.2 Hz, 1H), 6.89 - 6.81 (m, 3H), 6.70 - 6.67 (m, 2H), 3.51 (s, 3H). ^{13}C NMR (100 MHz, DMSO-d$_6$): δ 178.7, 152.4, 141.3, 136.9, 134.5, 132.1, 127.8, 126.0, 125.7, 125.0, 124.9, 124.3, 124.2, 121.4, 120.9, 120.7, 118.2, 112.0, 111.5, 110.4, 109.5, 103.3, 55.1, 52.5. IR (KBr): υ 3371, 1737, 1676, 1620, 1577, 1472, 1418, 1372, 1339, 1291, 1208, 1174, 1129, 1101, 1053, 1015, 958, 930, 889, 837, 796, 746, 679, 637 cm^{-1}. HRMS: Calcd for C$_{25}$H$_{19}$O$_2$N$_3$ [M + Na]$^+$ 416.1369; found 416.1383.

3-1-Methyl(5-methoxy-1H-indol-3-yl)-3-(1H-indol-3-yl)indolin-2-one (2e)

White solid; mp: 280°C - 282°C. ^{1}H NMR (400 MHz, DMSO-d$_6$): δ 10.97 (s, 1H), 10.82 (s, 1H), 7.37 - 7.32 (m, 2H), 7.28 - 7.23 (m, 2H), 7.19 - 7.16 (m, 2H), 7.04 - 7.00 (m, 2H), 6.89 (d, J = 2.4 Hz, 1 H), 6.82 - 6.77 (m, 2H), 6.69 (dd, J = 2.4 Hz, J = 8.8 Hz, 1H), 6.58 (d, J = 2.4 Hz, 1H), 3.50 (s, 3H), 3.26 (s, 3H). ^{13}C NMR (100 MHz, DMSO-d$_6$): δ 177.0, 152.5, 142.8, 137.0, 133.7, 132.2, 128.1, 126.0, 125.7, 125.1, 124.7, 124.5, 122.3, 121.0, 120.7, 118.4, 113.9, 113.6, 112.2, 111.7, 110.6, 108.6, 103.0, 55.1, 52.2, 26.3. IR (KBr): υ 3396, 3345, 3314, 2924, 2854, 2363, 2337, 1734, 1692, 1608, 1462, 1423, 1350, 1292, 1212, 1172, 1123, 1086, 1038, 1016, 913, 891, 844, 795, 743, 692, 650 cm^{-1}. HRMS: Calcd for C$_{26}$H$_{21}$O$_2$N$_3$ [M + Na]$^+$ 430.1531; found 430.1538.

3-1-Allyl(5-methoxy-1H-indol-3-yl)-3-(1H-indol-3-yl)indolin-2-one (3e)

White solid; mp: 266 - 268. ^{1}H NMR (400 MHz, DMSO-d$_6$): δ 10.99 (s, 1H), 10.83 (s, 1H), 7.36 (d, J = 8.4 Hz, 1 H), 7.33 - 7.29 (m, 2H), 7.24 (d, J = 8.4 Hz, 1H), 7.20 (d, J = 8 Hz, 1H), 7.11 (d, J = 7.6 Hz, 1H), 7.02 (t, J = 7.6 Hz, J = 14.8 Hz, 2H), 6.89 (d, J = 2.4 Hz, 1H), 6.82 (d, J = 2.8 Hz, 1H), 6.79 (d, J = 7.2 Hz, 1H), 6.68 (dd, J = 2.4 Hz, J = 8.4 Hz, 1H), 6.55 (d, J = 2.4 Hz, 1H), 5.95 - 5.86 (m, 1H), 5.21 (d, J = 1.6 Hz, 1H), 5.18 - 5.15 (m, 1H), 4.43 (d, J = 4.8 Hz, 2H), 3.48 (s, 3H). ^{13}C NMR (100 MHz, DMSO-d$_6$): δ 176.8, 152.6, 141.8, 137.0, 133.8, 132.2, 128.0, 126.0, 125.7, 125.2, 124.8, 124.4, 122.3, 121.1, 120.8, 118.4, 117.2, 114.0, 113.6, 112.3, 111.8, 110.8, 109.3, 102.9, 55.1, 52.2, 41.8. IR (KBr): υ 3754, 3708, 3345, 3010, 1736, 1669, 1606, 1540, 1478, 1456, 1358, 1293, 1208, 1171, 1093, 1018, 926, 841, 798, 744, 702, 632 cm^{-1}. HRMS: Calcd for C$_{28}$H$_{23}$O$_2$N$_3$ [M + Na]$^+$ 456.1688; found 456.1701.

3-1-Benzyl(5-methoxy-1H-indol-3-yl)-3-(1H-Indol-3-yl)indolin-2-one (4e)

White solid; mp: 260°C - 262°C. ^{1}H NMR (400 MHz, DMSO-d$_6$): δ 11.00 (s, 1H), 10.9 (s, 1H), 7.38 - 7.37 (m, 3H), 7.32 - 7.25 (m, 6H), 7.15 - 7.09 (m, 2H), 7.04 - 6.98 (m, 2H), 6.93 (d, J = 2 Hz, 1H), 6.85 (d, J = 2.4 Hz, 1H), 6.75 (t, J = 7.6 Hz, J = 15.2 Hz, 1H), 6.69 (dd, J = 2 Hz, J = 8.8 Hz, 1H), 6.56 (d, J = 2 Hz, 1H), 5.02 (s, 2H), 3.42 (s, 3H).^{13}C NMR (100 MHz, DMSO-d$_6$): δ 177.2, 152.5, 141.8, 137.0, 133.8, 132.2, 128.6, 127.9, 127.5, 126.0, 125.6, 125.2, 124.9, 124.5, 122.4, 121.1, 120.7, 118.3, 113.9, 113.4, 112.2, 111.7, 110.8, 109.3, 103.0, 55.0, 52.3, 43.0. IR (KBr): υ 3347, 3025, 2850, 1699, 1676, 1607, 1535, 1483, 1458, 1358, 1291, 1209, 1172, 1133, 1099, 1016, 930, 847, 799, 745, 698, 63 cm^{-1}. HRMS: Calcd for C$_{32}$H$_{25}$O$_2$N$_3$ [M + Na]$^+$ 506.1839 found 506.1848.

3-1-(Prop-2-ynyl)(5-methoxy-1H-indol-3-yl)-3-(1H-indol-3-yl)indolin-2-one (5e)

White solid; mp: 266°C - 268°C. ^{1}H NMR (400 MHz, DMSO-d$_6$): δ 10.99 (s, 1H), 10.85 (s, 1H), 7.38 - 7.36 (m, 2H), 7.27 - 7.25 (m, 4H), 7.08 - 7.01 (m, 2H), 6.87 (m, 1H), 6.81 - 6.79 (m, 2H), 6.71 - 6.69 (m, 1H), 6.63 (s, 1H), 4.65 (s, 2H), 3.51 (s, 3H), 3.31 (s, 1H). ^{13}C NMR (100 MHz, DMSO-d$_6$): δ 176.3, 152.6, 140.9, 137.0, 133.6, 132.2, 128.0, 125.9, 125.6, 125.3, 124.9, 124.6, 122.7, 121.2, 120.9, 118.4, 113.8, 113.2, 112.3, 111.7, 111.0, 109.4, 103.0, 78.2, 74.5, 55.2, 52.3, 29.0. IR (KBr): υ 3388, 3326, 3267, 1689, 1605, 1580, 1482, 1460, 1427, 1378, 1356, 1338, 1295, 1250, 1209, 1173, 1128, 1100, 1037, 1013, 932, 911, 888, 857, 800, 748, 703, 660 cm^{-1}. HRMS: Calcd for C$_{28}$H$_{21}$O$_2$N$_3$ [M + Na]$^+$ 454.1531; found 454.1538.

CONCLUSION

A simple and green alternative protocol for the synthesis 3-indolyl-3-hydroxy oxindoles in moderate to excellent yields is reported here. The unsymmetrical 3,3-di(indolyl)indolin-2-ones are also obtained in moderate to good yields under microwave irradiation. The highlights of the method are that no solvent, no catalysts are needed, and the reaction times are very short, i.e. five minutes for 3-indolyl-3-hydroxy oxindoles and ten minutes for unsymmetrical 3,3-di(indolyl)indolin-2-ones. Hence, this methodology can be conveniently used to synthesize the hybrid molecules of isatin and indoles in a short reaction time.

ACKNOWLEDGEMENTS

We thank the Department of Biotechnology and IIT-Madras for infrastructure.

REFERENCES

1. Paira, P., Hazra, A., Kumar, S., Paira, R., Sahu, K.B., Naskar, S., Saha, P., Mondal, S., Maity, A., Banerjee, S. and Mondal, N.B. (2009) Efficient Synthesis of 3,3-Diheteroaromatic Oxindole Analogues and Their in Vitro Evaluation for Spermicidal Potential. Bioorganic & Medicinal Chemistry Letters, 19, 4786-4789.http://dx.doi.org/10.1016/j.bmcl.2009.06.049

2. Subba Reddy, B.V., Rajeswari, N., Sarangapani, M., Prashanthi, Y., Ganji, R.J. and Addlagatta, A. (2012) Iodine-Cat- alyzed Condensation of Isatin with Indoles: A Facile Synthesis of Di(indolyl)indolin-2-ones and Evaluation of Their Cytotoxicity. Bioorganic & Medicinal Chemistry Letters, 22, 2460-2463.http://dx.doi.org/10.1016/j.bmcl.2012.02.011

3. Pajouhesh, H., Parson, R. and Popp, F.D. (1983) Potential Anticonvulsants VI: Condensation of Isatins with Cyclohexanone and Other Cyclic Ketones. Journal of Pharmaceutical Sciences, 72, 318-321. http://dx.doi.org/10.1002/jps.2600720330

4. DeLorbe, J.E., Jabri, S.Y., Mennen, S.M., Overman, L.E. and Zhang, F.L. (2011) Enantioselective Total Synthesis of (+)-Gliocladine C: Convergent Construction of Cyclotryptamine-Fused Polyoxopiperazines and a General Approach for Preparing Epidithiodioxopiperazines from Trioxopiperazine Precursors. Journal of the American Chemical Society, 133, 6549-6552. http://dx.doi.org/10.1021/ja201789v

5. Guo, C., Song, J., Huang, J.-Z., Chen, P.-H., Luo, S.-W. and Gong, L.-Z. (2012) Core-Structure-Oriented Asymmetric Organocatalytic Substitution of 3-Hydroxyoxindoles: Application in the Enantioselective Total Synthesis of (+)-Foli- canthine. Angewandte Chemie-International Edition, 51, 1046-1050. http://dx.doi.org/10.1002/anie.201107079

6. Berens, U., Brown, J.M., Long, J. and Selke, R.D. (1996) Synthesis and Resolution of 2,2'-Bis-diphenylphosphino [3,3']biindolyl, a New Atropisomeric Ligand for Transition Metal Catalysis. Tetrahedron: Asymmetry, 7, 285-292. http://dx.doi.org/10.1016/0957-4166(95)00447-5

7. Kumar, V.P., Reddy, V.P., Sridhar, R., Srinivas, B., Narender, M. and Rao, K.R. (2008) Supramolecular Synthesis of 3-Indolyl-3-hydroxy Oxindoles under Neutral Conditions in Water. Journal of Organic Chemistry, 73, 1646-1648.http://dx.doi.org/10.1021/jo702496s

8. Jafarpour, M., Rezaeifard, A., Gazkar, S. and Danehchin, M. (2011) Catalytic Activity of a Zirconium (IV) Schiff Base Complex in Facile and Highly Efficient Synthesis of Indole Derivatives. Transition Metal Chemistry, 36, 685-690. http://dx.doi.org/10.1007/s11243-011-9519-6

9. Sarrafi, Y., Alimohammadi, K., Sadatshahabi, M. and Norozipoor, N. (2012) An Improved Catalytic Method for the Synthesis of 3,3-Di(indolyl) oxindoles Using Amberlyst 15 as a Heterogeneous and Reusable Catalyst in Water. Monatshefte für Chemie, 143, 1519-1522.http://dx.doi. org/10.1007/s00706-012-0723-7

10. Jafarpour, M., Rezaeifard, A. and Gorzin, G. (2011) Enhanced Catalytic Activity of Zr(IV) Complex with Simple Tetradentate Schiff Base Ligand in the Clean Synthesis of Indole Derivatives. Inorganic Chemistry Communications, 14, 1732-1736.http://dx.doi.org/10.1016/j. inoche.2011.07.017

11. Kamal, A., Srikanth, Y.V.V., Khan, M.N.A., Shaik, T.B. and Ashraf, M. (2010) Synthesis of 3,3-Diindolyl Oxyindoles Efficiently Catalysed by FeCl$_3$ and Their in Vitro Evaluation for Anticancer Activity. Bioorganic & Medicinal Chemistry Letters, 20, 5229-5231.http://dx.doi.org/10.1016/j. bmcl.2010.06.152

12. Azizian, J., Mohammadi, A.A., Karimi, N., Mohammadizadeh, M.R. and Karimi, A.R. (2006) Silica Sulfuric Acid a Novel and Heterogeneous Catalyst for the Synthesis of Some New Oxindole Derivatives. Catalysis Communications, 7, 752-755.http://dx.doi.org/10.1016/j. catcom.2006.01.026

13. Alinezhad, H., Haghighi, A.H. and Salehian, F. (2010) A Green Method for the Synthesis of Bis-Indolylmethanes and 3,3'-Indolyloxindole Derivatives Using Cellulose Sulfuric Acid under Solvent-Free Conditions. Chinese Chemical Letters, 21, 183-186.http://dx.doi.org/10.1016/j. cclet.2009.09.001

14. Saffar-Teluri, A. (2014) Boron Trifluoride Supported on Nano-SiO$_2$: An Efficient and Reusable Heterogeneous Catalyst for the Synthesis of Bis(indolyl)methanes and Oxindole Derivatives. Research on Chemical Intermediates, 40, 1061-1067.http://dx.doi.org/10.1007/s11164-013-1021-7

15. Sarrafi, Y., Alimohammadi, K., Sadatshahabi, M. and Norozipoor, N. (2012) An Improved Catalytic Method for the Synthesis of 3,3-Di(indolyl) oxindoles Using Amberlyst 15 as a Heterogeneous and Reusable Catalyst in Water. Monatshefte für Chemie—Chemical Monthly, 143, 1519-1522. http://dx.doi.org/10.1007/s00706-012-0723-7

16. Karimi1, N., Oskooi1, H., Heravi, M., Saeedi, M., Zakeri, M. and Tavakoli, N. (2011) On Water: Bronsted Acidic Ionic Liquid [(CH$_2$)$_4$SO$_3$HMIM] [HSO$_4$] Catalysed Synthesis of Oxindoles Derivatives. Chinese Journal of Chemistry, 29, 321-323.http://dx.doi.org/10.1002/cjoc.201190085

17. Azizian, J., Mohammadi, A.A., Karimi, A.R. and Mohammadizadeh, M.R. (2004) KAl(SO$_4$)$_2$ · 12H$_2$O as a Recyclable Lewis Acid Catalyst for Synthesis of Some New Oxindoles in Aqueous Media. Journal of Chemical Research, 2004, 424-426.http://dx.doi.org/10.3184/0308234041423600

18. Yadav, J.S., SubbaReddy, B.V., Uma, G.K., Meraj, S. and Prasad, A.R. (2006) Bismuth (III) Triflate Catalyzed Condensation of Isatin with Indoles and Pyrroles: A Facile Synthesis of 3,3-Diindolyl- and 3,3-Dipyrrolyl Oxindoles. Synthesis, 2006, 4121-4123.http://dx.doi.org/10.1055/s-2006-950373

19. Feng, G.L., Geng, L.J. and Zhang, H.L. (2009) Facile Synthesis of 3,3-Di(indolyl)indolin-2-one Derivatives Catalyzed by ZrO$_2$/S$_2$O$_8$$^{2-}$ Solid Superacid under Grinding Condition. Chemical Journal on Internet, 11, Article ID: 111001pe.

20. Chakrabarty, M., Sarkar, S. and Harigaya, Y. (2005) A Facile Clay-Mediated Synthesis of 3,3-Diindolyl-2-indolinones from Isatins. Journal of Chemical Research, 8, 540-542.http://dx.doi.org/10.3184/030823405774663264

21. Deb, M.L. and Bhuyan, P.J. (2009) Water-Promoted Synthesis of 3,3'-Di(indolyl)oxindoles. Synthetic Communications, 39, 2240-2243. http://dx.doi.org/10.1080/00397910802654690

22. Praveen, C., Ayyanar, A. and Perumal, P.T. (2011) Practical Synthesis, Anticonvulsant, and Antimicrobial Activity of N-Allyl and N-Propargyl Di(indolyl)indolin-2-ones. Bioorganic & Medicinal Chemistry Letters, 21, 4072-4077.http://dx.doi.org/10.1016/j.bmcl.2011.04.117

23. Wang, S.Y. and Ji, S.J. (2006) Facile Synthesis of 3,3-Di(heteroaryl) indolin-2-one Derivatives Catalyzed by Ceric Ammonium Nitrate (CAN) under Ultrasound Irradiation. Tetrahedron, 62, 1527-1535. http://dx.doi.org/10.1016/j.tet.2005.11.011

24. Moghadam, K.R., Kiasaraie, M.S. and Amlashi, H.T. (2010) Synthesis of Symmetrical and Unsymmetrical 3,3-Di(indolyl)- indolin-2-ones under Controlled Catalysis of Ionic Liquids. Tetrahedron, 66, 2316-2321. http://dx.doi.org/10.1016/j.tet.2010.02.017

25. Nikpassand, M., Mamaghani, M., Tabatabaeian, K. and Samimi, H.A. (2010) An Efficient and Clean Synthesis of Symmetrical and Unsymmetrical 3,3-Di(indolyl)indolin-2-ones Using KSF. Synthetic Communications,

40, 3552-3560.http://dx.doi.org/10.1080/00397910903457399

26. Shanthi, G., Lakshmi, N.V. and Perumal, P.T. (2009) A Simple and Eco-Friendly Synthesis of 3-Indolyl-3-hydroxy Oxindoles and 11-Indolyl-11H-indeno[1,2-b]quinoxalin-11-ols in Aqueous Media. ARKIVOC, 2009, 121-130.http://dx.doi.org/10.3998/ark.5550190.0010.a12

27. Meshram, H.M., Kumar, D.A., Goud, P.R. and Reddy, B.C. (2010) ChemInform Abstract: Triton B Assisted, Efficient, and Convenient Synthesis of 3-Indolyl-3-hydroxy Oxindoles in Aqueous Medium. Synthetic Communications, 40, 39- 45.http://dx.doi.org/10.1002/chin.201025093

28. Hosseini-Sarvari, M. and Tavakolian, M. (2012) Preparation, Characterization, and Catalysis Application of Nano-Rods Zinc Oxide in the Synthesis of 3-Indolyl-3-hydroxy Oxindoles in Water. Applied Catalysis A: General, 441-442, 65-71.http://dx.doi.org/10.1016/j.apcata.2012.07.009

29. Khorshidi, A. and Tabatabaeian, K.J. (2011) An Ultrasound-Promoted Green Approach for the Synthesis of 3-(Indol- 3-yl)-3-hydroxyindolin-2-ones Catalyzed by Fe(III). Journal of the Serbian Chemical Society, 76, 1347-1353.http://dx.doi.org/10.2298/JSC110420120K

30. Makarem, S., Fakhari, A.R. and Mohammadi, A.A. (2012) Electro-Organic Synthesis of Nanosized Particles of 3-Hydroxy- 3-(1H-indol-3-yl)indolin-2-one Derivatives. Monatshefte für Chemie—Chemical Monthly, 143, 1157-1160. http://dx.doi.org/10.1007/s00706-011-0693-1

31. Srihari, G. and Murthy, M.M. (2011) Kaolin/KOH Is an Efficient Heterogeneous Catalyst for the Synthesis of 3-Hydroxy- 3-indolyl Oxindoles. Synthetic Communications, 41, 2684-2692. http://dx.doi.org/10.1080/00397911.2010.515342

32. Deng, J., Zhang, S., Ding, P., Jiang, H., Wang, W. and Li, J. (2010) Facile Creation of 3-Indolyl-3-hydroxy-2-oxindoles by an Organocatalytic Enantioselective Friedel-Crafts Reaction of Indoles with Isatins. Advanced Synthesis & Catalysis, 352, 833-838.http://dx.doi.org/10.1002/adsc.200900851

33. Hanhan, N.V., Sahin, A.H., Chang, T.W., Fettinger, J.C. and Franz, A.K. (2010) Catalytic Asymmetric Synthesis of Substituted 3-Hydroxy-2-oxindoles. Angewandte Chemie International Edition, 49, 744-747. http://dx.doi.org/10.1002/anie.200904393

34. Prathima, P.S., Rajesh, P., Rao, J.V., Kailash, U.S., Sridhar, B. and Rao, M.M. (2014) "On Water" Expedient Synthesis of 3-Indolyl-3-hydroxy Oxindole Derivatives and Their Anticancer Activity in Vitro. European

Journal of Medicinal Chemistry, 84, 155-159.http://dx.doi.org/10.1016/j. ejmech.2014.07.004

35. Lakhdar, S., Westermaier, M., Terrier, F., Goumont, R., Boubaker, T., Ofial, A.R. and Mayr, H. (2006) Nucleophilic Reactivities of Indoles. The Journal of Organic Chemistry, 71, 9088-9095. http://dx.doi.org/10.1021/ jo0614339

Chapter 3

NEW SCHIFF BASES FROM 6,6'-DIFORMYL-2,2'-BIPYRIDYL WITH AMINES CONTAINING O, S, N AND F: SYNTHESIS AND CHARACTERIZATION

Md. Razzak[1], Mohammad R. Karim[1], Md. Rejaul Hoq[1], Aminul H. Mirza[2]

[1]Department of Chemistry, Tennessee State University, Nashville, USA

[2]Department of Chemistry, Faculty of Science, Universiti Brunei Darussalam, Bandar Seri Begawan, Brunei Darussalam

ABSTRACT

Eight novel Schiff Bases, from 6,6'-diformyl-2,2'-bipyridyl (1a, 43%)with O, N, S and F containing amines: Thiosemicarbazide (2a, 70%), 4-Ethyl-3-thiosemicabazide (2b, 75%), 4,4-Dimethyl-3-thio- semicarbazide (2c, 75%), S-benzyldithiocarbazide, SBDTC (2d, 80%), (Trifluromethyl) phenylhydrazine (2e, 80%), 4-Phenyl-3-thiosemcarbazide (2f, 80%), Thiocarbazide (2g, 70%), 2-Amino- thiophenol (2h, 65%), have been synthesized. The conventional method of synthesis of the Schiff bases involves refluxing the reaction mixture containing the diformyls and amines for 1 hour. The solid products that had formed were filtered off using suction filtration. In few reactions, 2 - 3 drops of conc. sulfuric acid were used to obtain high yield. The structures of all eight novel synthesized compounds have fully been characterized by spectroscopic (IR, NMR, MS) methods.

INTRODUCTION

Previously we have reported synthesis of several Schiff Bases derived from 1, 10-phenanthroline-2,9-dicarbal- dehyde [1] . Phenanthroline backbone rather produced structurally rigid, less flexible Schiff Bases. We therefore, decided to add more flexibility to the molecule by synthesizing similar Schiff Bases derived form 6,6'-difor- myl-2,2'-bipyridyl, thereby eliminating structural rigidity. Moreover, design and synthesis of organic chelating agents containing nitrogen and sulfur as donor atoms is our major focus. In this case, bi-dentate N,N chelating agent such as 2,2'-bipyridyl is expected to play a vital role in building many mixed-ligand complexes for their desired predictable co-ordination behavior and their electrochemical and photo-physical properties [2] - [4] . The 2,2'-bipyridyl and ligands derived from it also extensively used in different areas, such as molecular scaffolding, supramolecular assemblies, catalysis, biochemistry, electrochemistry, ring-opening metathesis polymerization and biochemistry [5] - [9] , biologically photoredox reactions [10] , synthetic, medicinal chemistry, biotechnology [11] and solar cell [12] [13] . Being more flexible, these Schiff bases are expected to form stable complexes with a wide variety of metal ions, which are expected to show interesting properties as observed in the literature for similar metal complexes [14] - [16] . In view of the importance of Schiff bases derived from 6,6'-formyl-2,2'-bipyridyl and O, S, N and F-containing amines in different fields, we report here synthesis and characterization of eight new Schiff bases from 6,6'-diformyl-2,2'-bipyridyl. We also plan to continue studies like anti-cancer, antibacterial, and antitumor activities with novel Schiff bases derived from 6,6'-formyl-2,2'-bipyridyl with O, S, N and F-containing amines.

GENERAL METHOD AND PROCEDURES

Experimental

HPLC grade solvents were used in all the reactions. All reagents were purchased from Sigma-Aldrich Chemical Co. (St. Louis, MO, USA) and ACROS (drive Pittsburgh, PA, USA) and were used without further purification. Routine thin-layer chromatography (TLC) was performed on aluminum-backed Whatman, Sigma-Aldrich Chemical Co. (St. Louis, MO, USA). The conventional method of synthesis of the Schiff bases involves refluxing the reaction mixture containing the diformyls and amines for 1 hour. The solid product that had formed was filtered off using suction filtration. In some of the reactions, 2 - 3 drops of conc. sulfuric acid were used to obtain high yield. To obtain NMR spectra, all compounds were dissolved in DMSO-d6 and recorded on a Bruker Ascend 400 M Hz NMR spectrometer using TMS as an internal standard. To obtain Mass spectra, samples were dissolved in $CH_3CN:H_2O:AcOH$ (50%:50%:0.1%) and injected by a direct infusion method. Data were recorded on a LTQ XL linear Ion Trap Mass spectrometer with electrospray ionization (ESI) mode. MS spectrophotometer was purchased from Thermo Scientific. All infra-red (IR) data were recorded (υ_{max} in cm^{-1}) on Smart iRT purchased from Thermo Scientific.

Synthesis of 6,6′-Diformyl-2,2′-Bipyridyl (1a) from 6,6′-Dimethyl-2,2′-Bipyridyl

6,6′-Diformyl-2,2′-bipyridyl (245 mg, 43%) 1a was synthesized from 6,6′-dimethyl-2,2′-bipyridyl (500 mg, 0.003 mol) through direct oxidation by SeO_2 (3 g, 0.03 mol) in presence of glacial acetic acid (40 ml) following a previously reported procedure [17] . ^{1}H NMR and ^{13}C NMR data were consistent with the literature value (Scheme 1).

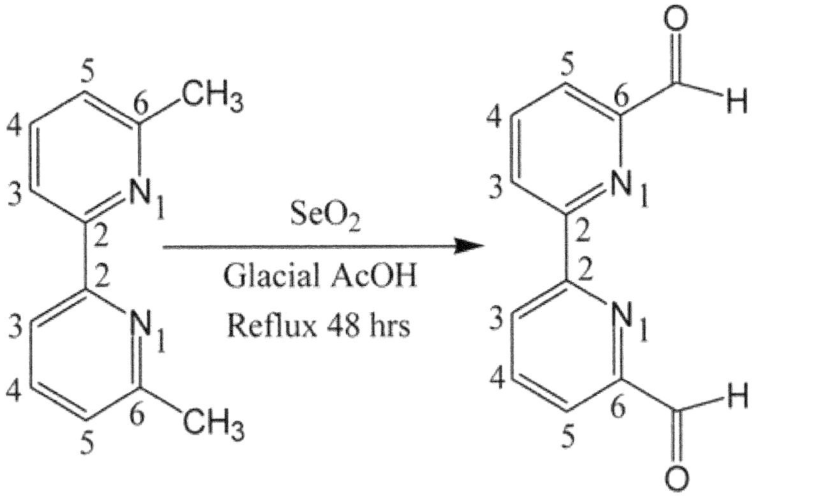

6, 6′-dimethyl-2, 2′-bipyridyl 1a, 43%

Scheme 1. Synthesis of 6,6′-diformyl-2,2′-bipyridyl (1a) from 6,6′-dimethyl-2,2′-bipyridyl.

General Procedure for Synthesis of Schiff Bases (2a-2h) from 6,6′-Diformyl-2,2′-Bipyridyl (1a)

3 equivalents of O, N, S and F containing amines were added to a solution of 1 equivalent of 6,6'-diformyl-2,2'- bipyridyl (1a) in 30 ml of MeOH. The solution was refluxed for 1 hour and then allowed to cool to room temperature. The solid product formed was filtered off and washed with methanol and dried under vacuum. For synthesizing compound 2d, S-benzyldithiosemicarbazide was prepared following a previously reported procedure [18] and ^{1}H NMR and ^{13}C NMR data were consistent with the literature value. 2 - 3 drops of conc. H_2SO_4 were added in the reaction mixture to obtain high yield of compound 2e (Scheme 2). The spectral data to confirm the structures of all eight desired Schiff bases have been shown in Table 1.

Table 1: Spectral data of the synthesized compounds.

Compound	Spectral Data
2a	IR, v (cm^{-1}): 3433, 3240 and 3154 (NH, stretching), 2990, 2850 (=CH, stretching), 1522 (C=N, imine), 1595 (C=C, aromatic), 1114 (C=S). ^{1}H-NMR (DMSO-d$_6$, 400 M Hz, δ ppm): δ_H = 11.77 (s, 2H-N), 8.45 - 8.30 (m, 6H, pyridine ring and imine protons), 8.27 (s, 2N-H), 8.18 (s, 2N-H), 8.01 (t, $J_1 = J_2 = 7.6$ Hz, 2H, pyridine ring protons). ^{13}C-NMR (DMSO-d$_6$, 100 M Hz, δ ppm): δ_C = 178, 154, 153, 152, 138, 121.2, 121.1. LC-MS (m/z): 359 (M+H), 381 (M + Na) (100%).
2b	IR, v (cm^{-1}): 3314 and 3155 (NH, stretching), 2960 and 2810 (>CH, stretching), 1515 (C=N, imine), 1573 (C=C, aromatic), 1076 (C=S). ^{1}H-NMR (DMSO-d$_6$, 400 M Hz, δ ppm): δ_H = 11.80 (s, 2N-H), 8.79 (t, $J_1 = J_2 = 4$ Hz, 2H, pyridine ring protons), 8.40 - 8.3 (m, 4H, pyridine ring and imine protons), 8.20 (s, 2N-H), 8.03 (t, $J_1 = J_2 = 8$ Hz, 2H, pyridine ring protons), 3.65 (m, 4H, methylene bridge), 1.2 (t, $J_1 = J_2 = 7$ Hz, 6H, methyl group). ^{13}C-NMR (DMSO-d$_6$, 100 M Hz, δ ppm): δ_C = 177, 155,153, 142, 138, 121.1, 121.0, 49, 14. LC-MS (m/z): 415 (M + H), 437 (M+Na) (100%).
2c	IR, v (cm^{-1}): 3415, 3345 and 3220 (N-H, stretching), 2910 and 2870 (CH, stretching), 1470 (C=N, imine), 1557 (C=C, aromatic), 911 (C=S). ^{1}H-NMR (DMSO-d$_6$, 400 M Hz, δ ppm): δ_H = 11.40 (s, 2N-H), 8.40 - 8.3 (m, pyridine ring and imine protons), 8.10 - 7.90 (m, 4H, pyridine ring protons), 3.35 (s, 12H, methyl protons). ^{13}C-NMR (DMSO-d$_6$, 100 M Hz, δ ppm): δ_C = 180, 155, 153, 144,138, 120.9, 120.4, 42. LC-MS (m/z): 415 (M + H), 437 (M + Na) (100%).
2d	IR, v (cm^{-1}): 3115 (NH, stretching), 2988 and 2849 (CH, stretching), 1510 (C=N, imine), 1580 (C=C, aromatic), 1041 (C=S). ^{1}H-NMR (DMSO-d$_6$, 400 M Hz, δppm): δ_H = 13.70 (s, 2N-H), 8.40 (d, J = 8 Hz, 2H), 8.35 (s, 2H, imine proton), 8.03 (t, $J_1 = J_2 = 8$ Hz, 2H, pyridine ring protons), 7.98 (d, J = 8 Hz, 2H, pyridine ring protons), 7.45 (d, J = 8 Hz, 2H, benzene ring protons), 7.35 (t, $J_1 = J_2 = 8$ Hz, 4H, benzene ring protons), 7.30 (t, $J_1 = J_2 = 8$ Hz, 4H, benzene ring protons). ^{13}C-NMR (DMSO-d$_6$, 100 M Hz, δ ppm): δ_C = 198, 155, 152, 146,138,136, 129.7, 129.0, 127, 122, 121, 38. LC-MS (m/z): 572.92 (M + H), 594.83 (M + Na) (100%).
2e	IR, v (cm^{-1}): 3334 and 3089 (NH, stretching), 2940 (CH, stretching), 1505 (C=N, imine), 1615 (C=C, aromatic). ^{1}H-NMR (DMSO-d$_6$, 400 M Hz, δ ppm): δ_H = 11.20 (s, 2N-H), 8.35 (d, J = 6 Hz, 2H, pyridine ring protons), 8.11 - 8.05 (m, 4H, pyridine ring protons), 8.02 (t, $J_1 = J_2 = 6$ Hz, 2H, pyridine ring protons), 7.60 (d, $J_1 = 6$ Hz, 4H, benzene ring protons), 7.30 (d, J = 6 Hz, 4H, benzene ring protons). ^{13}C-NMR (DMSO-d$_6$, 100 M Hz, δ ppm): δ_C = 154.4, 154.1, 148,139,138, 127, 126, 124, 120.3, 120.1, 112. LC-MS (m/z): 529.01 (M + H), (100%).
2f	IR, v (cm^{-1}): 3290, 3149 and 3000 (NH, stretching), 2940 (CH, stretching), 1550 (C=N, imine), 1595 (C=C, aromatic), 1163 (C=S). ^{1}H-NMR (DMSO-d$_6$, 400 M Hz, δ ppm): δ_H = 12.20 (s, 2N-H), 10.30(s, 2N-H), 8.55 (d, J = 8 Hz, 2H, pyridine ring protons), 8.40 (d, J = 8 Hz, 2H, pyridine ring protons), 8.3 (d, $J_1 = 8$ Hz, 2H, imine protons), 8.11 (t, $J_1 = J_2 = 8$ Hz, 2H, pyridine ring proton), 7.55 (d, $J_1 = 8$ Hz, 4H, benzene ring protons), 7.40 (t, $J_1 = J_2 = 8$ Hz, 4H, benzene ring protons), 7.25 (t, $J_1 = J_2 = 8$ Hz, 2H, benzene ring protons). ^{13}C-NMR (DMSO-d$_6$, 100 M Hz, δ ppm): δ_C = 176, 155, 153, 143, 139, 138, 128, 126.6, 126.1, 121.5, 121.4. LC-MS (m/z): 533 (M + Na), 551(M + K) (100%).
2g	IR, v (cm^{-1}): 3264, 3120 (NH, stretching), 2946 (CH, stretching), 1506 (C=N, imine), 1556 (C=C, aromatic), 1042 (C=S). ^{1}H-NMR (DMSO-d$_6$, 400 M Hz, δ ppm): δ_H = 11.80 (s, 2N-H), 10.10 (s, 2N-H), 8.48 (d, J = 8 Hz, 2H, pyridine ring protons), 8.33 (d, J = 8 Hz, 2H, imine protons), 8.14 (d, J = 8 Hz, 2H, pyridine ring protons), 8.00 (t, $J_1 = J_2 = 8$ Hz, 2H, pyridine ring protons), 5.10 (s, broad, 4H, NH$_2$ group). ^{13}C-NMR (DMSO-d$_6$, 100 M Hz, δ ppm): δ_C = 176, 154, 153, 142, 138, 126.2, 121. LC-MS (m/z): 389 (M+H).
2h	IR, v (cm^{-1}): 3360, 3053 (NH, stretching), 2920 (CH, stretching), 1562 (C=C, aromatic). ^{1}H-NMR (DMSO-d$_6$, 400 M Hz, δ ppm): δ_H = 8.52 (d, J = 4 Hz, 1H, pyridine ring proton), 8.38 - 8.5 (m, 2H, thiazolyl C-H proton), 8.32 (s, 1H, thiazolyl C-H proton), 8.30 - 8.08 (m, 3H, pyridine ring and thiazolyl C-H proton protons), 7.50 (d, J = 4 Hz, 1H, benzene ring proton), 7.65 - 7.40 (m, 3H, benzene ring protons),7.25 (s, broad, 1 N-H thiazolyl ring proton), 7.10 (d, J = 4 Hz, 1H, benzene ring proton), 6.90 (t, $J_1 = J_2 = 4$ Hz, 1H, benzene ring proton), 6.75 (d, J = 4 Hz, 1H, benzene ring proton), 6.67 (t, $J_1 = J_2 = 4$ Hz, 1H, benzene ring proton), 6.5 (s, 1H, N-H proton). ^{13}C-NMR (DMSO-d$_6$, 100 M Hz, δppm): δ_C = 169, 162, 155, 154, 153, 150, 148, 139.7, 139.3, 135, 127, 126.6, 126.0, 124, 123.8, 123.1, 121.7, 121.2, 121.1, 120, 119, 109. LC-MS (m/z): 449 (M + Na).

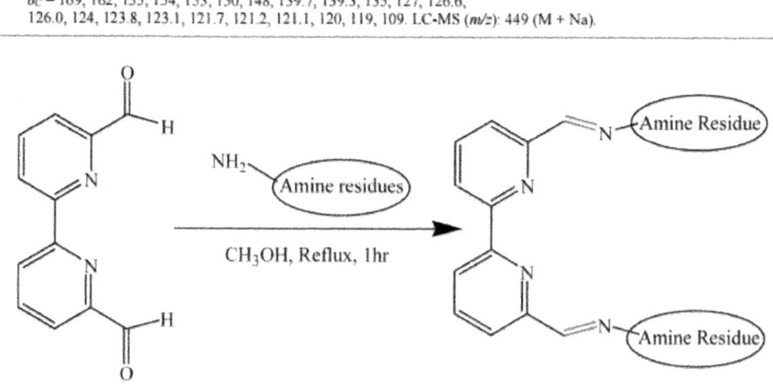

1a → 2a-2h

Scheme 2: Synthesis of Schiff Bases from 1a with O, N, S and F containing amines. 2a, amine: Thiosemicarbazide (Y 70%); 2b, amine: 4-Ethyl-3-thiosemicarbazide (Y

75%); 2c, amine: 4,4-Di- methyl-3-thiosemicarbazide (Y 75%); 2d, amine: 4 with S-Benzyldithiosemicarbazide (Y 80%); 2e, amine: (Trifluromethyl) phenylhydrazine, (Y 80%); 2f, amine: 4-Phenyl-3-thiosemicarbazide (Y 80%); 2g, amine: Thiocarbazide (Y 70%); 2h, amine: 2-Aminothiophenol (Y 65%).

RESULTS AND DISCUSSIONS

The synthesis of novel Schiff Bases were outlined in Scheme 3. All compounds were obtained from 6, 6'-diformyl-2, 2'-bipyridyl and confirmed by the help of infrared, ^{1}H-NMR, ^{13}C-NMR and Massspectrometry. In IR spectrum of 2a, bands appeared at 3433 cm^{-1}, 3240 cm^{-1} and 3154 cm^{-1} were identified as stretching frequencies for the presence of NH and NH$_2$ group. The IR spectrum of 2a revealed C=N stretching band of imine group at 1522 cm^{-1}. Band at 1595 cm^{-1} was for aromatic C=C bond. Band at 1114 cm^{-1} was due to the presence of C=S bond in amine residue.^{1}H-NMR spectrum of compound 2a showed peak at 11.77 ppm (highly deshielded) as singlet for the presence of secondary amine >N-H which confirms the formation of imine C=N bond. 6 pyridyl protons appeared between 8.45 - 8.30 ppm as multiplet. Total number of ^{13}C-NMR peaks were 7 in which peak at 178 ppm was for the presence of C=S group. Other peaks appeared between 154 and 121. In LC-MS, the molecular ion as base peak appeared at m/z 381 (M+Na).

IR spectrum of compound 2b showed two bands 3314 cm^{-1} and 3155 cm^{-1} due to NH group. Bands at 2960 cm^{-1} and 2810 cm^{-1} stretching frequencies were for >CH group. The imine C=N bond showed IR band at 1515 cm^{-1} whereas 1573 cm^{-1} for aromatic C=C bond. Thiol (C=S) band appeared at 1076 cm^{-1}. In ^{1}H-NMR for the compound 2b, two singlet peaks appeared at 11.80 ppm and 8.20 ppm due to 4 N-H groups. 2 pyridine protons showed triplet at 8.79 ppm with coupling constant value of 4 Hz. 4 pyridine ring and 2 imine protons showed peaks as multiplet between 8.40 - 8.3 ppm. Other 2 pyridine ring protons showed triplet at 8.03 ppm with coupling constant, J_1 = J_2 = 8 Hz. 4 methylene protons appeared as multiplet at 3.65 ppm whereas 6 methyl protons at 1.2 ppm with coupling constant of 7 Hz. 9 peaks were observed for compound 2b in ^{13}C-NMR spectrum with 177 ppm for thiol carbon and remaining peaks appeared for pyridine and aliphatic carbons. Molecular ion peak appeared at m/z 437 (M + Na) and M+H ion peaks at m/z 415 for compound 2b.

Compound 2c, IR spectrum showed bands at 3415 cm^{-1}, 3345 cm^{-1} and 3220 cm^{-1} for the group N-H's stretching frequencies. >CH group's stretching bands appeared at 2910 cm^{-1} and 2870 cm^{-1}. Imine band appeared at 1470 cm^{-1}. The bands at 1557 cm^{-1} and 911 cm^{-1} were for C=C bond (aromatic) and thiol bond C=S, respectively. In compound 2c, the indication of imine bond formation in ¹H-NMR spectrum was found at 11.40 ppm due to two N-H protons which is in highly deshielded region. 6 pyridine ring protons and 2 imine proton peaks appeared between 8.40 - 8.3 ppm and 8.10 - 7.90 ppm as multiplet. 12 methyl protons peak appeared at 3.35 ppm as singlet. Number of peaks in ¹³C-NMR spectrum for compound 2c were 8, including thiol carbon at 180 ppm. The molecular ion peak for compound 2c appeared at m/z 437 (M + Na).

Compound 2d in IR spectrum showed only 1 band at 3115 cm⁻¹ for the group NH. The stretching frequencies at 2988 cm⁻¹ and 2849 cm⁻¹showed the presence of =CH groups. The imine and thiol bands appeared at 1510 cm⁻¹ and 1041 cm⁻¹ whereas aromatic C=C bond frequency appeared at 1580 cm⁻¹. Peak at 13.70 ppm in ¹H-NMR spectrum appeared for two N-H groups as singlet which is highly deshielded region and implies the imine bond formation. Two doublets at 8.40 ppm and 7.98 ppm with J = 8 Hz appeared for 4 pyridine ring protons. The imine proton peak appeared at 8.35 ppm as singlet for compound 2d. Another peak at 8.03 ppm as triplet showed for 2 pyridine ring protons. 8 benzene protons appeared at 7.30 ppm and 7.35 ppm as triplet with coupling constant $J_1 = J_2 = 8$ Hz where as other two at 7.45 ppm as doublet with J = 8 Hz. There were twelve

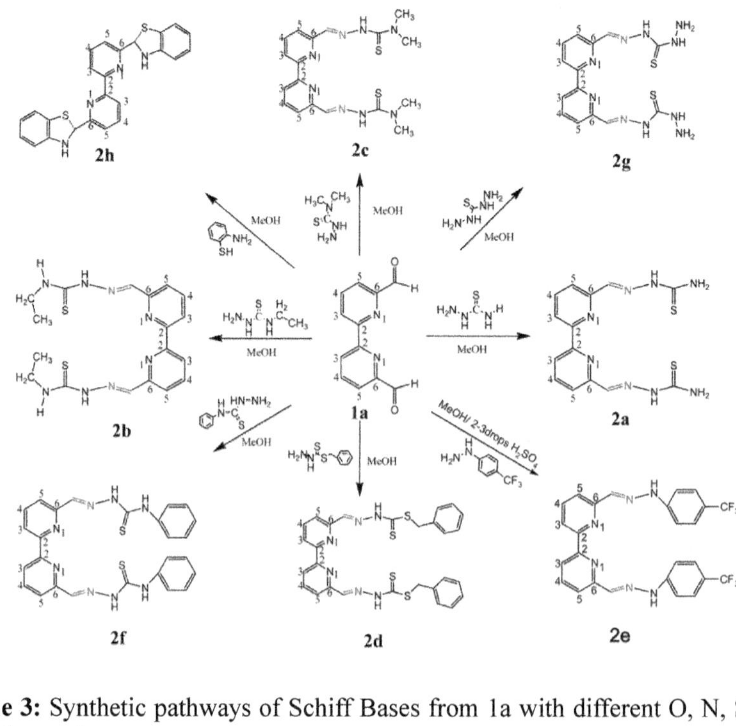

Scheme 3: Synthetic pathways of Schiff Bases from 1a with different O, N, S and F containing amines.

^{13}C-NMR peaks appeared in 2d compound spectrum with thiol peak at 198 ppm. Other peaks were for aliphatic, pyridine and benzene rings. The molecular ion peak for compound 2d was shown at m/z 594.83 in LC-MS.

There were 2 IR bands at 3334 cm^{-1} and 3089 cm^{-1} for the presence of NH for compound 2e. 1 stretching frequency observed at 2940 cm^{-1} due to =CH. For imine bond and aromatic C=C bond, bands appeared at 1505 cm^{-1} and 1615 cm^{-1} respectively. A singlet in ^{1}H-NMR spectrum in highly deshielded region at 11.20 ppm of two N-H groups helps to determine the formation of imine bond. 2 doublets appeared at 7.60 ppm and 7.30 ppm of 8 benzene ring protons with J value 6 Hz for compound 2e. Between 8.11 - 8.05 ppm, peak of 4 of pyridine ring protons appeared as multiplet whereas 2 pyridine ring protons at 8.35 ppm appeared as doublet with J = 6 Hz and remaining 2 pyridine ring protons appeared at 8.02 ppm as triplet with J$_1$ = J$_2$ = 6 Hz. ^{13}C-NMR spectrum for compound 2e showed 11 peaks all with pyridine, benzene and aliphatic regions. In LC-MS spectrum, the molecular ion peak for the compound 2e appeared at m/z: 529.01 (M + H).

There were 3 IR stretching frequencies 3290 cm^{-1}, 3149 cm^{-1} and 3000 cm^{-1} appeared for NH group for compound 2f. 1 stretching frequency at 2940 cm^{-1} appeared for =CH group whereas imine, aromatic and thiol bond frequencies were observed at 1550 cm^{-1} , 1595 cm^{-1} and 1163 cm^{-1} respectively. For compound 2f, a singlet in highly deshielded region peat appeared at 12.20 ppm due to 2N-Hgroups which helps to identify the formation of imine bond. Peak at 10.30 ppm appeared for 2N-H groups as singlet. 4 doublets appeared at 8.55 ppm, 8.40 ppm, 8.3 ppm and 7.55 ppm with coupling constant value 8 Hz in which first 2 of them were pyridine ring protons, imine protons and benzene ring protons respectively. 3 triplets appeared at 8.11 ppm, 7.40 and 7.25 ppm with coupling constant value J$_1$ = J$_2$ = 8 Hz for 2 pyridine ring protons, 4 benzene ring protons and 2 benzene ring protons respectively. In total 11 peaks appeared in ^{13}C-NMR spectrum for the compound 2f with all in the pyridine carbon and benzene carbon regions. LC-MS spectrum showed molecular ion as base peak at m/z: 551 (M + K).

2 stretching frequencies observed in IR spectrum for the compound 2g at 3264 and 3120 cm^{-1} due to the presence of NH group. 2 stretching frequencies at 2946 cm^{-1} and 1556 appeared for =C-H and aromatic C=C groups. The imine and thiol bands appeared at 1505 cm^{-1} and 1042 cm^{-1}respectively. Compound 2g in ^{1}H-NMR spectrum showed peak at 11.80 ppm as singlet due to 2N-H groups which is in highly deshielded region and thus indicated the imine bond formation. Other 2N-H groups' singlet peaks appeared at 10.10 ppm which was also relatively deshielded region. 3 doublet peaks observed at 8.48 ppm, 8.33 ppm and 8.14 ppm for 6 pyridine ring protons and imine protons with the

value of coupling constant 8 Hz whereas 1 triplet peak appeared at 8.00 ppm for 2 pyridine protons with J value 8 Hz equally. There was a broad singlet observed at 5.10 ppm due to presence of 2 NH_2 groups in compound 2g. In ^{13}C-NMR spectrum for compound 2g, 7 peaks were observed with thiol carbon at 176 ppm. Mass analyses showed base peak at 389 (M + H).

For compound 2h, 2 bands appeared for N-H group at 3360 cm^{-1} and 3053 cm^{-1} due to stretching vibrations. 1 band appeared at 2920 for the presence of =CH group. Aromatic C=C bond frequency appeared at 1562 cm^{-1}. Compound 2h showed twisted geometry that appeared in both ^{1}H-NMR and ^{13}C-NMR spectrums. In ^{1}H-NMR spectrum of compound 2h, 3 singlet peak appeared: at 8.32 ppm for 1 thiazolyl ring C-H proton, at 6.5 ppm as singlet for 1 thiazolyl ring N-H proton and 7.25 ppm for another thiazolyl ring N-H proton. 4 doublet peaks observed at 8.52 ppm, 7.70 ppm, 7.05 ppm and 6.75 ppm for 1 pyridine ring proton and later 3 peaks for benzene ring protons. 2 triplet peaks appeared at 6.90 ppm and 6.67 ppm due to 2 benzene ring protons. 2 multiplet peaks were observed between 8.38 - 8.5 ppm for 2 pyridine ring protons and between 8.30 - 8.08 ppm for 3 pyridine protons and 1 thiazolyl C-H protons. 1 multiplet peak between 7.65 - 7.50 ppm due to 3 benzene ring protons. In ^{1}H-NMR spectrum of compound 2h, all coupling constant value were observed at J = 4 Hz. In ^{13}C-NMR spectrum, 22 peaks of carbon were observed due to twisted geometry of the product which made 2h carbons diastereotopic in nature. All carbons appeared in pyridine ring, benzene ring regions. Molecular ion peak appeared in LC-MS spectrum at m/z: 449 (M + Na).

CONCLUSION

Eight novel Schiff bases of 6,6'-diformyl-2,2'-Bipyridyl with O, S, N and F containing amines have been successfully synthesized. Conc. sulfuric addition has been found to significantly enhance the yields of the products. In some cases, sulfuric acid was not added to avoid salt formation with nitrogen and amine group. However, it was observed that the yield increased significantly when the reaction was carried out under mild acidic conditions. This is due to the fact that protonation of the carbonyl group (C=O) enhances the nucleophilic attack-NH_2 group of the amine.

ACKNOWLEDGEMENTS

We thank the Department of Chemistry at Tennessee State University for providing the necessary support to carry out the research. We also thank the Department of Education, Title III funds for providing instrumental support.

CITE THIS PAPER

Md.Razzak,Mohammad R.Karim,Md. RejaulHoq,Aminul H.Mirza, (2015) New Schiff Bases from 6,6'-Diformyl-2,2'-Bipyridyl with Amines Containing O, S, N and F: Synthesis and Characterization.*International Journal of Organic Chemistry*,**05**,264-270. doi: 10.4236/ijoc.2015.54026

REFERENCES

1. Jaman, Z., Karim, M.R., Siddiquee, T.A., Mirza, A.H. and Ali, M.A. (2013) Synthesis of 5-Substituted 2, 9-Dimethyl-1,10-Phenanthroline Dialdehydes and Their Schiff Bases with Sulfur-Containing Amines. International Journal of Organic Chemistry, 3, 214-219.http://dx.doi.org/10.4236/ijoc.2013.33029

2. Siebert, R., Winter, A., Dietzek, B., Schubert, U.S. and Popp, J. (2010) Dual Emission from Highly Conjugated 2,2':6':2"-Terpyridine Complexes—A Potential Route to White Emitters. Macromolecular Rapid Communications, 31, 883-888. http://dx.doi.org/10.1002/marc.200900894

3. Siebert, R., Winter, A., Schubert, U.S., Dietzek, B. and Popp, J. (2010) Excited-State Planarization as Free Barrierless Motion in a π-Conjugated Terpyridine. The Journal of Physical Chemistry C, 114, 6841-6848. http://dx.doi.org/10.1021/jp100313x

4. Siebert, R., Winter, A., Schubert, U.S., Dietzek, B. and Popp, J. (2011) The Molecular Mechanism of Dual Emission in Terpyridine Transition Metal Complexes—Ultrafast Investigations of Photoinduced Dynamics. Physical Chemistry Chemical Physics, 13, 1606-1617. http://dx.doi.org/10.1039/C0CP01134G

5. Binnemans, K., Lenaerts, P., Driesen, K. and Gorller-Walrand, C. (2004) A Luminescent Tris(2-thenoyltrifluoroace- tonato)europium(III) Complex Covalently Linked to a 1,10-Phenanthroline-Functionalised Sol-Gel Glass. Journal of Materials Chemistry, 14, 191-195. http://dx.doi.org/10.1039/b311128h

6. Larsson, K. and Ohrstrom, L. (2004) X-Ray and NMR Study of the Fate of the Co(1,10-phenanthroline-5,6-diketone)33+ Ion in Aqueous Solution: Supramolecular Motifs in the Packing of 1,10-Phenanthroline-5,6-Diketone and 1,10-Phe- nanthroline-5,6-Diol Complexes. Inorganica Chimica Acta, 357, 657-664. http://dx.doi.org/10.1016/j.ica.2003.07.001

7. Steed, J.W. and Atwood, J.L. (2000) Supramolecular Chemistry. Wiley, Chichester.

8. Chow, C.S. and Bogdan, F.M. (1997) A Structural Basis for RNA—Ligand Interactions. Chemical Reviews, 97, 1489-1514. http://dx.doi.org/10.1021/cr960415w

9. Sammes, P.G. and Yahioglu, G. (1994) 1,10-Phenanthroline: A Versatile Ligand. Chemical Society Reviews, 23, 327-334. http://dx.doi.org/10.1039/cs9942300327

10. Balzani, V., Juris, A., Venturi, M., Campagna, S. and Serroni, S. (1996) Luminescent and Redox-Active PolynuclearTransition Metal Complexes. Chemical Reviews, 96, 759-834. http://dx.doi.org/10.1021/cr941154y

11. Daniel, S. and Gnana Raj, G.A. (2013) Photoinduced Electron-Transfer Reactions of Tris(4,4›-dinonyl-2,2›-bipyridyl) Ruthenium(II) Cation with Phenolate Ions in Aqueous Acetonitrile. Journal of Chemical and Pharmaceutical Research, 5, 220-227.

12. Smith, N.A. and Sadler, P.J. (2013) Photoactivatable Metal Complexes: From Theory to Applications in Biotechnology and Medicine. Philosophical Transactions of the Royal Society A, 373, Article ID: 20120519. http://dx.doi.org/10.1098/rsta.2012.0519

13. Monat, J.E., Rodriguez, J.H. and McCusker, J.K. (2002) Ground- and Excited-State Electronic Structures of the Solar Cell Sensitizer Bis(4,4›-dicarboxylato-2,2›bipyridine)bis(isothiocyanato)ruthenium(II). Journal of Physical Chemistry A, 106, 7399-7406. http://dx.doi.org/10.1021/jp020927g

14. Gomathi, V., Selvameena, R., Subbalakshmi, R. and Valarmathy, G. (2013) Synthesis, Spectral Characterization and Antimicrobial Screening of Mn(ll) and Zn(ll) Complexes Derived from (E)-1-((p-tolylimino) methyl)naphthalene-2-ol. Oriental Journal of Chemistry, 29, 533-538. http://dx.doi.org/10.13005/ojc/290220

15. Wang, J., Onions, S., Pilkington, M., Stoeckli-Evans, H., Halfpenny, J.C. and Wallis, J.D. (2007) Metal Catalyzed Rearrangement of a 2,2-Bipyridine Schiff-Base Ligand to a Quaterpyridine-Type Complex. Chemical Communications, 2007, 3628-3630. http://dx.doi.org/10.1039/b705555b

16. Hodacova, J. and Budesmsky, M. (2007) New Synthetic Path to 2,2›-Bipyridine-5,5›-Dicarbaldehyde and Its Use in the [3+3] Cyclocondensation with Trans-1,2-Diaminocyclohexane. Organic Letters, 9, 5641-5643. http://dx.doi.org/10.1021/ol702612t

17. Newkome, G.R. and Lee, H.-W. (1983) 18[(2,6) Pyridin6coronand-6: "Sex-ipyridine". Journal of the American Chemical Society, 105, 5956-5957. http://dx.doi.org/10.1021/ja00356a061

18. Mughrabi, F.F., Hashim, H., Ameen, M., Khaledi, H. and Ali, H.M. (2011) Cytoprotective Effect of Benzyl N'-(indol- 3-ylmethylidene)-Hydrazinecarbodithioate against Ethanol-Induced Gastric Mucosal Injury in Rats. African Journal of Pure and Applied Chemistry, 5, 34-42. |http://www.academicjournals.org/AJPAC.

Chapter 4

UTILITY OF STYRYLPYRAZOLOFORMIMIDATE IN THE SYNTHESIS OF FUSED HETEROCYCLIC COMPOUNDS

Hamdi M. Hassaneen[*], Zakaria Ahmed Gomaa

Department of Chemistry, Faculty of Science, Cairo University, Giza, Egypt

ABSTRACT

Refluxing of (E)-5-amino-1-phenyl-3-styryl-1H-pyrazole-4-carbonitrile 2 with triethylorthoformate in acetic anhydride afforded the corresponding formimidate 3. Treatment of 3 with hydrazine hydrate in ethanol afforded amino imino compound 4. Reaction of 4 with diethyl dicarbonate at reflux gave (E)-7-phenyl-9-styryl-7H-pyrazolo[4,3-e][1,2,4]triazolo[1,5-c]pyrimidine 7. Refluxing of 4 with hydrazine hydrate afforded (E)-4-hydrazinyl-1-phenyl-3-styryl-1H-pyrazolo[3,4-d] pyri- midine 8. Treatment of the latter compound 8 with aldehydes in boiling ethanol in the presence of acetic acid afforded the corresponding hydrazone 10. Oxidative cyclization of the hydrazone 10 led to the formation of pyrazolo[4,3-e][1,2,4]triazolo[4,3-c]pyrimidine 11. The latter products re- arranged to pyrazolo[4,3-e][1,2,4]triazolo[1,5-c]pyrimidines 13. The structures of the new pro- ducts were established on the basis of elemental analysis and spectral data.

INTRODUCTION

The chemistry of hydrazonoyl halides has attracted the interest of many research groups as they have proved to be useful organic synthesis [1] - [10] . In continuation of our long standing interest for the utility of nitrilimines derived from hydrazonoyl halides in the synthesis of heterocycles [11] - [14] , we are interested in (Z)-N'-phenyl- cinnamohydrazonoyl chloride 1 to study the effect of C=C double bond on the cycloaddition reactions [15] - [17] . We wish to report herein a simple and convenient route for the synthesis of pyrazolo[3,4-d]pyrimidine, pyrazolo[4,3-e][1,2,4]triazolo[4,3-c]pyrimidines and its isomeric pyrazolo[4,3-e][1,2,4]triazolo[1,5-c]pyrimidine derivative via Dimroth rearrangement. Such compounds have been used as a new pharmacological test for characterization of human A_3 adenosine receptors [18] - [20] .

RESULTS AND DISCUSSION

Compound (E)-5-amino-1-phenyl-3-styryl-1H-pyrazole-4-carbonitrile 2 was prepared from our laboratory via reaction of (Z)-N'-phenylcinnamohydrazonoyl chloride 1 with malononitrile in ethanolic sodium ethoxide solution (Scheme 1) [21] . Refluxing of compound 2 with triethylorthoformate in acetic anhydride afforded ethyl N-(4-cyano-1-phenyl-3-((E)-styryl)-1H-pyrazol-5-yl)formimidate 3 (Scheme 1). The structure of compound 3 was established on the basis of elemental analysis and spectral data. The IR spectrum of 3 revealed the absence of amino group, while it showed a characteristic band at υ 2215 cm^{-1} assignable to cyano group. Its ^{1}H NMR data showed signals at δ, 1.29 (t, 3H, CH$_3$), 4.32 (q, 2H, CH$_2$), 7.15 - 7.69 (m, 12H, Ar H), and 8.60 (s, 1H, NCH). Also, its ^{13}C-NMR spectrum showed 17 carbon atoms. Moreover, the mass spectrum showed molecular ion peak as a base peak at m/z 342 (100%). Reaction of 3 with hydrazine hydrate in ethanol at room temperature yielded a product 4 which analyzed correctly for $C_{19}H_{16}N_6$ (Scheme 1). The IR spectrum of 4 showed the absence of cyano group and it showed bands at υ 3351, 3309, 3177 cm^{-1} assignable to amino and imino groups. Also, the mass spectrum revealed a base peak at m/z 328 (100%) corresponding to its molecular ion peak. On the basis of elemental analysis and spectral data, the product is (E)-4-imino-1-phenyl-3-styryl-1H-pyrazolo [3,4-d]pyrimidin-5(4H)-amine 4.

When compound 4 was refluxed with diethyl dicarbonate a single product was obtained, its mass spectrum and elemental analysis are consistent with the molecular formula $C_{23}H_{18}N_6O_2$ (Scheme 2). Two possible structures were proposed for the isolated product 5a and 5b. The identity of the isolated product was confirmed to be (E)-ethyl-7-phenyl-9-styryl-7H-pyrazolo[4,3-e][1,2,4] triazolo[1,5-c]pyrimidine-2-carboxylate 5b. Thus, sapo- nification of the reaction product obtained 5b gave the intermediate acid 6 which decarboxylated to a pro- duct identical to all respects (m.p., mixed m.p., IR) with (E)-7-phenyl-9-styryl-7H-pyrazolo[4,3-e][1,2,4]triazo- lo[1,5-c]pyrimidine 7. The latter product 7 was also confirmed via its alternative synthesis by treatment of compound 4 with triethylorthoformate or formic acid. Structure of 5a was accordingly discarded. In addition, structure 5b was further substantiated by IR and ^{1}H NMR spectra. Its IR spectrum exhibits a carbonyl band at υ 1743 cm^{-1} and ^{1}H NMR spectrum showed signals at: δ 1.42 (t, J = 7 Hz, 3H), 4.49 (q, J = 7 Hz, 2H), 7.37-8.78 (m, J = 7 Hz, 12H), and 9.84 (s, 1H, pyrimidine-CH).

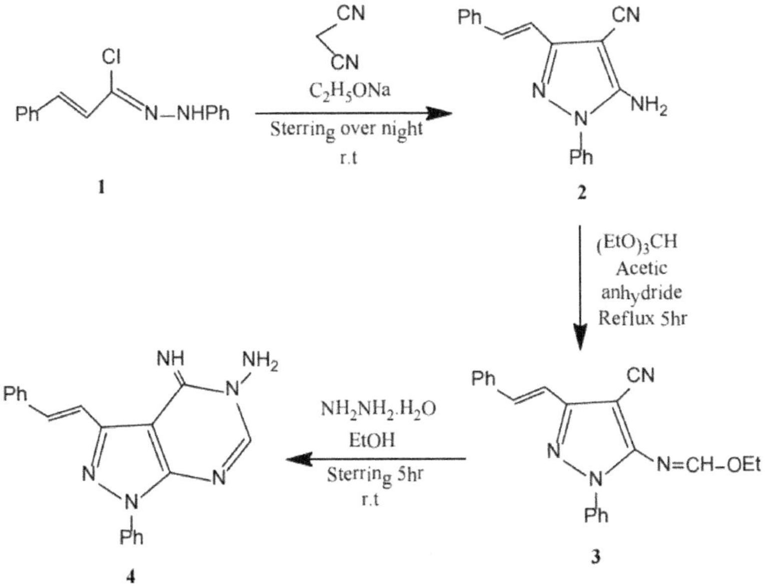

Scheme 1: Synthesis of formimidate 3 and amino imino compound 4.

Refluxing of compound 4 with hydrazine hydrate in ethanol, gave (E)-4-hydrazinyl-1-phenyl-3-styryl-1H- pyrazolo[3,4-d]pyrimidine 8, via Dimroth type rearrangement, which has not been reported hitherto (Scheme 3 and Scheme 4). This is consistent with a similar rearrangement that was reported recently [22] . The structure of 8 was confirmed by elemental and spectral data (see experimental) and its reaction described below.

Thus, treatment of hydrazine derivative 8 with the appropriate aldehydes 9a-i in refluxing ethanol in the presence of acetic acid led to the formation of the new condensation products, 4-(2-arylhydrazinyl)-1-phenyl -3-styryl-1H-pyrazolo[3,4-d]pyrimidine 10a-i. The structures of 10a-i were confirmed by their elemental analysis and spectral data. For example their IR spectra showed the characteristic band for NH at υ 3199 - 3352 cm^{-1}. Also, their ^{1}H NMR spectra revealed in each case, a signal in the region 11.97 - 12.13 assignable to NH proton which disappeared upon shaking its DMSO solution with D_2O.

Oxidative cyclization of hydrazone 10a-h led to the formation of pyrazolo[4,3-e][1,2,4]triazolo[4,3-c]pyri- midine derivatives 11a-h (Scheme 3 and Scheme 5). Thus, stirring of 10a-h with 4 equivalent of Fe(III) chloride in ethanol overnight gave, in each case, a single product as evidenced by TLC analysis. Mass spectra revealed that, each product has 2 hydrogen atoms less than that of the respective hydrazone. Also, IR and ^{1}H NMR revealed the absence of NH band and CH=N proton, respectively.

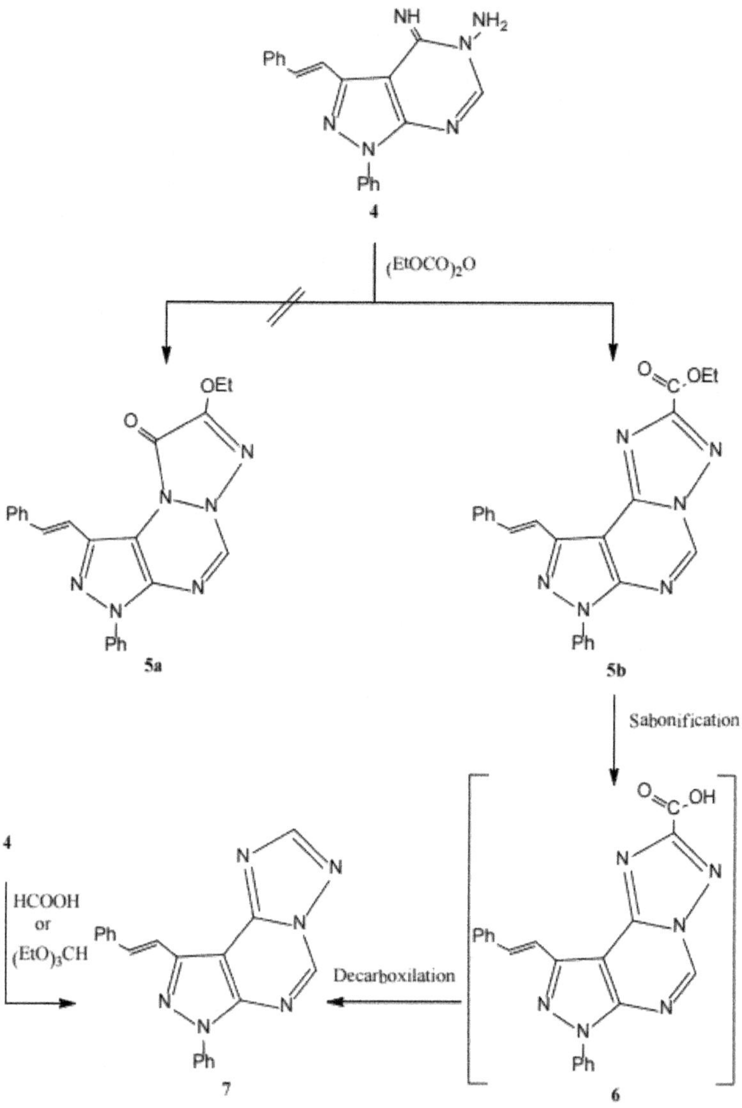

Scheme 2: Synthesis of pyrazolotriazolopyrimidine 7.

Compounds 11a,e were isomerized to the thermodynamically more stable pyrazolo[4,3-e][1,2,4]triazo- lo[1,5-c]pyrimidine derivatives 13a,e through tandem ring opening and ring closure reactions via heating of 11a,e in ethanol in the presence of sodium acetate (Scheme 3). This rearrangement is consistent with those reported in earlier reports [23] . The structures of 13a,e were established by elemental and spectral analysis (see experimental). Also, the structures of 13a,e were confirmed via their alternative synthesis. Thus, treatment of 4 with acid chlorides 12a,e in refluxing pyridine gave a products identical in all respects (m.p., mixed m.p., IR and ^{1}H NMR spectra) with those obtained above from base-catalyzed rearrangement of 11a,e (Scheme 3). Also,

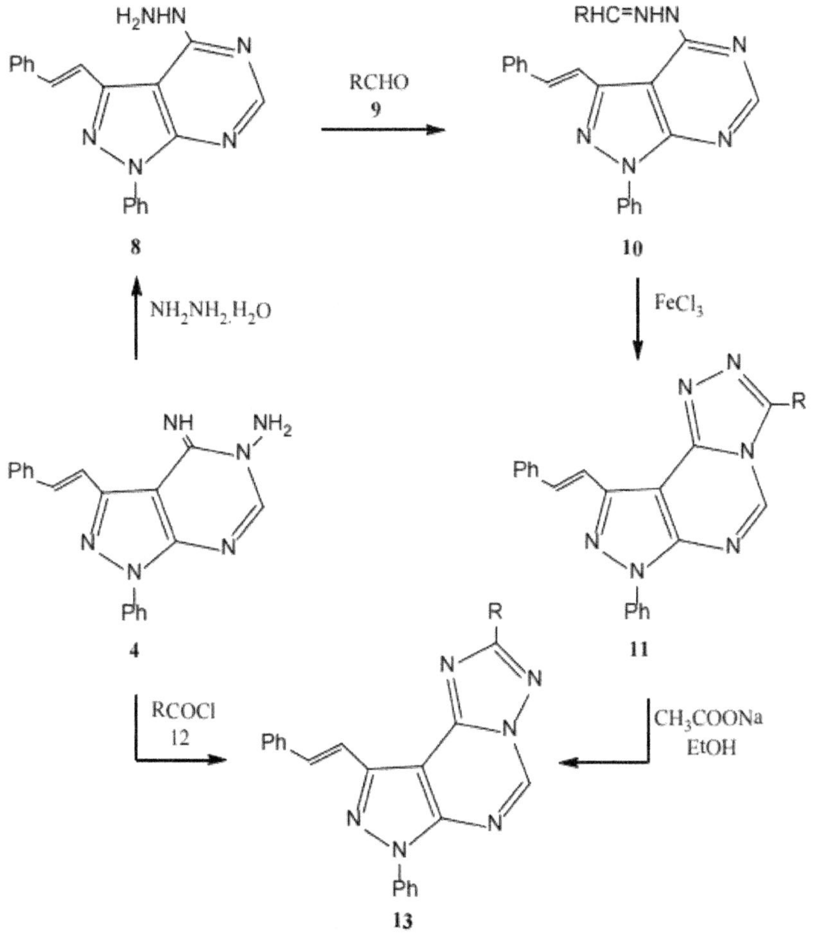

R: a) = Ph; b) = 3-CH$_3$C$_6$H$_4$; c) = 4-(CH$_3$)$_2$NC$_6$H$_4$, d) = 4-OHC$_6$H$_4$; e) = 4-ClC$_6$H$_4$; f) = 4-FC$_6$H$_4$; g) = 2-thienyl; h) = 4-pyridyl; i) = 3,4-methylenedioxyphenyl; j) = CH$_3$; K) = 4-BrC$_6$H$_4$; l) 2-furyl.

Scheme 3: Synthesis of hydrazinylpyrazolo pyrimidine 8, arylhydrazinylpyrazolopy-rimidine 10 and pyrazolotriazolopyrimidine derivatives 11, 13.

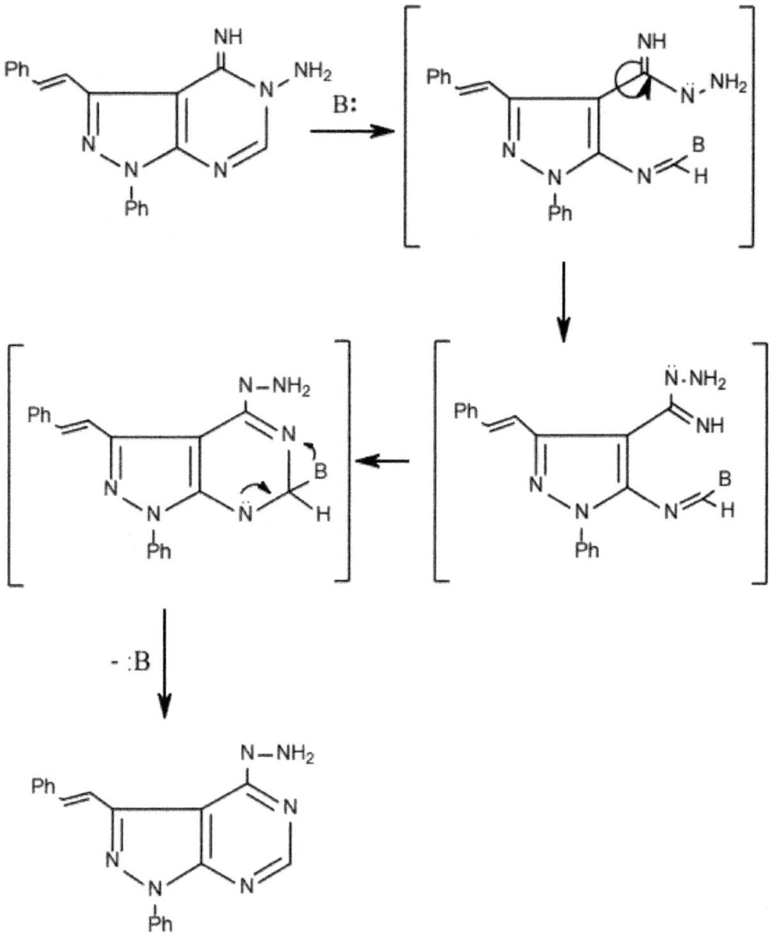

Scheme 4: Synthesis of (E)-4-hydrazinyl-1-phenyl-3-styryl-1H-pyrazolo [3,4-d]py-rimidine.

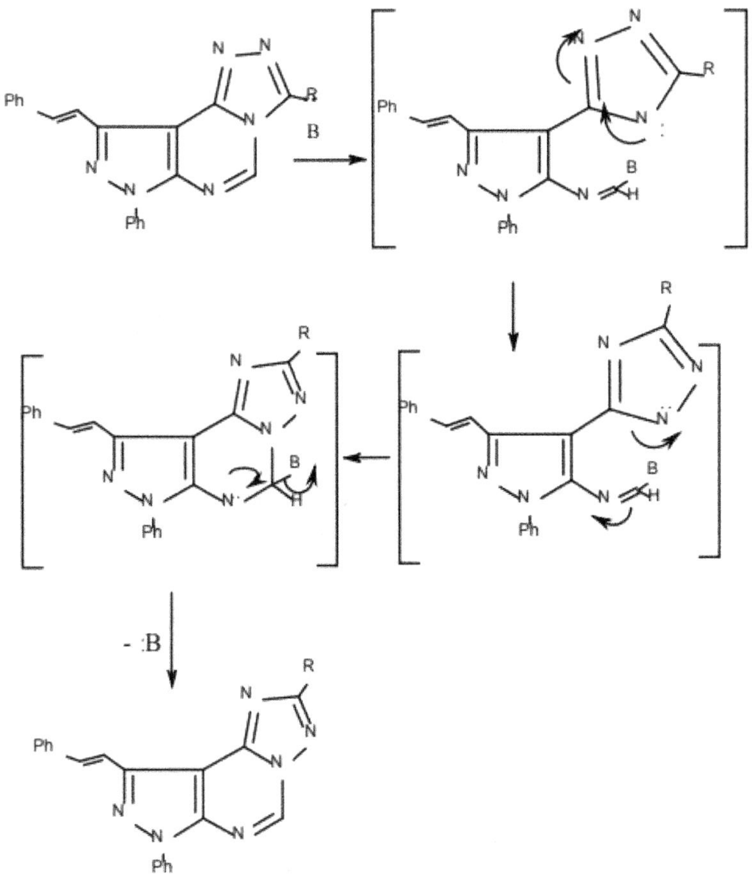

Scheme 5: Synthesis of (E)-2-alkyl-7-phenyl-9-styryl-7H-pyrazolo[4,3-e][1,2,4] triazolo[1,5-c]pyrimidine.

compound 4 reacted with 12j,k,l to give the corresponding 13j,k,l. The latter compounds 13j,k,l were confirmed by elemental and spectral analysis.

EXPERIMENTAL

General

All melting points were determined on an electrothermal GallenKamp melting point apparatus and are uncorrected. The IR spectra were recorded as KBr Pellets on a Jasco FTIR-460 plus Fourier transform infrared spectrophotometer. ^{1}H and ^{13}C NMR spectra were recorded at (300 MHz) and (75 MHz) respectively on Varian EM-300 MHz spectrometer. Chemical shifts (δ) are given from

TMS (ppm) as internal standard for ^{1}H NMR and ^{13}C NMR. Mass spectra were recorded on AEI MS 30 mass spectrometer operating at 70 eV. The elemental analyses were performed at the Microanalytical Center of Cairo University. Compound 2 was prepared as previously described [18] .

Preparation of Ethyl N-(4-Cyano-1-Phenyl-3-((E)-Styryl)-1H-Pyrazol-5-yl)Formimidate 3

To a solution of the compound (E)-5-amino-1-phenyl-3-styryl-1H-pyrazole-4-carbonitrile 2 (1.43 g, 5 mmol) in acetic anhydride (5 mL), triethylorthoformate (0.74 g, 5 mmol) was added. The reaction mixture was refluxed for 5 h and the solvent was evaporated under reduced pressure. The solid product was collected and crystallized from acetonitrile to afford the compound 3. Yellow crystals; m.p: 170°C - 171°C; yield (82%); IR (KBr): υ = 2215 (CN) cm^{-1}; ^{1}H NMR (DMSO-d6): δ = 1.29 (t, 3H, CH$_2$CH$_3$), 4.32 (q, 2H, CH$_3$CH$_2$), 7.15 - 7.69 (m, 12H, Ar H), 8.60 (s, 1H, NCH); ^{13}C NMR (DMSO-d6): δ = 13.74, 64.07, 79.46, 114.46, 117.78, 123.78, 126.88, 128.00, 128.74, 128.85, 128.97, 132.88, 135.58, 137.46, 149.57, 151.51, 162.46. MS: m/z (%) = 342 (M$^+$, 100), 313 (31), 285 (45), 77 (41). Anal. for C$_{21}$H$_{18}$N$_4$O: Calcd. C, 73.67; H, 5.30; N, 16.36. found C, 73.31; H, 5.03; N, 16.11.

Preparation of (E)-4-Imino-1-Phenyl-3-Styryl-1H-Pyrazolo[3,4-d] Pyrimidin-5(4H)-Amine 4

To a solution of the compound 3 (17.1 g, 50 mmol) in ethanol (250 mL), hydrazine hydrate (2.5 mL, 50 mmol) was added. The reaction mixture was stirred for 5 h at room temperature; the solid product was collected and crystallized from dioxane to afford the compound 4. Yellow crystals; m.p. 200°C - 202°C; yield (91%); IR (KBr): υ = 3351 & 3309 (NH$_2$), 3177 (NH) cm^{-1}. MS: m/z (%) = 328 (M$^+$, 100), 313 (29), 77 (30). Anal. for C$_{19}$H$_{16}$N$_6$: Calcd. C, 69.50; H, 4.91; N, 25.59. found C, 69.07; H, 4.72; N, 25.39.

Preparation of (E)-9-Styryl-7H-Pyrazolo[4,3-e][1,2,4]Triazolo[1,5-c]Pyrimidine-2-Carboxylate Ethyl 7-Phenyl 5b

A solution of the compound 4 (1.64 g, 5 mmol) in diethyl dicarbonate (10 mL) was refluxed for 4 h. and then cooled. The solid product was filtered off, dried and finally crystallized from acetic acid to give 5b. Yellow crystals; m.p. 212°C - 214°C; yield (88%); IR (KBr): υ = 1743 (CO) cm^{-1}; ^{1}H NMR (DMSO-d6): δ = 1.42 (t, 3H, CH$_2$CH$_3$), 4.49 (q, 2H, CH$_3$CH$_2$), 7.37 - 8.78 (m, 12H, Ar H), 9.84 (s, 1H, NCH). MS: m/z (%) = 410 (M+, 96), 409 (50),

77 (100). Anal. for $C_{23}H_{18}N_6O_2$: Calcd. C, 67.31; H, 4.42; N, 20.48. found C, 67.02; H, 4.37; N, 20.25.

Preparation of (E)-7-Phenyl-9-Styryl-7H-Pyrazolo[4,3-e][1,2,4] Triazolo[1,5-c]Pyrimidine 7

A solution of the compound 4 (1.64 g, 5 mmol) in triethylorthoformate or formic acid (10 mL) was refluxed for 4 h. left to cool and the solid product was filtered, dried and finally crystallized from acetic acid to afford 7. Pale yellow crystals; m.p. 198°C - 200°C (acetic acid); yield (86%); ^{1}H NMR (DMSO-d6): δ = 6.98 - 8.56 (m, 12H, Ar H), 9.77 (s, 1H, NCH), 9.79 (s, 1H, NCH). MS: m/z (%) = 338 (M$^+$, 98), 337 (100), 77 (30). Anal. for $C_{20}H_{14}N_6$: Calcd. C, 70.99; H, 4.17; N, 24.84. found C, 70.65; H, 4.12; N, 24.61.

Preparation of (E)-4-Hydrazinyl-1-Phenyl-3-Styryl-1H-Pyrazolo[3,4-d]Pyrimidine 8

A solution of the compound 4 (16.4 g, 50 mmol) in ethanol (250 mL) and hydrazine hydrate (10 mL) was refluxed for 5 h. The solvent was evaporated; the solid product was collected, dried and finally crystallized from acetonitrile to afford 8. Yellow crystals; m.p. 202°C - 204°C; yield (90%); IR (KBr): υ = 3310 (NH), 3351 & 3309 (NH$_2$) cm^{-1}; ^{1}H NMR (DMSO-d6): δ = 4.86 (s, 2H, NH$_2$), 7.31-8.41 (m, 13H, Ar H), 9.25 (s, 1H, NH). MS: m/z (%) = (M$^+$, 38), 313 (30), 297 (80), 77 (100). Anal. for $C_{19}H_{16}N_6$: Calcd. C, 69.50; H, 4.91; N, 25.59. found C, 69.22; H, 4.81; N, 25.33.

General Method for Preparation (E)-4-(2-Arylhydrazinyl)-1-Phenyl-3-Styryl-1H-Pyrazolo[3,4-d]Pyrimidine 10a-i

To a mixture of compound 8 (1.64 g, 5 mmol) in ethanol (30 mL), the appropriate aldehyde 9a-i (5 mmol) was added. The reaction mixture was refluxed for 2 h and the solvent was evaporated. The solid product formed was collected, dried and finally crystallized from suitable solvent to afford 10a-i. 4-((Z)-2-Benzylidenehydrazinyl)-1-phenyl-3-((E)-styryl)-1H-pyrazolo[3,4-d] pyrimidine 10a:

Yellow crystals; m.p. 188°C - 189°C (acetonitrile); yield (87%); IR (KBr): υ = 3205 (NH) cm^{-1}, ^{1}H NMR (DMSO-d6): δ = 7.15 - 8.58 (m, 19H, Ar H), 12.04 (s, 1H, NH). MS: m/z (%) = 416 (M$^+$, 14), 415 (42), 313 (43), 236 (41), 77 (100). Anal. for $C_{26}H_{20}N_6$: Calcd. C, 74.98; H, 4.84; N, 20.18. found C, 74.37; H, 4.71; N, 19.73.

4-((Z)-2-(3-Methylbenzylidene)hydrazinyl)-1-phenyl-3-((E)-styryl)-1H-pyrazolo[3,4-d]pyrimidine 10b:

Yellow crystals; m.p. 192°C - 194°C (acetonitrile); yield (81%); IR (KBr): υ = 3352 (NH) cm^{-1}, ^{1}H NMR (DMSO-d6): δ = 2.36 (s, 3H, CH$_3$), 7.24-8.52 (m, 18H, Ar H), 12.01 (s, 1H, NH); ^{13}C NMR (DMSO-d6): δ = 20.83, 99.47, 119.11, 121.83, 125.20, 126.783, 128.18, 128.41, 128.76, 129.01, 130.57, 132.80, 136.52, 137.80, 138.19, 147.89, 153.98. MS: m/z (%) = 430 (M$^+$, 79), 339 (52), 312 (100), 236 (79), 77 (80). Anal. for C$_{27}$H$_{22}$N$_6$: Calcd. C, 75.33; H, 5.15; N, 19.52. found C, 75.06; H, 4.92; N, 19.36.

N,N-Dimethyl-4-((Z)-(2-(1-phenyl-3-((E)-styryl)-1H-pyrazolo[3,4-d] pyrimidin-4-yl)hydrazono)methyl)aniline 10c:

Yellow crystals; m.p. 242°C - 244°C (dimethylformamide); yield (77%); IR (KBr): υ = 3199 (NH) cm^{-1}; ^{1}H NMR (DMSO-d6): δ = 2.99 (s, 6H, N(CH$_3$)$_2$), 6.75-8.43 (m, 18H, Ar H), 11.87 (s, 1H, NH). MS: m/z (%) = 459 (M$^+$, 34), 457 (62), 312 (100), 236 (65), 145 (60), 77 (91). Anal. for C$_{28}$H$_{25}$N$_7$: Calcd. C, 73.18; H, 5.48; N, 21.34. found C, 72.55; H, 5.23; N, 20.84.

4-((Z)-(2-(1-Phenyl-3-((E)-styryl)-1H-pyrazolo[3,4-d]pyrimidin-4-yl) hydrazono)methyl)phenol 10d:

Yellow crystals; m.p. 270°C - 272°C (acetonitrile); yield (83%); IR (KBr): υ = 3324 (NH) cm^{-1}; ^{1}H NMR (DMSO-d6): δ = 6.99 - 8.68 (m, 18H, Ar H), 11.98 (s, 1H, NH), 13.02 (s, 1H, OH). MS: m/z (%) = 432 (M$^+$, 82), 312 (100), 77 (93). Anal. for C$_{26}$H$_{20}$N$_6$O: Calcd. C, 72.21; H, 4.66; N, 19.43. found C, 71.76; H, 4.51; N, 19.11.

4-((Z)-2-(4-Chlorobenzylidene)hydrazinyl)-1-phenyl-3-((E)-styryl)-1H-pyrazolo[3,4-d]pyrimidine 10e:

Yellow crystals; m.p. 230°C - 232°C (dimethylformamide); yield (84%); IR (KBr): υ = 3205 (NH) cm^{-1}; ^{1}H NMR (DMSO-d6): δ = 7.02 - 8.70 (m, 18H, Ar H), 12.11 (s, 1H, NH). MS: m/z (%) = 450 (M$^+$, 100), 77 (72). Anal. for C$_{26}$H$_{19}$ClN$_6$: Calcd. C, 69.25; H, 4.25; Cl, 7.86; N, 18.64. found C, 68.76; H, 4.03; Cl, 7.65; N, 18.23.

4-((Z)-2-(4-Fluorobenzylidene)hydrazinyl)-1-phenyl-3-((E)-styryl)-1H-pyrazolo[3,4-d]pyrimidine 10f:

Yellow crystals; m.p. 192°C - 194°C (acetonitrile); yield (87%); IR (KBr): υ = 3199 (NH) cm^{-1}; ^{1}H NMR (DMSO-d6): δ = 7.04 - 8.71 (m, 18H, Ar H), 12.13 (s, 1H, NH). MS: m/z (%) = 434 (M$^+$, 87), 339 (32), 312 (100), 236 (79), 77 (56). Anal. for C$_{26}$H$_{19}$FN$_6$: Calcd. C, 71.88; H, 4.41; F, 4.37; N, 19.34. found C, 71.51; H, 4.30; F, 4.29; N, 19.03.

1-Phenyl-3-((E)-styryl)-4-((Z)-2-(thiophen-2-ylmethylene)hydrazinyl)-1H-pyrazolo[3,4-d]pyrimidine 10g:

Yellow crystals; m.p. 186°C - 188°C (acetonitrile); yield (83%); IR (KBr): υ = 3337 (NH) cm^{-1}; ^{1}H NMR (DMSO-d6): δ = 7.01 - 8.71 (m, 17H, Ar H), 12.01 (s, 1H, NH). MS: m/z (%) = 422 (M$^+$, 100), 312 (88), 236 (61), 77 (42). Anal. for $C_{24}H_{18}N_6S$: Calcd. C, 68.23; H, 4.29; N, 19.89; S, 7.59. found C, 67.81; H, 4.20; N, 19.64; S, 7.50.

1-Phenyl-4-((Z)-2-(pyridin-4-ylmethylene)hydrazinyl)-3-((E)-styryl)-1H-pyrazolo[3,4-d]pyrimidine 10h:

Yellow crystals; m.p. 284°C - 286°C (dimethylformamide); yield (84%); IR (KBr): υ = 3322 (NH) cm^{-1}; ^{1}H NMR (DMSO-d6): δ = 6.89 - 8.64 (m, 18H, Ar H), 11.97 (s, 1H, NH). MS: m/z (%) = 417 (M$^+$, 54), 312 (71), 77 (100). Anal. for $C_{25}H_{19}N_7$: Calcd. C, 71.93; H, 4.59; N, 23.49. found C, 71.58; H, 4.24; N, 23.11.

4-((Z)-2-(Benzo[d][1,3]dioxol-5-ylmethylene)hydrazinyl)-1-phenyl-3-((E)-styryl)-1H-pyrazolo[3,4-d]pyrimidine 10i:

Yellow crystals; m.p. 214°C - 216°C (acetonitrile); yield (82%); IR (KBr): υ = 3341 (NH) cm^{-1}; ^{1}H NMR (DMSO-d6): δ = 6.09 (s, 2H, CH$_2$), 6.97 - 8.47 (m, 17H, Ar H), 12.01 (s, 1H, NH). MS: m/z (%) = 460 (M$^+$, 54), 312 (100), 236 (63), 77 (92). Anal. for $C_{27}H_{20}N_6O_2$: Calcd. C, 70.42; H, 4.38; N, 18.25. found C, 70.01; H, 4.30; N, 18.11.

General Method for Preparation of 7H-Pyrazolo[4,3-e][1,2,4] Triazolo[4,3-c]Pyrimidine 11a-h

To a solution of an appropriate arylhydrazinyl compound 10a-h (5 mmol) in ethanol (20 mL), ferric chloride (4 mL, 2 M) was added, and the reaction mixture was stirred for 24 h. The solid that separated was collected, dried and finally crystallized from dimethylformamide to afford the corresponding 7H-pyrazolo[4,3-e][1,2,4]triazo- lo[4,3-c]pyrimidine 11a-h.

(E)-3,7-Diphenyl-9-styryl-7H-pyrazolo[4,3-e][1,2,4]triazolo[4,3-c] pyrimidine 11a:

White crystals; m.p. 276°C - 278°C; yield (87%); ^{1}H NMR (DMSO-d6): δ = 7.37 - 8.81 (m, 17H, Ar H), 9.32 (s, 1H, NCH). MS: m/z (%) = 414 (M$^+$, 100), 77 (80). Anal. for $C_{26}H_{18}N_6$: Calcd. C, 75.35; H, 4.38; N, 20.28. found C, 75.03; H, 4.32; N, 20.01.

(E)-7-Phenyl-9-styryl-3-(m-tolyl)-7H-pyrazolo[4,3-e][1,2,4] triazolo[4,3-c]pyrimidine 11b:

White crystals; m.p. 182°C - 184°C; yield (86%); ^{1}H NMR (DMSO-d6): δ = 2.23 (s, 3H, CH$_3$), 7.31 - 8.47 (m, 16H, Ar H), 9.08 (s, 1H, NCH). MS: m/z

(%) = 428 (M⁺, 100), 77 (26). Anal. for $C_{27}H_{20}N_6$: Calcd. C, 75.68; H, 4.70; N, 19.61. found C, 75.22; H, 4.410; N, 19.15.

(E)-N,N-Dimethyl-4-(7-phenyl-9-styryl-7H-pyrazolo[4,3-e][1,2,4] triazolo[4,3-c]pyrimidin-3-yl)aniline 11c:

Yellow crystals; m.p. 222°C - 224°C; yield (85%); ¹H NMR (DMSO-d6): δ = 3.02 (s, 6H, N(CH₃)₂), 6.89 - 8.41 (m, 16H, Ar H), 9.24 (s, 1H, NCH). MS: m/z (%) = 457 (M⁺, 47), 77 (100). Anal. for $C_{28}H_{23}N_7$: Calcd. C, 73.50; H, 5.07; N, 21.43. found C, 73.21; H, 4.98; N, 21.24.

(E)-4-(7-Phenyl-9-styryl-7H-pyrazolo[4,3-e][1,2,4]triazolo[4,3-c] pyrimidin-3-yl)phenol 11d:

Pale green crystals; m.p. 284°C - 286°C; yield (88%); ¹H NMR (DMSO-d6): δ = 6.93 - 8.45 (m, 16H, Ar H), 9.28 (s, 1H, NCH), 13.11 (s, 1H, OH). MS: m/z (%) = 430 (M⁺, 18), 77 (100). Anal. for $C_{26}H_{18}N_6O$: Calcd. C, 72.55; H, 4.21; N, 19.52. found C, 72.31; H, 4.16; N, 19.39.

(E)-3-(4-Chlorophenyl)-7-phenyl-9-styryl-7H-pyrazolo[4,3-e][1,2,4] triazolo[4,3-c]pyrimidine 11e:

Yellow crystals; m.p. 250°C - 252°C; yield (80%); ¹H NMR (DMSO-d6): δ = 7.33 - 8.12 (m, 16H, Ar H), 8.66 (s, 1H, NCH). MS: m/z (%) = 448 (M⁺, 100), 77 (48). Anal. for $C_{26}H_{17}ClN_6$: Calcd. C, 69.56; H, 3.82; Cl, 7.90; N, 18.72. found C, 69.14; H, 3.75; Cl, 7.78; N, 18.59.

(E)-3-(4-Fluorophenyl)-7-phenyl-9-styryl-7H-pyrazolo[4,3-e][1,2,4] triazolo[4,3-c]pyrimidine 11f:

White crystals; m.p. 228°C - 230°C; yield (81%); ¹H NMR (DMSO-d6): δ = 7.34 - 8.14 (m, 16H, Ar H), 8.81 (s, 1H, NCH). MS: m/z (%) = 432 (M⁺, 100), 77 (21). Anal. for $C_{26}H_{17}FN_6$: Calcd. C, 72.21; H, 3.96; F, 4.39; N, 19.43. found C, 71.43; H, 3.90; F, 4.32; N, 19.31.

(E)-7-Phenyl-9-styryl-3-(thiophen-2-yl)-7H-pyrazolo[4,3-e][1,2,4] triazolo[4,3-c]pyrimidine 11 g:

Green crystals; m.p. 262°C - 264°C; yield (84%); ¹H NMR (DMSO-d6): δ = 7.33 - 8.73 (m, 15H, Ar H), 9.48 (s, 1H, NCH); ¹³C NMR (DMSO-d6): δ = 98.57, 118.90, 122.25, 126.39, 126.81, 127.59, 128.51, 128.88, 129.24, 129.72, 135.62, 136.38, 137.73, 138.69, 142.28, 144.87, 145.56. MS: m/z (%) = 420 (M⁺, 100), 77 (18). Anal. for $C_{24}H_{16}N_6S$: Calcd. C, 68.55; H, 3.84; N, 19.99; S, 7.63. found C, 68.01; H, 3.80; N, 19.51; S, 7.52.

(E)-7-Phenyl-3-(pyridin-4-yl)-9-styryl-7H-pyrazolo[4,3-e][1,2,4] triazolo[4,3-c]pyrimidine 11 h:

Orange crystals; m.p. 268°C - 270°C; yield (83%); ¹H NMR (DMSO-d6): δ = 6.91 - 8.68 (m, 16H, Ar H), 9.36 (s, 1H, NCH). MS: m/z (%) = 415 (M⁺,

100), 77 (41). Anal. for $C_{25}H_{17}N_7$: Calcd. C, 72.28; H, 4.12; N, 23.60. found C, 71.74; H, 4.06; N, 23.31.

General Methods for Preparation of Pyrazolo[4,3-e][1,2,4] Triazolo[1,5-c]Pyrimidine Derivatives 13

Method A: To a solution of the appropriate 11a,e (1 mmol) in absolute ethanol (30 mL), sodium acetate (0.16 g, 2 mmol) was added and the mixture was refluxed for 8 h. The precipitated solid after cooling was filtered, washed with water, dried and finally crystallized from suitable solvent to give the respective products 13a,e.

Method B: To a solution of the compound 4 (1.64 g, 5 mmol) in pyridine (20 mL), the appropriate acid chloride 12a,e,j,k,l (5 mmol) was added. The reaction mixture was refluxed for 4 h, then cooled and poured over crushed ice containing hydrochloric acid (10%) with stirring. The solid product was filtered, washed with water, dried and finally crystallized from suitable solvent to afford 13a,e,j,k,l.

(E)-2,7-Diphenyl-9-styryl-7H-pyrazolo[4,3-e][1,2,4]triazolo[1,5-c] pyrimidine 13a:

Pale brown crystals; m.p. 210°C - 212°C (dimethylformamide); yield (80%); ^{1}H NMR (DMSO-d6): δ = 6.66 - 8.75 (m, 17H, Ar H), 9.78 (s, 1H, NCH). MS: m/z (%) = 414 (M$^+$, 100), 77 (28). Anal. for $C_{26}H_{18}N_6$: Calcd C, 75.35; H, 4.38; N, 20.28. found C, 74.98; H, 4.18; N, 19.89.

(E)-2-(4-Chlorophenyl)-7-phenyl-9-styryl-7H-pyrazolo[4,3-e][1,2,4] triazolo[1,5-c]pyrimidine 13e:

Pale yellow crystals; m.p. 268°C - 270°C (dimethylformamide); yield (87%); ^{1}H NMR (DMSO-d6): δ = 6.68 - 8.84 (m, 16H, Ar H), 9.81 (s, 1H, NCH). MS: m/z (%) = 448 (M$^+$, 100), 447 (68), 77 (54). Anal. for $C_{26}H_{17}ClN_6$: Calcd C, 69.56; H, 3.82; Cl, 7.90; N, 18.72. found C, 69.31; H, 3.66; Cl, 7.69; N, 18.40.

(E)-2-Methyl-7-phenyl-9-styryl-7H-pyrazolo[4,3-e][1,2,4]triazolo[1,5-c] pyrimidine 13j:

White crystals; m.p. 186°C - 187°C (acetic acid); yield (92%); ^{1}H NMR (DMSO-d6): δ = 2.57 (s, 3H, CH$_3$), 7.32 - 8.59 (m, 12H, Ar H), 9.50 (s, 1H, NCH); ^{13}C NMR (DMSO-d6): δ = 14.28, 118.81, 122.05, 126.82, 127.37, 128.54, 128.88, 129.22, 135.60, 136.37, 137.87, 140.16, 144.17, 164.41. MS: m/z (%) = 352 (M$^+$, 98), 351 (100), 77 (35). Anal. for $C_{21}H_{16}N_6$: Calcd. C, 71.58; H, 4.58; N, 23.85. found C, 71.01; H, 4.33; N, 23.52.

(E)-2-(4-Bromophenyl)-7-phenyl-9-styryl-7H-pyrazolo[4,3-e][1,2,4] triazolo[1,5-c]pyrimidine 13k:

Pale yellow crystals; m.p. 290°C - 292°C (dimethylformamide); yield (85%); ^{1}H NMR (DMSO-d6): δ = 6.71 - 8.80 (m, 16H, Ar H), 9.84 (s, 1H, NCH). MS: m/z (%) = 492 (M$^+$, 100), 491 (75), 206 (30), 102 (29), 77 (81). Anal. for $C_{26}H_{17}BrN_6$: Calcd C, 63.30; H, 3.47; Br, 16.20; N, 17.03. found C, 63.01; H, 3.39; Br, 16.03; N, 16.86.

(E)-2-(Furan-2-yl)-7-phenyl-9-styryl-7H-pyrazolo[4,3-e][1,2,4] triazolo[1,5-c]pyrimidine 13l:

Pale yellow crystals; m.p. 264°C - 266°C (dimethylformamide); yield (86%); ^{1}H NMR (DMSO-d6): δ = 6.75 - 8.77 (m, 15H, Ar H), 9.62 (s, 1H, NCH). MS: m/z (%) = 404 (M$^+$, 100), 403 (65), 77 (53). Anal. for $C_{24}H_{16}N_6O$: Calcd C, 71.28; H, 3.99; N, 20.78. found C, 70.79; H, 3.63; N, 20.47.

CITE THIS PAPER

Hamdi M.Hassaneen,Zakaria AhmedGomaa, (2015) Utility of Styrylpyrazoloformimidate in the Synthesis of Fused Heterocyclic Compounds. *International Journal of Organic Chemistry*,**05**,213-222. doi: 10.4236/ijoc.2015.54021

REFERENCES

1. Wolkoff, P. (1975) A New Method of Preparing Hydrazonoyl Halides. Canadian Journal of Chemistry, 53, 1333-1335. http://dx.doi.org/10.1139/v75-183

2. Shawali, A.S. and Abdel Hamid, A.O. (1976) Reaction of Dimethyiphenacyl-sulfonium Bromide with N-Nitrosoace-tarylamides and Reactions of the Products with Nucleophiles. Bulletin of the Chemical Society of Japan, 49, 321. http://dx.doi.org/10.1246/bcsj.49.321

3. Farag, A.M. and Algharib, M.S. (1988) Synthesis and Reactions of C-(2-Thenoyl)-N-Arylformhydrazidoyl Bromides. Organic Preparations and Procedures International: The New Journal for Organic Synthesis, 20, 521-526. http://dx.doi.org/10.1080/00304948809356298

4. Eweiss, N.F. and Osman, A. (1980) Synthesis of Heterocycles. Part II. New Routes to Acetylthiadiazolines and Alkylazothiazoles. Journal of Heterocyclic Chemistry, 17, 1713-1717. http://dx.doi.org/10.1002/jhet.5570170814

5. Hassaneen, H.M., Shawali, A.S., Elwan, N.M. and Abounada, N.M. (1992) Reaction of 1-(2-Naphthoyl)methyl-2-di-methylsulfonium

Bromide with N-Nitroso-N-arylacetamides and Reactions of the Products with Some Nucleophiles. Sulfur Letters, 13, 273-285.

6.   Shawali, A.S. and Parkanyi, C. (1980) Hydrazidoyl Halides in the Synthesis of Heterocycles. Journal of Heterocyclic Chemistry, 17, 833-854. http://dx.doi.org/10.1002/jhet.5570170501

7.   Shawali, A.S. and Sherif, S.M. (2007) The Chemistry of Hydrazonates. Current Organic Chemistry, 11, 773. http://dx.doi.org/10.2174/138527207780831747

8.   Shawali, A.S. and Hassaneen, H.M. (1977) ArylAlkanehydrazonates and Their Thio Analogs. Synthesis of Some 2-Alkyl Derivatives of 4H-1,3,4-Benzoxadiazines and 4H-1,3,4-Benzothiadiazines. Bulletin of the Chemical Society of Japan, 50, 2827. http://dx.doi.org/10.1246/bcsj.50.2827

9.   Gouma, S.M., Salah, T.A. and Abdel Hamid, A.O. (2015) Synthesis and Cytotoxicity Evaluation of Some Novel Thiazoles, Thiadiazoles and Pyrido[2,3-d][1,2,4]triazolo[4,3-a]pyrimidin-5(1H)-ones Incorporating Triazole. Monatshefte für Chemie, 146, 149-158.

10.   Gouma, S.M., Ahmed, S.A. and Abdel Hamid, A.O. (2015) Synthesis, Characterization and Pharmacological Evaluation of Some Novel Thiadiazoles and Thiazoles Incorporating Pyrazole Moity As Anticancer Agents. Molecules, 20, 1357-1376.

11.   Hassaneen, H.M., Shawali, A.S., Abdallah, T.A. and Saleh, F.M. (2014) Synthesis and Cycloaddition Reactions of 4,4-dimethyl-2,6-dioxocyclohexane-thiocarboxamides with Nitrilimines. ARKIVOC, 2014, 155-169.

12.   Hassaneen, H.M., Shetta, A.H., Elwan, N.M. and Shawali, A.S. (1982) Reaction of Phenyltrichloromethane with Semicarbazide and Thiosemicarbazide Derivatives. Heterocycles, 19, 1477-1482. http://dx.doi.org/10.3987/R-1982-08-1477

13.   Hassaneen, H.M., Mousa, H.A.H. and Shawali, A.S. (1987) Chemistry of C-Heteroarylnitrilimines. Synthesis and Cycloaddition Reactions of N-phenyl-C-(2-thienyl)nitrilimine. Journal of Heterocyclic Chemistry, 24, 1665-1668. http://dx.doi.org/10.1002/jhet.5570240633

14.   Hassaneen, H.M. and Mousa, H.A.H. (1988) C-(2-Thienyl)-N-Arylformohydrazidoyl Chlorides in the Synthesis of Selenadiazolines and Thiadiazolines. Sulfur Letters, 7, 93.

15.   Hassaneen, H.M., Shawali, A.S. and Elwan, N.M. (1990) Synthesis and Cycloaddition Reaction of N-Phenyl-C-Sty- rylmethanohydrazonyl

Bromide. Heterocycles, 31, 247-253. http://dx.doi.org/10.3987/COM-89-5045

16. Hassaneen, H.M., Shawali, A.S. and Elwan, N.M. (1990) A Convenient Synthesis of 3,5'-Bipyrazolyl Derivatives via Hydrazonyl Halides. Heterocycles, 31, 1041-1047. http://dx.doi.org/10.3987/COM-90-5310

17. Hassaneen, H.M., Hilal, R.H., Elwan, N.M., Harhash, A. and Shawali, A.S. (1984) The Regioselectivity in the Formation of Pyrazolines and Pyrazoles from Nitrile Imines. Journal of Heterocyclic Chemistry, 21, 1013-1016. http://dx.doi.org/10.1002/jhet.5570210417

18. Baraldi, P.G., Elkashef, H., Farghaly, A., Vanelle, P. and Fruttarolo, F. (2004) Synthesis of New Pyrazolo[4,3-e]- 1,2,4-triazolo[1,5-c] pyrimidines and Related Heter-ocycles. Tetrahedron, 60, 5093-5104. http://dx.doi.org/10.1016/j.tet.2004.04.010

19. Nagamatsu, T. and Fujita, T. (1999) Facile and General Syntheses of 3- and/or 5-Substituted 7H-pyrazolo[4,3-e]-1,2,4-triazolo[4,3-c] pyrimidines as a New Class of Potential Xanthine Oxidase Inhibitors. Chemical Communications, 16, 1461.

20. Seifi, N., Niaki, M.H.Z., Barzegari, M.R., Davoodnia, A., Zhiani, R. and Kaju, A.A. (2006) Synthesis of 8-aryl-1H- pyrazolo[4,3-e][1,2,4] triazolo[4,3-a] Pyrimidine-4(5H)-imine by Using the Preyssler's Anion [NaP5W30O110]$^{14-}$ as a Green and Eco-Friendly Catalyst. Journal of Molecular Catalysis A: Chemical, 260, 77-81. http://dx.doi.org/10.1016/j.molcata.2006.06.043

21. Elwan, N.M. (1989) Synthesis and Reactivity of Some Nitrogen Heterocyclic System via Dipolar Cycloadditions. Ph.D. Thesis, Cairo University, Giza City.

22. Rashad, A.E., Heikel, O.A., El-Nezhawy, A.O.H. and Abdel-Megeid, F.M.E. (2004) Synthesis and Isomerization of Thienotriazolopyrimidine and Thienotetrazolopyrimidine Derivatives with Potential Anti-Inflammatory Activity. Heteroatom Chemistry, 16, 226-234. http://dx.doi.org/10.1002/hc.20114

23. Abdelfattah, B., Kandeel, M.M., Abdel-Hakeem, M. and Fahmy, Z.M. (2006) Synthesis of Certain Fused Thienopyrimidines of Biological Interest. Journal of the Chinese Chemical Society, 53, 403-412. http://dx.doi.org/10.1002/jccs.200600051

Chapter 5

SYNTHESIS OF FLUORINATED HETEROBICYCLIC NITROGEN SYSTEMS CONTAINING 1,2,4-TRIAZINE MOIETY AS CDK2 INHIBITION AGENTS

Mohammed Saleh Tawfik Makki, Reda Mohammdy Abdel-Rahman, Faisal Mohammed Aqlan

Department of Chemistry, Faculty of Science, King Abdulaziz University, Jeddah, Saudi Arabia

ABSTRACT

New fluorine substituted heterobicyclic nitrogen system as imidozolopyrimidines (2,3), pyrimido- 1,2,4-triazinones (4-7), 1,2,4-triazinyl-1,2,4-triazine (12-16), 1,2,4-triazinyl-1,2,4-triazinones (14- 17) and substituted thiobarbituric acids (19-20), have been synthesized using the reaction of 3- amino-5,6-di (4¢-fluorophenyl)-1,2,4-triazine (1) with α,β?bifunctional compounds. Structures of the title compounds were characterized by UV, IR, $^1H/^{13}C$-NMR and mass spectrometric method. The studied compounds were tested for CDK2 inhibiting activity in DNA damage, as well as in vitro anti-tumor activity.

INTRODUCTION

1,2,4-triazines and their condensed analogues have shown to display wide-ranging applications in medicinal and pharmaceutical chemistry. The NCNN group of 1,2,4-triazine ring is an essential part in various biological activities. Tirapazamine as antitumor [1] , Lamotrigine as anti-epileptic drug [2] , and fused 1,2,4-triazines as antimicrobial [3] -[5] , anti HIV [6] antimycobacterial [7] , antiviral [8] [9] , anxiolytic [10] and antidepressant [11] agents are already reported in literature. Heterobicyclic nitrogen systems containing 1,2,4-triazine moiety have also shown anti-HIV and anticancer activities [12] -[16] . However, introduction of fluorine atoms to these bioactive heterocycles often improves their pharmacological properties mainly due to increased membrane permeability, enhanced hydrophobic bonding and stability against metabolic transformation owing to the strength of the C-F bond [17] -[21]

. Also, introduction of fluorine in these heterobicyclic systems exhibited enhanced anti-tumor activity [20] . The search for novel cancer treatment has however entered a new post-genomic era and the emphasis now is on influencing cell signaling mechanisms such as those triggered by kinases [22] where selective inhibition of the Kinome constituents and related signaling proteins has opened up a diversity of new drug targets.

Cyclin-Dependent Kinases (CDKs) are families of serine/threonine kinases that play a well-established role in the regulation of the eukaryotic cell division cycle and have also been implicated in the control of gene transcription and other processes [23] . Dysregulation of cell cycle in cancer cells has provided a rationale for the development of small molecule inhibitors of CDK1 as novel anticancer drugs. It was believed that CDK2 was the master regulator of S phase entry. Gene knockout mouse studies of cell cycle regulators revealed that CDK2 is dispensable for S phase inhibition and progression whereby CDK1 can compensate for the loss of CDK2 and the latter was found to be involved in cell cycle independent functions such as DNA damage repair [24] .

Rational approaches to the design of fluorine-containing potential inhibitors associated with the strategic placement of fluorine in these molecules led us to design CDK2 inhibitors as possible anti-cancer agents. The aim of the present work is to synthesize and develop fluorinated fused and/or isolated heterobicyclic nitrogen systems, derived from the interaction between 3-amino,6-(4'-fluorophenyl)-1,2,4-triazines (1) and α, β-bifunc- tional oxygen, sulfur, halogen and nitrogen compounds. In this work, we evaluate the protein kinase inhibiting and cytotoxic activity of fluorinated heterobicyclic systems containing 1,2,4-triazine moiety.

EXPERIMENTAL

Chemicals and Methods

Melting points were determined using an electrothermal Bibby Stuart Scientific melting point apparatus and are uncorrected. The infrared (IR) spectra were recorded on a Perkin-Elmer FT-IR infrared spectrophotometer using the KBr pellet technique. Electronic absorption spectra were recorded in DMF using a Shimadzu UV-Visible 1650 PC spectrophotometer. ^{1}H and ^{13}C NMR spectra were recorded on a Bruker DPX-400 FT NMR spectrometer using tetramethylsilane (internal standard DMSO-d$_6$) as a solvent (Chemical shifts in δ, ppm). ^{19}F NMR Spectra were determined at 84.25 MHz using hexafluorobenzene as a solvent. Splitting patterns were designated as follows: s: singlet; m: multiplet. Mass spectra were measured on a GCMS-Q 1000 Ex spectrometer. Elemental analyses were performed on a 2400 Perkin Elmer

Series 2 analyzer and the found values were within ±0.4% of the theoretical values. Follow up of the reactions and checking the homogeneity of the compounds were made by TLC on silica gel-protected aluminum sheets (Type 60 F254, Merck) and the spots were detected by exposure to UV-lamp.

3-Amino-5,6-di(4'-fluorophenyl)-1,2,4-triazine (1)

A mixture of 4,4'-bifluorobenzil (0.01 mol) and aminoguanidine bicarbonate (0.01 mol) in n-butanol [28] was refluxed for 4 h. The solid obtained after cooling the mixture was filtered and crystallized from ethanol to give orange crystals, yield 75%, m.p. 208°C - 209°C. IR (n/cm^{-1}): 3200 (NH$_2$), 1620 (deformation, NH$_2$), 1600 (C=N), 1250 (C-F) and 880, 820 (p-substituted phenyl). ^{1}H-NMR (DMSO d$_6$) δ: 2.89 - 2.51 (br. s, 2H, NH$_2$), 7.31 - 7.39, 7.39 - 7.44 (each m, 8H, aromatic protons), 7.0 - 6.61 (m). ^{13}C-NMR (DMSO d$_6$) δ: 164,156, 148, 136 - 131, 129.8 - 127.7. Anal. Calcd: C, 63.15; H, 3.40; N, 21.67; F, 11.76% for C$_{17}$H$_5$N$_5$F$_2$ (323). Found: C, 62.98; H, 3.25; N, 21.50; and F, 11.6%.

3-Amino-5,7-di(4'-fluorophenyl)imidazo[3,2-b][1,2,4-]triazine (2)

A mixture of 1 (0.01 mol) and chloroacetonitrile (0.01 mol) in DMF was refluxed for 4 h. The solid obtained on cooling the reaction mixture was filtered and crystallized from dioxan to give 2 as yellowish crystals, yield 70%, m.p. 141°C - 142°C. IR (n/cm^{-1}): 3350, 3228 (NH$_2$, NH), 31.86 (aromatic CH), 1623 (deformation NH$_2$), 1601, 1517 (C=N), 1350 (NCN), 1226 (C-F). ^{1}H-NMR (DMSO d$_6$) δ: 7.89, 7.4, 7.37, 7.28, 7.24, 7.166, 7.00 and 6.94 (m, 8H, aromatic protons), 6.89, 6.717 (d, d of H adjacent of F), 4.01 - 3.97 (s, 2H, NH$_2$). ^{13}C-NMR (DMSO d$_6$) δ: 164, 155, 147, 132, 131 - 130.4, 115, 114, 77.88 - 77.46, 59.68. M/S (Int. %): 323 (8.89), 282 (33.01), 214 (100). Anal. Calcd: C, 62.96; H, 3.08; N, 17.28; F, 11.72%; for C$_{17}$H$_{10}$N$_4$F$_2$O (324). Found: C, 62.79; H, 2.88; N, 17.15; and F, 11.55%.

6,7-Di(4'-fluorophenyl)-2,3-dihydro-3-oxo-imidazo[3,2-b][1,2,4-]triazine(3)

A mixture of 1 (0.01 mol) and monochloroacetic acid (0.01 mol) in DMF was heated for 15 min. and then left to cool at room temperature. The solid obtained was filtered and crystallized from dioxan to give 3 as faint yellowish crystals, yield 65%, m.p. 193°C - 194°C. IR (n/cm^{-1}): 3090 (ArCH), 1640 (C=O ↔ C-OH), 1480 (deformation CH$_2$), 2950 (aliphatic CH), 1350 (NCN), and 1227 (C-F). ^{1}H-NMR (DMSO d$_6$) δ: 7.61, 7.41, 7.39, 7.38, 7.32, 7.31, 7.306 and 7.301 (m, 8H, aromatic protons), 7.29 (s, 1H, CH=), 6.98 - 6.97, 6.95 and 6.54 (d, d of H adjacent of F), 2.52 - 2.51 (s, 2H, CH$_2$). ^{13}C-NMR (DMSO d$_6$) δ:

164, 155, 147, 132.1, 131 - 130.4, 115, 114 and 77.77 - 77.34. Anal. Calcd: C, 62.96; H, 3.08; N, 17.28; F, 11.72%; for $C_{17}H_{10}N_4F_2O$ (324). Found: C, 62.76; H, 2.91; N, 17.01; and F, 11.55%.

7,8-Di(4'-fluorophenyl)-2,3-dihydro-pyrimido[3,2-b][1,2,4-]triazin-2,4-dione (4)

A mixture of 1 (0.01 mol) and diethylmalonate (0.01 mol) in THF was refluxed for 4 h. On cooling the reaction mixture, a yellowish solid separated which was crystallized from THF to give 4 as yellow crystals, yield 72%, m.p. 199°C - 200°C. UV (EtOH): λ_{max} 346 nm. IR (n/cm^{-1}): 3091 (Ar-CH), 2980 (aliphatic CH), 1680, 1650 (2C=O), 1576 (C=N) 1492 (deformation CH$_2$), 1381 (NCN), 1225 (C-F), 919, 867 and 815 (p-substituted phenyl). ^{1}H-NMR (DMSO d$_6$) δ: 7.69 - 7.29 (m, 8H, aromatic protons), 6.98 - 6.67 (d, d, s, H adjacent F), 3.65 - 3.613 (t, 2H, CH$_2$). ^{13}C-NMR (DMSO d$_6$) δ: 164.01, 162.93, 155, 147.86, 132.18, 131.83 - 131.29, 130.54, 130.49, 115.03, 114.88, 77.97 - 77.54, 67.22 and 40.03-39.33. Anal. Calcd: C, 61.36; H, 2.84; N, 15.90; and F, 10.79%; for $C_{18}H_{10}N_4F_2O_2$ (324). Found: C, 61.86; H, 2.56; N, 15.78; and F, 10.58%.

4-Amino-2-oxo-7,8-di(4'-fluorophenyl)pyrimido[3,2-b][1,2,4-]triazine (6)

A mixture of 1 (0.01 mol) and ethyl cyanoacetate (0.01 mol) in THF was refluxed for 4 h. The solid obtained after cooling the reaction mixture was filtered and crystallized from ethanol to give deep-yellow crystals, yield 65%, m.p. 198°C - 200°C. UV (EtOH): λ_{max} 346 nm. IR (n/cm^{-1}): 3300 (NH$_2$), 1670, 1610 (deformation NH$_2$), 1229 (C-F), 828 and 810 (p-substituted ring). ^{1}H-NMR (DMSO d$_6$) δ: 7.40 - 7.29 (m, 8H, aromatic protons), 6.971 (s, 1H, =CH cyclic), 6.96 - 6.69 (d, d, s, H adjacent of F), 3.62 (s, 2H, NH$_2$). ^{13}C-NMR (DMSO d$_6$) δ: 164.01, 154.9, 147.8, 132.16, 131.81 - 130.48, 115, 114.8, 77.78 - 77.57 and 67.21. M/S (Int. %): 351 (8.18), 307 (28.13), 214 (100). Anal. Calcd: C, 61.53; H, 3.13; N, 19.94; and F, 10.82%; for $C_{18}H_{11}N_5F_2O$ (351). Found: C, 61.33; H, 3.02; N, 19.58; and F, 10.69%.

7,8-Di(4'-fluorophenyl)-3-(arylidene)pyrimido[3,2-b][1,2,4-]triazin-2,4-dione (5)

A mixture of 4 (0.01 mol) and p-chlorobenzaldehyde (0.01mol) in EtOH was refluxed for 1h and then allowed to cool at room temperature. The solid thus obtained was filtered and crystallized from ethanol to give 5 as yellowish crystals, yield 78%, m.p. 208°C - 209°C. IR (n/cm^{-1}): 3024 (Ar-CH), 1680, 1660 (2C=O), 1610 (C=C), 1576 (C=N), 1382 (NCN), 1227 (C-F), 820 and

815 (p-substituted ring). ^{1}H-NMR (DMSO d$_6$) δ: 9.9 (s, 1H, CH=), 7.76 - 7.75, 7.70 - 7.60 (m, 8H, aromatic protons), 7.35 - 7.20 (m, 4H, aryl protons), and 6.95 - 6.37 (d, d, s, H of adjacent F). Anal. Calcd: C, 63.15; H, 2.73; N, 11.78; F, 8.0; and Cl, 7.57%; for $C_{25}H_{13}N_4F_2ClO_2$ (475). Found: C, 62.88; H, 2.60; N, 11.58; F, 7.69; and Cl, 7.49%.

7,8-Di(4'-fluorophenyl)-4-arylidino-pyrimido[3,2-b][1,2,4-] triazin-2-one(7)

A mixture of 6 (0.01 mol) and p-chlorobenzaldehyde (0.01 mol) in EtOH was refluxed for 1h and subsequently left to cool at room temperature. A solid appeared which was filtered and crystallized from ethanol to give yellow crystals, yield 82%, m.p. 212°C - 214°C. IR (n/cm^{-1}): 3092 (Ar-CH), 1668 (C=O), 1602 (C=N), 1226 (C-F), 919, 858 and 815 (p-substituted ring). ^{1}H-NMR (DMSO d$_6$) δ: 7.85 - 7.81 (m, 4H, aromatic H), 7.40 - 7.28 (m, 8H, aromatic H), 6.99 - 6.97 (d, d, s, H of adjacent F), 6.838 (1H, CH=N). ^{13}C-NMR (DMSO d$_6$) δ: 163.96, 161.24, 154.92, 147.68, 132.32, 131.95, 130.58 - 130.52, 115.03, 115.02, 114.89, 114.88, 78.25 - 77.81 and 40.16. Anal. Calcd: C, 63.29; H, 2.95; N, 14.76; F, 8.01; and Cl, 7.59%; for $C_{25}H_{14}N_5F_2ClO$ (474). Found: C, 62.98; H, 2.75; N, 14.49; F, 7.78; and Cl, 7.38%.

Ethyl 3-(carboxyamino)-5,6-di(4'-fluorophenyl)-1,2,4-triazine (8)

A mixture of 1 (0.01 mol) and ethyl chloromethanoate (0.01 mol) in benzene-TEA mixture was refluxed for an hour. The solid obtained after cooling the reaction mixture was filtered and crystallized from THF to give 8 as faint yellow crystals, yield 60%, m.p. 178°C - 180°C. IR (n/cm^{-1}): 3300 (NH), 3080 (Ar-CH), 2980 (R-CH), 1676 (C=O), 1483 (deformation CH$_3$), 1381 (NCN), 1226 (C-F), 1010 (O-C-O), 840, and 815 (p-substituted ring). ^{1}H-NMR (DMSO d$_6$) δ: 8.87 (s, 1H, NH), 7.68 - 7.40, 7.39 - 7.29 (each m, 8H, aromatic H), 7.03 - 7.00, 6.94 - 6.82 (d, d, s, H adjacent F), 4.20 (t, 2H, CH$_2$O), and 1.4 - 1.2 (s, 3H, CH$_3$). ^{13}C-NMR (DMSO d$_6$) δ: 164.87, 161.81, 160.98, 157.05, 147.64, 146.91, 141.26, 132.14, 132.08, 130.86 - 127.10, 115.40, 115.25, 114.96, 114.45, 78.46 - 78.03, 39.98, and 14.19. Anal. Calcd: C, 60.67; H, 3.93; N, 15.73 and F, 10.67%; for $C_{18}H_{14}N_4F_2O_2$ (356). Found: C, 60.41; H, 3.64; N, 15.61 and F, 10.55%.

N^4-(5'-6'-diaryl-1,2,4-triazin-3'-yl)semicarbazide (10)

A mixture of 8 (0.01 mol) and hydrazine hydrate (0.01 mol) in EtOH was refluxed for 2 h, and then cooled at room temperature. The solid obtained was filtered and crystallized from ethanol to give 10 as faint yellow crystals, yield 60%, m.p. 198°C - 200°C.

IR (n, cm⁻¹): 3334, 3269 (NH, NH₂), 3092(Ar-CH), 1660 (C=O), 1228 (C-F), 860 and 810 (p-substituted ring). ¹H-NMR (DMSo-d₆): 8.8 (s, 1H, NH), 7.72-7.25 (m, 8H aromatic H), 6.98 - 6.67 (m, d, d, s, H, adjacent F) and 3.16 (s, 2H, NH₂). ¹³C-NMR(DMSO-d₆): 164.02, 161.29, 155.02, 147.86, 132.17, 131,82, 131.35, 131.29, 130.54 - 130.49, 115.13, 115.02, 114.88, 77.94 - 77.53 and 40.14 - 39.19. Analy. Calcd.: C, 56.14; H, 3.50; N, 24.26 and F, 11.11%; for $C_{16}H_{12}N_6F_2O$ (342). Found: C, 55.89; H, 3.35; N, 24.39; and F, 11.01%.

N⁴-(5'-6'-Di(4'-fluorophenyl)-1,2,4-triazin-3'-yl)thiosemicarbazide (11)

A mixture of 1 (0.01 mol) and carbon disulfide (0.01 mol) in 1% KOH solution was stirred for 2 h at room temperature to give the intermediate 9. Now a mixture of 9 (0.01mol) and hydrazine hydrate (0.01mol) in EtOH was refluxed for 2 h and subsequently cooled in an ice-bath for few minutes to form a brownish yellow solid. The solid was filtered and crystallized from ethanol to give 11 as orange crystals, yield 78%, m.p. 203°C - 204°C. IR (n/cm⁻¹): 3314, 3229, 3192 (NH, NH, NH). 1381 (NCN), 1225 (C-F), 1157 (C=S) and 850 (p-substituted ring). ¹H-NMR (DMSO-d₆): 772 - 7.29 (m, 8H, aromatic H), 6.98 - 6.72 (m, d, d, s, H, adjacent F) and 3.14 (s, 2H, NH₂). ¹³ C-NMR (DMSO-d₆): 164.0, 162.2, 161.28, 154.98, 147.81, 132.21, 131.85 - 130.50, 115.03, 114.88, 78.04 - 77.61 and 40.02-39.19. Analy. Calcd: C, 53.63; H, 3.35; N, 23.46; and F, 10.61; S, 8.93%, for $C_{16}H_{12}N_6F_2S$ (358), Found: C, 53.38; H, 3.20; N, 23.41; and F, 10.38; S, 8.78%.

5,6-Di(4'-fluorophenyl)-3-(2',3'-dihydro-3'-oxo-1',2',4'-triazol-4'-yl)-1,2,4-triazine (12)

A mixture of 10 (0.5 g) and triethyl orthoformate (10 ml) was refluxed for 2h and then cooled at room temperature. The produced solid was filtered and crystallized from dioxan to give 12 as faint yellowish crystals, yield 65%, m.p. 190°C - 192°C. IR (n/cm⁻¹): 3192 (NH), 1679 (C=O), 1572(C=N), 1382 (NCN),1227 (C-F), 880 and 816 (p-substituted ring). ¹H-NMR (DMSO-d₆): 8.81 (s, 1H, NH), 7.62 - 7.27 (m, 8H, aromatic H), 6.966 - 6.92 (d, d, s, H, adjacent F) and 6.60 (s, 1H,CH=N):¹³CNMR (DMSO-d₆): 164.02, 161.40, 154.9, 132.16, 131.28, 130.53 - 130.48, 115.02, 114.88 and 77.93 - 77.49. Analy. Calcd: C, 57.95; H, 2.84; N, 23.86; and F.10.79%, for $C_{17}H_{10}F_2O$ (352). Found: C, 57.55; H, 2.70; N, 23.59; and F, 10.61%.

5,6-Di(4'-fluorophenyl)-3-(3'-mercapto-5'-hydroxy-1',2',4'-triazol-4'-yl)-1,2,4-triazine (13)

A mixture of 11 (0.01 mol) and carbon disulfide (0.01 mol) in dimethyl formamide was refluxed for 2 h. The reaction mixture was cooled in ice-bath to form a solid which was filtered and crystallized from ethanol to give 13 as orange crystals, yield 65%, m.p. 210°C - 212°C. IR (n/cm⁻¹): 3259, 3192 (NH, NH), 1658 (C=O), 1389 (NCN), 1225 (C-F), 1156 (C=S), 880 and 829 (p-substituted ring). ¹H-NMR (DMSO-d₆): 13.3, 12.8 (each s, NH, NH), 7.91 - 7.28 (m, 8H, aromatic H) and 6.98 - 6.69 (m, d, d, s, H adjacent F). ¹³CNMR (DMSO-d₆):195 (164.0), 161.28, 154.99, 147.83, 132.20 131.84, 131.29 - 130.49, 115.03, 114.88 and 78.01 - 77.58. Analy. Calcd: C, 53.12; H, 2.60; N, 21.87: F. 9.89; and S, 8.33%, for $C_{17}H_{10}N_2SO$ (384), Found: C, 52.89; H, 2.39; N, 21.55; F, 9.69; and S, 8.01%.

5,6-Di(4'-fluorophenyl)-3-(6'-methyl-3'-hydroxy-5'-oxo-1',2',4'-triazin-4'-yl)-1,2,4-triazine (14)

Equimolar mixture of 10 and sodium pyruvate in 5% NaOH solution (50 ml) was refluxed for 2 h, cooled then poured onto ice-HCl. The solid obtained was filtered and crystallized from ethanol to give 14 as faint yellow crystals, yield 65%, m.p. 213°C - 214°C. IR (n/cm⁻¹): 3372, 3297 (OH+NH), 1700, 1626 (CO, CONH), 1567 (C=N), 1483 (deformation CH₃) 1379 (NCN), 1224 (C-F), 1155 (C=S), 829 and 810 (p-substituted ring). ¹H-NMR (DMSO-d₆): 13.481 (s, 1H, NH), 7.65, 7.41, 7.39, 7.22, 7.20, 7.19, 6.985, 6.978 (8H, aromatic H), 6.971 - 6.964, 6.957 - 6.950 (m, d, d, s, H of F) and 1.903 (s, 3H, CH₃).¹³C-NMR (DMSO-d₆): 195.63, 164.73, 163.06, 161.66, 153.43, 140.91, 131.69, 131.11, 131.09 - 129.99, 115.21, 115.06, 114.95, 77.88 - 77.44, 40.02, and 39.19. Analy. Calcd: C, 57.86; H, 3.04; N, 12.13; and F, 9.64%, for $C_{19}H_{12}N_6 F_2O_2$ (394). Found: C, 57.49; H, 2.89; N, 21.01; and F, 9.33%.

5,6-Di(4'-fluorophenyl)-3-(3'-mercapto-1,2,4-triazol-4'-yl)-1,2,4-triazine (15)

A mixture of 11 (0.01 mol) and triethyl orthoformate (0.01 mol) was refluxed for 2 h and then cooled at room temperature. The solid obtained was filtered and crystallized from dioxan to give 15 as yellow crystals, yield 72%, m.p. 183°C - 184°C. IR (n/cm-¹): 3298, 3191 (NH), 1610, 1576 (C=N), 1381 (NCN), 1225 (C-F), 1157 (C=S), 840 and 810 (p-substituted ring). ¹H-NMR (DMSO-d₆): 10.03 (S, 1H, NH=S), 7.60 - 511, 7.48 - 7.27 and 7.0 - 6.55 (m, 9H, aromatic H). Analy. Calcd: C, 55.43; H, 2.71; N, 22.82; F.10.32 and S,

8.69%, for $C_{17}H_{10}N_6F_2S$ (368). Found: C, 55.21; H, 2.33; N, 22.55; F.10.01 and S, 8.49%.

5,6-Di(4'-fluorophenyl)-3-(3',5'-dimercapto-1',2',4'-triazol-4'-yl)-1,2,4-triazine (16)

A mixture of 11 (0.01 mol) and carbon disulfide (0.01 mol) in DMF was refluxed for 2 h and then cooled at room temperature. The solid obtained was filtered and crystallized from methanol to give 16 as orange crystals, yield 80%, m.p. 203°C - 204°C. IR (n/cm^{-1}): 3197 (NH), 3150 (NH), 1595 (C=N), 1564 (C=N), 1382 (NCN), 1224 (C-F), 1151(C=S), 831 and 810 (p-substituted ring). ^{13}C-NMR (DMSO-d$_6$): 163.99162.90‹ ‹ ‹161.47 161.26 ‹ ‹154.96 147 ‹ 132.25 ‹ 132.23 ‹ 131.89 ‹ 131.87 131.31 130.56 ‹ ‹ 130.50 ‹ 115.02 ‹ 114.88 ‹ 77 and 77.69-78.12. Analy. Calcd: C, 51.0; H, 2.50; N, 21.0; F, 9.5; and S, 16.0%, for $C_{17}H_{10}N_6F_2S_2$ (400). Found: C, 50.85; H, 2.31; N, 20.88; F, 9.33; and S.15.79%.

5,6-Diaryl-3-(6'-methyl-3'-mercapto-6'-oxo-1',2',4'-triazin-4'-yl)-1,2,4-triazine (17)

A mixture of 11 (0.01 mol) and sodium pyruvate (0.01 mol) in 5% NaOH aq. (20 ml) was refluxed for 3h. The reaction mixture was cooled and then poured on ice cold HCl solution. The solid obtained was filtered and crystallized from ethanol to give 17 as yellow crystals, yield 60%, m.p. 220-222°C. IR (n/cm^{-1}): 3197 (NH), 2906, 2855(CH$_3$), 1666 (C=C), 1586 (C=N), 1472 (deformation CH$_3$), 1367 (NCN), 1224 (C-F), 1158 (C=S), 872 and 792 (p-substituted ring). ^{1}H-NMR (DMSO-d$_6$): 11.86 (S, 1H, NH), 7.75 - 7.38, 7.30 - 7.28 (each, m, 8H, aromatic H), 6.97 - 6.69 (d, d, s, H-CF), and 1.08 (S, 3H, CH$_3$). ^{13}C-NMR (DMSO-d$_6$): 180, 163.98, 162.91, 161.27, 154.96, 147.84, 132.21, 131.85, 131.36, 131.30, 130.55 - 130.02, 115.02, 114.88, 78.03 - 77.60 and 39.20. Analy. Calcd: C, 55.60; H, 2.92; N, 20.48; F, 9.26; and S.7.80%, for $C_{19}H_{12}N_6F_2SO$ (410). Found: C, 55.39; H, 2.80; N, 20.29; F.9.01 and S, 7.55%.

N,N'-[5,6-di(4¢fluorophenyl)-1,2,4-triazin-3¢-yl]thiourea (18)

Equimolar amount of compounds 1 and 9 in ethanol (50 ml) was refluxed for 2 h. The reaction mixture was cooled in an ice bath to form a solid which was filtered and crystallized from ethanol to give 18 as faint yellow crystals, yield 65%, m.p. 193°C - 194°C. IR (n, cm^{-1}): 3300, 3289, 3192 (C-F), 1157(C=S), 827 and 814 (p- substituted ring). Analy. Calcd: C, 60.98; H, 2.95; N, 18.36; F, 12.45; and S.5.24%, for $C_{31}H_{18}N_8F_4S$ (610). Found: C, 60.65; N, 2.80; F, 12.13; and S, 5.00%.

N,N'-Di[di(4'-fluorophenyl)1,2,4-triazin-3-yl]-thiobarbituric Acid (19)

A mixture of 18 (0.01 mol) and malonic acid (0.01 mol) in glacial acetic acid (50 ml) was refluxed for 4h. The reaction mixture was cooled and then poured onto ice. The solid obtained was filtered and crystallized from ethanol to give 19 as yellow crystals, yield 60%, m.p. 300°C - 302°C. UV (Ethanol): λ_{max} 334 nm. IR (n/cm^{-1}): 2980, 2868 (aliphatic CH), 1663 (C=O), 1595, 1561(C=N), 1366 (NCN), 1223 (C-F), 1156 (C=S), 854 and 820 (p-substituted ring). ^{1}H-NMR (DMSO-d$_6$): 13.52 (s. IH, OH), 7.82 - 7.00 (m, 16H, aromatic H), 6.98, 6.97, 6.85(d, d, S, H, of C-F) and 2.499 (S, CH$_2$). ^{13}C-NMR (DMSO-d$_6$):180, 165.68, 164.65, 163.27, 162.98, 161.62, 154.93, 140.95, 131.95 - 130.14, 115.22, 114.89, 78.28 - 77.85 and 24.88. Analy. Calcd: C, 60.15; H, 2.66; N, 16.49; F, 11.10; and S, 4.55%. for $C_{34}H_{18}N_8F_4SO_2$ (678). Found: C, 59.89; H, 2.51; N, 16.30; F, 10.89; and S, 4.40%.

N,N'-Di[di(4'-fluorophenyl)1,2,4-triazin-3-yl]-5-arylidene-thiobarbituric Acid (20)

Equimolar quantity of 19 and 4-chlorobenzaldehyde in ethanol (50 ml), piperidine (0.5 ml) was refluxed for 6h. The reaction mixture was cooled and neutralized with few drops of acetic acid. The solid precipitated was filtered and crystallized from ethanol to give 20 as yellow crystals, yield 60%, m.p. 310°C - 312°C. UV (EtOH): λ_{max} 346 nm. IR (n, cm^{-1}): 3080 (aromatic CH), 2920, (aliphatic CH), 1710, 1680(2C=O), 1385 (NCSN), 1250 (C-F), 1190 (C-S), 880 and 810 (p-substituted ring). ^{1}H-NMR (DMSO-d$_6$): 7.66 - 7.64, 7.38 - 7.36, 7.29 - 7.27, (each, m, 21H, aromatic H), 6.61, 6.96 and 6.92, (CH=, d, d, s, of H adjacent to the F atoms). ^{13}C-NMR (DMSO-d6): 180, 164, 162.94, 162.36, 161.30, 155.02, 147.89, 132.14, 131.79, 131.34, 131.28, 130.53, 130.48, 115.03, 114.88, 77.90 - 77.47, 66.42 and 40.12 - 39.17. Analy. Calcd: C, 61.42; H, 2.62; N, 13.48; F, 9.48, Cl, 4.49; and S, 3.99%, for $C_{41}H_{21}N_8FClSO_2$ (801). Found: C, 61.01; H, 2.55; N, 12.98; F, 9.03; Cl. 3.99 and S, 3.55%.

RESULTS AND DISCUSSION

Chemistry

3-Amino-1,2,4-triazines are important intermediates in the synthesis of heterobicyclic systems. Also, 5,6-di- phenyl-1,2,4-triazine containing functional groups at position-3, especially amino and/or hydrazine groups, are used as nucleophilic reagents (electron donors) towards π-acceptors carbonitriles [25]

. In addition, in most of these reactions, DMF is used as a strong polar aprotic solvent which accelerates the SN^2 reactions [26] .

Nucleophilic attack on exo C=N bonds of 1,2,4-trizaine provides a direct and convenient method for functionalization of 1,2,4-triazines, which makes it possible to introduce various substituents in a single reaction. Thus, it is quite probable that their reactions with bifunctional nucleophiles would result in polycyclic compounds via attack by the second nucleophilic center at another carbon atom of hetero-ring [26] .

The presence of hetero atoms results in significant changes in the cyclic molecular structure, due to the availability of unshared pairs of electrons on nitrogen atoms and the difference in electronegativity between heteroatoms and carbons within the closed system.

The starting material 3-amino5,6-di(4'-fluorophenyl)-1,2,4-triazine (1) was prepared from refluxing amino-guanidine bicarbonate with 4,4'-difluorobenzil in n-ButOH [27] (Scheme 1). Structure of compound 1 was deduced from elemental analysis and spectral data. IR spectrum showed an absorption bands at 3200, 1620 cm^{-1} for NH$_2$ group, and 1250 cm^{-1} for C-F. ^{1}H-NMR (DMSO-d$_6$) recorded δ at 2.89-1 (br, s, 2H, NH$_2$), 7.31 - 7.34, 7.39 - 7.44 ppm for 8 aromatic protons with δ at 7.0-6.61 ppm as a characteristic doublet and a signal for the adjacent H to the F atom. ^{13}C-NMR spectrum exhibited a resonance signals for carbons at δ 127.7 - 129.8 for aromatic carbons and four other peaks at δ 136, 148, 156, and 161 ppm for C=N, C-N, C-F and C-NH$_2$ respectively.

X-ray analysis of 1,2,4-triazine ring reveal that the ring is slightly distorted due to the asymmetry induced by the electronegativity of nitrogen atoms and two intermolecular hydrogen bonds that stabilized the 1,2,4-triazine structure [28] .

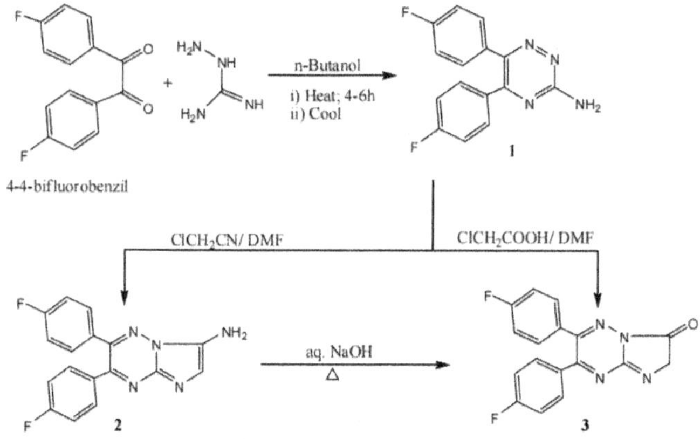

Scheme 1: Synthesis of 2 and 3.

The ring closure of 3-amino-5,6-di(4'-fluorophenyl)-1,2,4-triazine (1) with chloroacetonitrile in boiling DMF produced 3-amino-6,7-di(4'-fluorophenyl) imidazolo[3,2-b][1,2,4]triazine (2), while reaction of 1 with monochloro acetic acid in refluxing DMF yielded 2,3-dihydro-6,7-diaryl-imidazolo[3,2-b] [1,2,4]triazin-3-one (3). Moreover, compound 3 was also obtained via basic hydrolysis of 2 by warming in aqueous NaOH.

Structure of compound 2 showed a characteristic bands at 3350 and 1623 cm^{-1} attributed to NH_2 group while that of compound 4 had a characteristic band at 1660 cm^{-1} for C=O, in addition to bands at 2950 and 1480 cm^{-1} for CH_2 group and an additional absorption at 1226 cm^{-1} for C-F. Mass spectra of 3 showed a molecular ion peak at m/z 323, with a base peak at m/z 214 for 4,4'-difluorophenyl acetylene radical [29].

Ring closure reactions of compound 1 with diethyl malonate and/or ethyl cyanoacetate in refluxing THF afforded 1,2,3,4-tetrahydro-7,8-di(4'-fluorophenyl)-pyrimido[3,2-b][1,2,4]traizin-2,4-dione (4) and 4-amino-7, 8-di(4'-fluorophenyl)-pyrimido[3,2-b][1,2,4]traizin-2-one (6), respectively. The condensation of both 4 and 6 with aromatic aldehyde in boiling ethanol?piperidine gave the arylidene 5 and the Schiff base 7 (Scheme 2).

Structures of both compounds 4 and 6 were elucidated from their analytical and spectral data. IR spectrum of 4 showed absorption bands at 1680 and 1650 cm^{-1} for the two C=O groups, while that of 6 recorded absorption bands at 3300, 1670 cm^{-1} due to NH_2 and C=O groups, respectively. Mass spectrum of 6 recorded a molecular ion peak at 351 m/z and a base peak at 214 m/z. UV absorption spectrum of 5 showed a higher maximum (λ_{max} = 351 nm) than that of 4 (λ_{max} = 346 nm), due to the extended conjugation in the heterocyclic ring in the former. ^{1}H-NMR of 5 showed a prominent peak at δ 8.2 ppm due to C=CHAr. ^{13}C-NMR of compound 7 showed the resonance signal of exo N=CH in addition to an endo N=C of 1,2,4-triazine.

The treatment of compound 1 with ethyl chloroformate in benzene and triethylamine and/or CS_2/KOH led to the formation of ethyl carboxylate 8 and the intermediate 9, respectively. Hydrazinolysis of 8 and 9 by refluxing with hydrazine hydrate in ethanol afforded N^4-substituted semicarbazide/ thiosemicarbazide 10 and 11, respectively (Scheme 3). IR spectra of both compounds 10 and 11 showed C=O and C=S at 3400, 3100 cm^{-1} for NH, NH_2 groups, respectively.

Ring closure reactions of compound 10 with triethylorthoformate, carbondisulfide, dimethylformamide and/or sodium pyruvate (NaOH solution) led to the direct formation of 3-(2',3'- dihydro-3'-oxo-1,2,4-triazol-4'- yl)- 5,6- di(4'-fluorophenyl)-1,2,4-triazine(12), 3-(3'-oxo-1,2,3,4-tetrahydro-5'- thioxo-1,2,4-triazol-4'-yl)-5,6-di-(4'-flu- orophenyl)-1,2,4-triazine (13) and

3-(3′,4′-dihydro-3′,5′-dioxo-6′-methyl-1′,2′,4′-triazin-3′-yl)-5,6-di   (4′-fluoro-phenyl)-1,2,4-triazine (14) (Scheme 4).

IR spectra of 12 and 14 evealed absorption bands at 3150 - 3100 and 1680 - 1660 cm^{-1} for NH and C=O groups, respectively. ^{1}H-NMR spectrum of 14 showed resonance signals at δ 1.95, 13.48 and 8-6.6 ppm for CH$_3$, NH and aromatic protons respectively. However compound 13 recorded a resonance C=S at δ 195 ppm. Mass spectra of 13 exhibit a molecular ion peak at m/z 384 and a base peak at m/z 214 due to 4,4′-difluorophenyl acetylene radical [29] .

Under the same experimental conditions, ring closure reactions of compound 11 with triethylorthoformate, carbondisulfide and sodium pyruvate resulted in   3-(2′,3′-   dihydro-3′-thioxo-1′,2′,4′-triazol-4′-yl)-5,6-di(4′-fluorophenyl)-1,2,4-triazine (15), 3-(1′,2′,3′,4′-tetrahydro-3′,5′-thioxo-1′,2′,4′-triazol-4′-yl)-5,6-di(4′-fluorophenyl)- 1,2,4-triazine (16) and 3-(3′,4′-dihydro-3′-thioxo-6′-methyl-5′-oxo-1,2,4-triazol-4′-yl)-5,6-di(4′-fluorophenyl)- 1,2,4-triazine (17), respectively (Scheme 5).

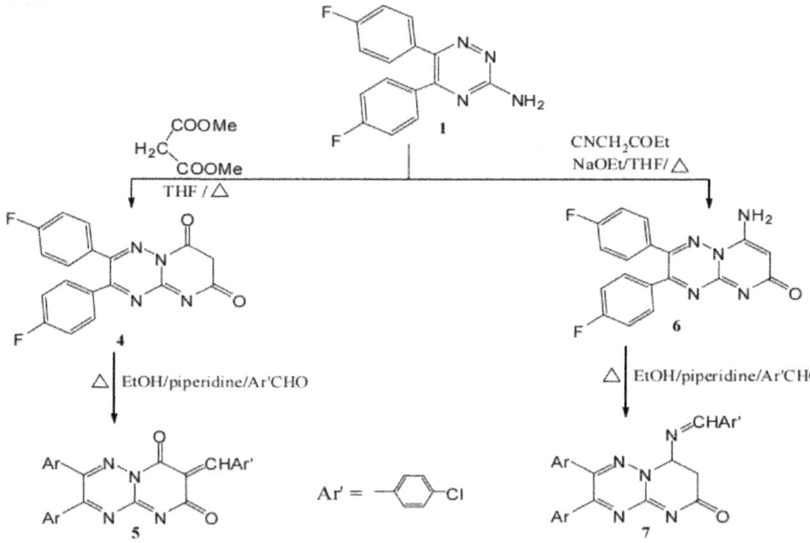

Scheme 2: Synthesis of 5 and 7.

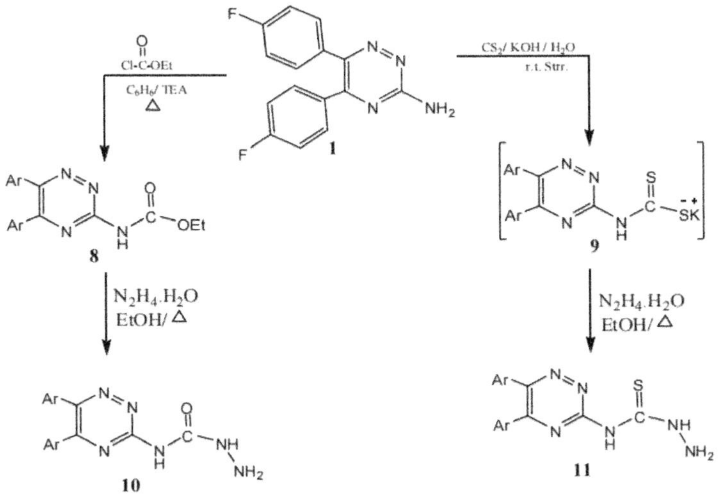

Scheme 3: Synthesis of 10 and 11.

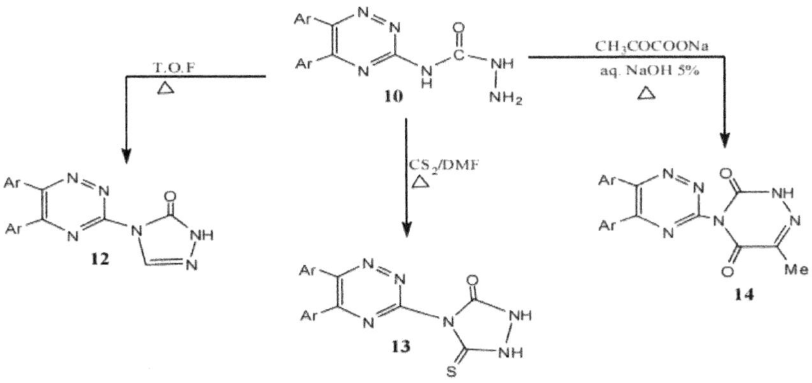

Scheme 4: Synthesis of 12-14.

IR spectra of compounds 15-17 showed an absorption bands at 3200-3100 and 1180-1150 cm^{-1} for NH and C=S groups, respectively. ^{1}H-NMR of 17 recorded resonance signals at 1.08 and 11.85 ppm for CH$_3$ and NH protons, respectively. Also, ^{13}C-NMR exhibited resonances for C=S, C=O, C=N and CH$_3$ at δ 180, 163 and 130 ppm respectively.

Fluorinated N,N-disubstituted-thiobarbituric acid 19 was prepared from refluxing compound 9 with 1 in ethanol to give N,N-disubstituted-thiourea 18. Ring closure reaction of 18 with malonic acid on warming with glacial acetic acid yielded compound 19 (Scheme 6). IR spectra of compound 18 showed absorption bands at 3300 - 3190 cm^{-1} for NH, NH of thiourea derivative which was not found in compounds 19 and 20. ^{1}H-NMR spectrum of compound 19 recorded a resonance signal at δ 2.44 ppm for CH$_2$ protons. Also, ^{13}C-NMR showed a signal at δ 24.88 ppm for aliphatic carbons. IR spectrum of 20 showed absorption bands at 1680, 1660 and 1611 cm^{-1} for the two C=O and CH=C groups, in addition to two bands at 1255 and 1185 cm^{-1} for C-F and C=S groups, respectively. ^{13}C-NMR of 20 exhibited resonance signals at δ 188, 166 and 39.48 ppm attributed to C=S, C=O and CH=C carbons, respectively. It is interesting to note that UV absorption of compound 20 recorded λ_{max} at 334 nm which is higher than the value for compound 19 (λ_{max} = 332), confirming ta presence of an α, β-unsaturated cyclic ketone system in the former.

Biological Evaluation

The CDK2 inhibitory activity of the synthesized compounds revealed that eleven out of the tested twenty compounds displayed variable inhibitory effects. However compounds 11, 13, 16 and 17 showed profound activity.

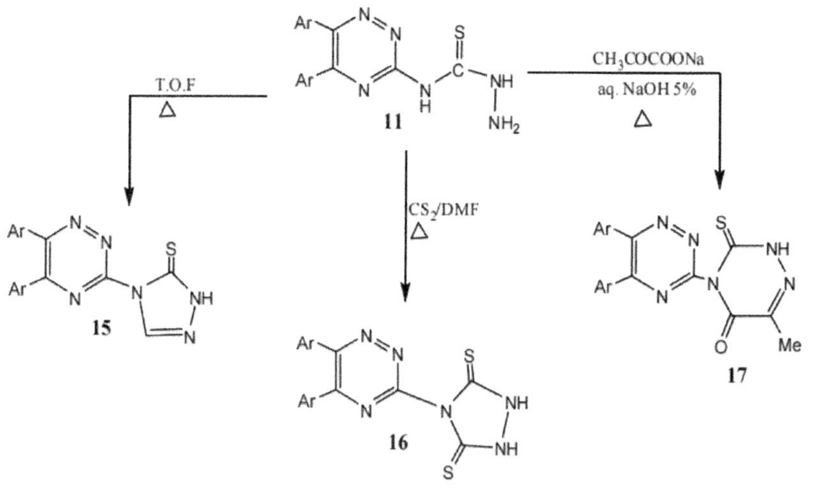

Scheme 5: Synthesis of 15 and 17.

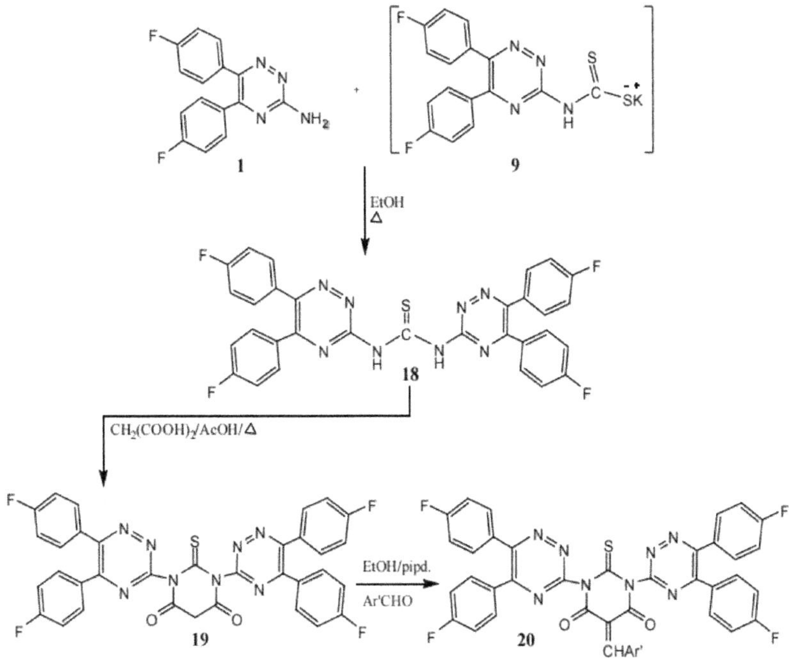

Scheme 6: Synthesis of 18-20.

Compound 16 was found to be equipotent to the positive reference Olomoucine.

Further examination of the structures of the active compounds revealed that all of them have almost similar N-(amino(hydrazinyl)methyl) hydrazinecarbothioamide scaffolding as displayed in Figure 1 by bold bonds. Though these structural features are present in other active compounds but other molecular framework variants rends them less active and less sensitive towards the CDK2 enzyme. Also, the presence of mercapto-groups in fluorinated heterobicyclic nitrogen systems (11, 13, 16, 17) exhibited good effects toward cell damage (Table 1).

However, in vitro anti-tumor testing of the highly active compounds (11, 13, 16, 17) were evaluated according to the described method, under different concentrations, a sulforhodamine B (SRB) protein assay was used to estimate cell viability or growth by determining GI_{50}, TGI and LC_{50}(Table 2).

The results obtained, indicated that all the tested compounds were effective towards some types of tumors. Compound 11 was active against non-small cell lung cancer, renal cancer and breast cancer cell lines, while compound 13 had activity towards leukemia, renal cancer, non-small cell lung cancer and breast cancer. Furthermore, compounds 16 and 17 exhibited significant degrees of activity towards non-small cell lung cancer and breast cancer (Table 2).

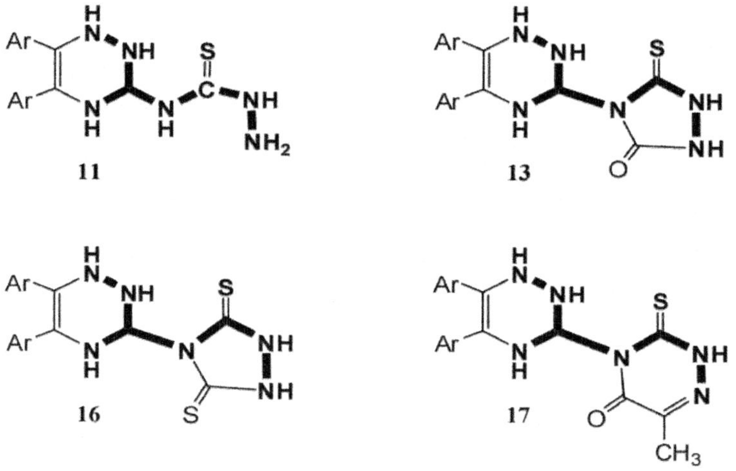

Figure 1: Common structural features of the active compounds.

Table 1: The CDK2 inhibitory activity of tested compounds (IC_{50} in µg/ml).

The data represent means values from three independent experiments plus the standard deviation (SD).

Compound. No.	IC_{50} CDK$_2$ ±SD µg/ml)
1	15+1.8
2	18+2.8
3	10+1.3
4	11+2.4
6	20+3.5
11	4.0+1.8
12	11+3.8
13	4.8+1.0
15	12+5.3
16	5.0+2.8
17	5.2+2.8
Olomoucine	5.0+1.0

Table 2: In vitro antitumor activity data of some active compounds.

Compound	Response Parameters* [mean log10]				Selectivity Analysis Subpanel Sensitivity
	**(Δ)		TGI	GI$_{50}$	
11	0.00 0.02 0.69-	Non small cell lung cancer, Renal cancer and Breast cancer	5.0 5.0 5.0	5.0 5.0 5.0	11
13	1.14	Leukemia, renal cancer, non-small cell lung cancer, Breast cancer	5.01	5.22	13
16	0.65	Non small cell lung cancer	5.00	5.06	16
17	0.37	Breast cancer	5.00	5.01	17

GI$_{50}$: concentration giving 50% inhibition; TGI: concentration giving total growth inhibition; LC$_{50}$: concentration having 50% lethal effect; Δ is considered low if 1, moderate if > 1 and high if ≥ 3; Subpanels showing a statistical measure of differential sensitivity with respect to the indicated response parameters.

CONCLUSION

The synthesized α, β-bifunctional oxygen, sulfur, halogen and nitrogen derivatives of 1,2,4-triazines have shown promising CDK2 activities. Out of eleven active compounds, four of them (11, 13, 16 and 17) have shown very good CDK2 enzyme inhibiting activity. In these compounds N-(amino(hydrazinyl) methyl)hydrazinecarbo- thioamide scaffold was found to be an essential molecular feature in CDK2 enzyme inhibition activity. However these four compounds have also shown significant activity against various tumor cell-lines in subpanel assay.

ACKNOWLEDGEMENTS

The authors are grateful to King Abdulaziz University for providing research facilities to perform the present research work.

CITE THIS PAPER

Mohammed Saleh TawfikMakki,Reda MohammdyAbdel-Rahman,Faisal MohammedAqlan, (2015) Synthesis of Fluorinated Heterobicyclic Nitrogen Systems Containing 1,2,4-Triazine Moiety as CDK2 Inhibition Agents. *International Journal of Organic Chemistry*,**05**,200-211. doi:10.4236/ijoc.2015.53020

REFERENCES

1. Fuchs, T., Chowdhury, G., Barnes, C.L. and Gates, K.S. (2001) 3-Amino-1,2,4-benzotriazine 4-Oxide: Characterization of a New Metabolite Arising from Bioreductive Processing of the Antitumor Agent 3-Amino-1,2,4-benzotriazine 1,4-Dioxide (Tirapazamine). The Journal of Organic Chemistry, 66, 107-114. http://dx.doi.org/10.1021/jo001232j

2. Leach, M., Marden, C. and Miller, A. (1986) Pharmacological Studies on Lamotrigine, a Novel Potential Anti-Epileptic Drug: II Neuro-chemical Studies on the Mechanism of Action. Epilepsia, 27, 490-497. http://dx.doi.org/10.1111/j.1528-1157.1986.tb03573.x

3. 3. Ibrahim, M.A., Abdel-Rahman, R.M., Abdel-Halim, A.M., Ibrahim, S.S. and Allimony, H.A. (2008) Synthesis and Antifungal Activity of Novel Polyheterocyclic Compounds Containing 1,2,4-Triazine Moiety. Arkivoc, Xvi, 202-215.

4. Abdel-Rahman, R.M., Makki, M.S.T., Ali, T.E. and Ibrahim, M.A. (2012) 1,2,4-Triazine Chemistry Part III: Synthetic Strategies to Functionalized Bridgehead Nitrogen Heteroannulated 1,2,4-Triazine Systems and Their Regiospecific and Pharmacological Properties, Current Org. Synthesis, 9, 1-25. The Merck Index & Co. Inc., White house Station, NJ, USA, (2006) 1953.

5. Abdel-Rahman, R.M. (2001) Rrole of Uncondensed 1,2,4-Triazine Compounds and Related Heterocyclic Systems as Therapeutic Agents. Pharmazie, 56, 18-30.

6. Abdel-Rahman, R.M. (1991) Synthesis and Anti Human Immune Virus Activity of Some New Fluorine Containing Substituted 3-Thioxo-1,2,4-triazin-5-ones, Farmaco, 46, 379-389.

7. Mamolo, M.G., Falagiani, V., Zampieri, D., Vio, L. and Banfi, E. (2000) Synthesis and Antimycobacterial Activity of Some 4H-1,2,4-Triazin-5-one Derivatives. Il Farmaco, 55, 590-595. http://dx.doi.org/10.1016/S0014-827X(00)00074-4

8. Abdel-Rahman, R.M., Morsy, J.M., Hanafy, F.I. and Amine, H.A. (1999) Synthesis of Heterobicyclic Nitrogen Systems Bearing the 1,2,4-Triazine Moiety as Anti-HIV and Anticancer Drugs; Part I. Pharmazie, 54, 347-351.

9. Liu, K.C.H., Shih, Bi-J. and Lee, Ch-Hs (1993) Synthesis of Representative 10-aryl-, 10-Aralkyl- and 10-Heteroaryl-9H-naphtha[1',2':4,5]thiazolo[3,2-b][1,2,4]triazin-9-ones as Potential Anti HIV Agents. Journal of Heterocyclic Chemistry, 10, 1331-1335. http://dx.doi.org/10.1002/jhet.5570300525

10. Russell, M.G., Carling, R.W., Street, L.J., Hallett, D.J., Goodacre, S., Mezzogori, E., Reader, M., Cook, S.M., Bromidge, F.A., Newman, R., Smith, A.J., Wafford, K.A., Marshall, G.R., Reynolds, D.S., Dias, R., Ferris, P., Stanley, J., Lincoln, R., Tye, S.J., Sheppard, W.F., Sohal, B., Pike, A., Dominguez, M., Atack, J.R. and Castro, J.L. (2006) Discovery of Imidazo[1,2-b][1,2,4]triazines as GABA(A) alpha2/3 Subtype Selective Agonists for the Treatment of Anxiety. Journal of Medical Chemistry, 49, 1235-1238. http://dx.doi.org/10.1021/jm051200u

11. Koek, W., Patoiseau, J.F., Assie, M.B., Cosi, C. and Kleven, M.S. Dupont-Passelaigue, E., Carilla-Durand, E., Palmier, C., Valentin, J.P., John, G., Pauwels, P. J., Tarayre, J.-P. and Colpaert, F.C. (1998) F 11440, a Potent, Selective, High Efficacy 5-HT1A Receptor Agonist with Marked Anxiolytic and Anti-Depressant Potential. Journal of Pharmacology and Experimental Therapeutics, 287, 266-283.

12. El-Gendy, Z., Morsy, J., Allimony, H.A., Abdel-Monem, W. R. and Abdel-Rahman R. M. (2001) Synthesis of Heterobicyclic Nitrogen Systems Bearing the 1,2,4-Triazine Moiety as anti HIV and Anticancer Drugs, Part III. Pharmazie, 56, 376-382.

13. El-Gendy, Z., Morsy, J.M., Allimony, H.A., Abdel-Monem, W.R. and Abdel-Rahman R.M. (2003) Synthesis of heterobicyclic nitrogen systems bearing a 1,2,4-triazine moiety as anticancer drugs, Part IV. Phosphorus, Sulfur, and Silicon and the Related Elements, 178, 2055-2071. http://dx.doi.org/10.1080/10426500390228738

14. Abdel-Rahman, R.M., Morsy, J., El-Edfawy, S. and Amine, H.A. (1999) Synthesis of Some Heterobicyclic Nitrogen Systems Bearing the 1,2,4-Triazine Moiety as Anti HIV and Anticancer Drugs, Part II. Pharmazie, 54, 667-671.

15. Mohared, R.M., Ibrahim, R.A. and Moustafa, H.E. (2010) Hydrazide-Hydrazones in the Synthesis of 1,3,4-Oxadiazine, 1,2,4-Triazine and Pyrazole Derivatives with Antitumor Activities. The Open Organic Chemistry Journal, 4, 8-14.

16. El-Nagger, S.A., El-Barbary, A.A., Mansour, M.A., Abdel-Shafy, F. and Talat S. (2011) Anti-Tumor Activity of Some 1,3,4-Thiadiazoles and 1,2,4-Triazine Derivatives against Ehrlichs Ascites Carcinoma. International Journal of Cancer Research, 7, 278-288. http://dx.doi.org/10.3923/ijcr.2011.278.288

17. Smart, B.E. (2001) Fluorine Substituent Effects (on Bioactivity). Journal of Fluorine Chemistry, 109, 3-11. http://dx.doi.org/10.1016/S0022-1139(01)00375-X

18. Ismail, F.M.D. (2002) Important Fluorinated Drugs in Experimental and Clinical Use. Journal of Fluorine Chemistry, 118, 27-33. http://dx.doi.org/10.1016/S0022-1139(02)00201-4

19. Dolbier Jr., W.R. (2005) Fluorine Chemistry at the Millennium. Journal of Fluorine Chemistry, 126, 157-163. http://dx.doi.org/10.1016/j.jfluchem.2004.09.033

20. Isanbor, C. and O'Hagan, D. (2006) Fluorine in Medicinal Chemistry: A Review of Anti-Cancer Agents. Journal of Fluorine Chemistry, 127, 303-319. http://dx.doi.org/10.1016/j.jfluchem.2006.01.011

21. Abdel-Rahman, R.M. (2001) Synthesis and Chemistry of Fluorine Containing 1,2,4-Triazine. Pharmazie, 56, 791-804.

22. Workman, P. (2005) Genomics and the Second Golden Era of Cancer Drug Development. Molecular BioSystems, 1, 17-26. http://dx.doi.org/10.1039/b501751n

23. Morgan, D.O. (1997) Cyclin-Dependent Kinases: Engines, Clocks, and Microprocessors. Annual Review of Cell and Developmental Biology, 13, 261-291. http://dx.doi.org/10.1146/annurev.cellbio.13.1.261

24. Satyanarayana, A. and Kaldis, P. (2009) A Dual Role of Cdk2 in DNA Damage Response. Cell Division, 4, 9. http://dx.doi.org/10.1186/1747-1028-4-9

25. Abdel-Rahman, R.M. (1999) Chemistry of Uncondensed 1,2,4-Triazines Part I, Chemical Reactivity of 5,6-Diphenyl- 1,2,4-triazine-5-yl Containing Active Groups. Trends in Heterocyclic Chemistry, 6, 126-133.

26. Abdel-Rahman, R.M. (2001) Chemistry of Uncondensed 1,2,4-Triazines Part IV. Synthesis and Chemistry of Bioactive 3-Amino-1,2,4-triazines and Related Compounds. Pharmaize, 56, 275-286.

27. Musator, D.M., Kurilor, D.V. and Rakisher, A.K. (2008) Efficient Synthesis of 3-Amino-5,6-diphenyl-1,2,4-triazine. Ukrainica Bioorganica Act, 1, 61-62.

28. Hwang, L.-C., Wu, R.-R., Jane, S.-Y. and Lee, G.-H. (2003) Crystals Structure of 3-Amino-1,2,4-triazine. Analytical Sciences, 19, x73-x74. http://dx.doi.org/10.2116/analscix.19.x73

29. Palmer, M.H., Preston, P.N. and Steven, M.F.G. (1971) The Mass Spectra of 1,2,4-Triazines and Related Compounds. Organic Mass Spectrometry, 5, 1085-1092. http://dx.doi.org/10.1002/oms.1210050908

Chapter 6

NEW METHOD FOR PREPARATION OF 1-AMIDOALKYL-2-NAPHTHOLS VIA MULTICOMPONENT CONDENSATION REACTION UTILIZING TETRACHLOROSILANE UNDER SOLVENT FREE CONDITIONS

Samy B. Said[1], Mohammad M. A. Mashaly[1], Ahmed M. Sheta[1], Saad S. Elmorsy[2]

[1]Chemistry Department, Faculty of Science, Damietta University, New Damietta, Egypt

[2]Chemistry Department, Faculty of Science, Mansoura University, Mansoura, Egypt

ABSTRACT

An efficient and direct procedure for the synthesis of amidoalkylnaphthol derivatives employing a multi-component and one-pot condensation reaction of 2-naphthol, aromatic aldehyde and acetonitrile in the presence of tetrachlorosilane (TCS). A binary reagent from (TCS)/ZnCl$_2$ **was used upon applying** benzonitrile.

INTRODUCTION

A recent increased attention witnessed to the use of tetrachlorosilane (TCS), a cheap industrial intermediate, in different areas of the organic chemistry has now reached significant levels, not only for the possibility to perform environmentally benign synthesis, but for the good yield. In continuation of our investigations on the development and applications of new in situ reagents derived from tetrachlorosilane in organic synthesis [1] - [9], we have developed an efficient, general, and convenient protocol for the multicomponent synthesis of 1-ami- doal-kyl-2-naphthols.

Multicomponent coupling reactions (MCRs) represent powerful time-, energy-, and material-saving synthetic protocols in modern chemistry in which molecular complexity could be generated in a single synthetic operation [10] [11] .

The preparation of 1-amidoalkyl-2-naphthols can be carried out by multi-component condensation of aryl aldehydes, β-naphthol and amide derivatives in the presence of Lewis or Bronsted acid catalysts such as $Ce(SO_4)_2$ [12] , montmorillonite K10 [13] , iodine [14] , cation exchanged resins [15] , $NaHSO_4$·H_2O [16] , $Fe(HSO_4)_3$ [17] , sulfamic acid/ultrasound [18] , $HClO_4/SiO_2$ [19] , cyanuric chloride [20] and $K_5CoW_{12}O_{40}$·$3H_2O$ [21] . Some of the reported methods suffer from one or more limitations such as high reaction temperature, lower product yield, tedious work up, and use of toxic reagents. They also lack general applicability to produce arrays of 1-ami- doalkyl-2-naphthols as they are restricted to only a few amides. Therefore, the development of a more general, cost-effective multicomponent coupling reaction (MCR) protocol for the synthesis of 1-amidoalkyl-2-naphthols remains a challenge. Therefore the introduction of new, efficient, and general methods involving various nitriles for this multicomponent reaction under milder conditions is still required.

EXPERIMENTAL

Infrared (IR) spectra were recorded on JASCO 410 spectrometer. Absorption maxima were recorded in cm-1. Nuclear Magnetic Resonance (NMR) spectra were run at Varian-Mercury (300 MHz) FTNMR spectrometer. Spectra were taken using DMSO solvent with chemical shifts quoted in parts per million (δ ppm) using TMS as internal standard. The Mass Spectra (M.S.) were recorded on GC-MS QP-2010 EX Schmiadzu (Japan) mass spectrometer. Melting points (uncorrected) were determined in an open capillary with a Griffin melting point apparatus. Column Chromatography was carried out by using Merck Kieselgel 60 GF-254 (230 - 400 mesh). Analytical TLC was performed on aluminum sheets (Merck, Kieselgel 60 F254, Thickness 0.2 mm). Tetrachlorosilane (TCS) was used as obtained from commercial sources. The solvents were distilled and dried before use. Acetonitrile was dried by refluxing over phosphorous pentoxide then distilled. Methylene chloride was dried by refluxing over anhydrous calcium chloride then distilled.

Typical Procedure

In 50 ml round bottom flask equipped with air condenser and magnetic stirring bar, a mixture of aromatic aldehyde (5 mmol), anhydrous naphthol (5 mmol), and tetrachlorosilane (1.8 ml, 15 mmol) was allowed to stir at room temperature for 10 minutes. To this reaction mixture, 5 mmol of nitrile [acetonitrile or benzonitrile] was added and the stirring process was continued for further time (monitored by TLC). The reaction mixture was quenched using ice cold water. The aqueous solution was extracted with chloroform (3 × 30 ml) and then the chloroform extract was dried over anhydrous Na_2SO_4 and concentrated over boiling water bath. The residual oil was purified using preparative thin layer chromatography using silica gel to give the products in pure form.

(1) N-((2-Hydroxynaphthalen-1-yl)(phenyl)methyl)acetamide:

mp 246°C (lit. [12] mp 241°C - 243°C); IR (KBr, cm^{-1}): 3399, 3245, 3062, 1639, 1581, 1517, 1369, 1334, 1101, 806, 740, 692, 617; 1H-NMR(300 MHz, DMSO-d$_6$): δ (ppm) 1.97 (s, 3H), 7.1 - 7.35 (m, 9H),7.74 - 7.85 (m, 3H), 8.38 (d, J = 9 Hz, 1H), 9.95 (s, 1H); ^{13}C NMR(75 MHz, DMSO-d$_6$): δ (ppm) 23.5, 41.3, 118.8, 120.2, 122.0, 123.9, 124.9, 125.7, 127.6, 128.1, 128.3, 128.5, 128.6, 134.2, 144.0, 152.6, 169.6; MS: m/z 231(100%), 232(75.7%), 233(13.16%), 291(20.72%)M$^+$.

(2) N-((4-chlorophenyl)(2-hydroxynaphthalen-1-yl)methyl)acetamide

mp 230°C (lit. [12] mp 224°C - 227°C); IR (KBr, cm^{-1}): 3394, 2960, 2705, 2609, 1631, 1581, 1521, 1436, 1371, 1328, 1272, 1240, 1170, 1091, 815, 748, 586, 499.

(3) N-((2-hydroxynaphthalen-1-yl)(p-tolyl)methyl)acetamide:

mp 224°C (lit. [17] mp 222°C - 223°C); IR (KBr, cm^{-1}): 3394, 2969, 2707, 2616, 1625, 1515, 1436, 1332, 1272, 1174, 1064, 811; ^{1}H-NMR(300 MHz, DMSO-d$_6$): δ (ppm) 2.02 (s, 3H), 2.24 (s, 3H), 7.05 - 7.40(m, 8H), 7.76 - 7.94 (m, 3H), 8.45 (d, J = 8.7 Hz, 1H), 10.25 (s, 1H); ^{13}C NMR(75 MHz, DMSO-d$_6$): 20.50, 22.68, 47.74, 118.53, 119.04, 122.33, 123.31, 126.03, 126.23, 128.52, 129.11, 132.35, 135.04, 139.56, 153.13, 169.18; MS: m/z 245(100%), 246(75.7%), 247(13.1%), 305(21%)M$^+$.

(4)N-((4-(dimethylamino)phenyl)(2-hydroxynaphthalen-1-yl)methyl) acetamide:

mp 215°C (lit. [12] mp 215°C - 217°C); IR (KBr, cm^{-1}): 3432, 3021, 2886, 2803, 1664, 1523, 1469, 748.

(5) N-((4-formylphenyl)(2-hydroxynaphthalen-1-yl)methyl)acetamide:

mp 232°C (lit. [22] mp 237°C - 240°C); IR (KBr, cm^{-1}): 3386, 3056, 1700, 1627, 1604, 1272, 817.

(6) N-((2-bromophenyl)(2-hydroxynaphthalen-1-yl)methyl)acetamide:

mp 210°C (lit. [23] mp 204°C); IR (KBr, cm^{-1}): 3421, 3122, 1656, 1517, 1434, 1058, 817, 752.

(7) N-((3-bromophenyl)(2-hydroxynaphthalen-1-yl)methyl)acetamide:

mp 230°C (lit. [24] mp 229°C - 230°C); IR (KBr, cm^{-1}): 3407, 3164, 1643, 1517, 1436, 1066, 811, 748.

(8) N-((2-nitrophenyl)(2-hydroxynaphthalen-1-yl)methyl)acetamide:

mp 215°C (lit. [25] mp 218°C - 219°C); IR (KBr, cm⁻¹): 3410, 3265, 1643, 1590, 1520, 1415, 1345, 1305, 1225, 1105, 1010, 950, 852, 745.

(9) N-((3-nitrophenyl)(2-hydroxynaphthalen-1-yl)methyl)acetamide:

mp 180°C (lit. [25] mp 256°C - 258°C); IR (KBr, cm⁻¹): 3241, 1625, 1596, 1525, 1436, 1346, 1286, 1211, 1091, 1014, 948, 856, 732; ¹H-NMR(300 MHz, DMSO-d₆): δ (ppm) 2.06 (s, 3H), 7.15 - 7.49 (m, 6H), 7.78 - 8.04 (m, 5H), 8.54 (d, J = 8.1 Hz, 1H), 10.12 (s, 1H); ¹³C NMR(75 MHz, DMSO-d₆): 25.55, 66.13, 108.61, 118.16, 120.28, 122.32, 123.94, 125.57, 127.46, 128.4, 129.20, 129.46, 130.84, 132.08, 134.82, 147.69, 148.04, 152.49, 191.67; MS: m/z 276(100%), 277(75.7%), 278(13.16%), 336(13%)M⁺.

(10) N-((4-methoxyphenyl)(2-hydroxynaphthalen-1-yl)methyl)acetamide:

mp 178°C (lit. [17] mp 183°C - 186°C); IR (KBr, cm⁻¹): 3399, 3245, 1639, 1511, 1438, 1365, 1303, 1247, 1172, 1029, 815, 748.

(11) N-((2-hydroxynaphthalen-1-yl)(phenyl)methyl)benzamide:

mp 230°C (lit. [26] mp 235°C - 237°C); IR (KBr, cm⁻¹): 3413, 3222, 1629, 1531, 1488, 1346, 752; ¹H-NMR (300 MHz, DMSO-d₆): δ (ppm) 7.23 - 8 (m, 17H), 9.01 (1H, d, J=9 Hz), 10.3 (1H, s); ¹³C NMR(75 MHz, DMSO-d₆): 167.30, 155.13, 142.59, 135.39, 132.59, 131.44, 130.27, 128.53, 128.42, 128.08, 127.98, 127.36, 126.99, 126.88, 126.56, 122.55, 117.26, 62.66; MS: m/z 233 (6.5%), 232 (42.25%), 231 (100%), 353 (4.74%) M⁺.

(12) N-((4-chlorophenyl)(2-hydroxynaphthalen-1-yl)methyl)benzamide:

mp 191°C (lit. [27] mp 185°C - 186°C); IR (KBr, cm⁻¹): 3424, 3116, 1633, 1569, 1535, 1484, 1434, 1344, 1270, 819, 707.

(13) N-((2-hydroxynaphthalen-1-yl)(p-tolyl)methyl)benzamide:

mp 213°C (lit. [27] mp 216°C - 218°C); IR (KBr, cm⁻¹): 3413, 1631, 1533, 1346, 1272, 1078, 817.

(14) N-((2-hydroxynaphthalen-1-yl)(4-methoxyphenyl)methyl)benzamide:

mp 196°C (lit. [28] mp 197°C - 199°C); IR (KBr, cm⁻¹): 3423, 1625, 1571, 1511, 1436, 1348, 1263, 1174, 1029, 933, 821; ¹H-NMR (300 MHz, DMSO-d₆): δ (ppm) 10.38 (1H, s), 9.05 (1H, d, J = 8.7 Hz), 8.11 (1H, d, J = 8.4 Hz),7.79 - 7.90 (3H,m), 7.45 - 7.55 (2H,m),7.25 - 7.32 (3H,m), 6.84 (2H, d, J = 8.7 Hz), 3.68 (3H, s); ¹³C NMR(75 MHz, DMSO-d₆): 48.98, 54.98, 113.61, 118.48, 118.76, 122.65, 126.69, 127.02, 127.73, 128.38, 128.47, 12858, 129.21, 131.33, 132.25, 133.92, 134.43, 153.07, 158.05, 165.54; MS: m/z 263 (16.5%), 262 (40.5%), 261 (100%), 383 (5.7%) M⁺.

(15)   N-((4-(dimethylamino)phenyl)(2-hydroxynaphthalen-1yl)methyl) benzamide:

mp 220°C (lit. [25] mp 220°C - 221°C); IR (KBr, cm^{-1}): 3405, 3273, 1631, 1573, 1521, 1434, 1340, 1180, 1029, 939, 811.

(16) N-((2-hydroxynaphthalen-1-yl)(2-nitrophenyl)methyl)benzamide:

mp 261°C (lit. [25] mp 266°C - 267°C); IR (KBr, cm^{-1}): 3440, 3145, 1649, 1510, 1490, 1441, 1347, 1249, 805, 712.

(17) N-((2-hydroxynaphthalen-1-yl)(3-nitrophenyl)methyl)benzamide:

mp 220°C (lit. [21] mp 216°C - 218°C); IR (KBr, cm^{-1}): 3413, 3180, 1644, 1522, 1468, 1428, 1356, 1272, 800, 712.

(18) N-((2-hydroxynaphthalen-1-yl)(4-nitrophenyl)methyl)benzamide:

mp 226°C (lit. [21] mp 228°C - 230°C); IR (KBr, cm^{-1}): 3444, 3155, 1649, 1510, 1493, 1447, 1347, 1244, 803, 713; ^{1}H-NMR(300 MHz, DMSO-d$_6$): δ (ppm) 10.24 (s, 1H, OH), 7.98 (d, J =7.8 Hz, 1H), 9.00 (d, J = 8.0 Hz, 1H), 7.10 - 7.25 (m, 4H, ArH), 7.46 - 7.66 (m, 7H, ArH), 7.76 - 7.98 (m, 4H, ArH); ^{13}C NMR(75 MHz, DMSO- d$_6$): 48.20, 116.29, 118.78, 119.94, 122.49, 125.42, 126.91, 128.54, 128.84, 129.55, 130.27, 131.48, 131.96, 132.69, 134.32, 141.78, 153.72, 165.44; MS: m/z 278(4.5%), 277(43.25%), 276(100%), 384(4.%) M$^+$.

RESULTS AND DISCUSSION

With the aim to develop more efficient synthetic processes and convenient protocol; for the one-pot synthesis of 1-amidoalkyl-2-naphthols, biologically active drug like molecules [29] - [31]. We herein describe a practical, inexpensive protocol for the preparation of 1-amidoalkyl-2-naphthols via multi component condensation reaction between various aromatic aldehydes, β-naphthols and nitriles including alkyl and aryl nitrile using readily available tetrachlorosilane (TCS) reagent at room temperature, and solvent-free conditions. To test the general scope and versatility of this procedure in the synthesis of a variety of substituted amidoalkyl naphthols, we examined a number of different substituted aromatic aldehydes, 2-naphthol and acetonitrile. To optimize the reaction condition, the reaction of benzaldehyde, β-naphthol and acetonitrile was selected as a model to investigate the effects of different amounts of reagent on the yield. The best result was obtained by carrying out the reaction with one molar amounts of aldehyde and β-naphthol, and two molar amounts of acetonitrile as shown in Table 1. We also examined the reaction in various solvents. Chlorinated solvents such as methylene chloride or 1,2-dichloroethane were found to be ineffective solvents yielding the reaction product in very low conversion. The donor solvents such as tetrahydrofuran (THF) and diethyl ether (DEE) were completely inhibited the reaction.

The reaction between benzaldehyde, acetonitrile and 2-naphthol was carried out without TCS and we found that no reaction takes place. To determine the optimum quantity of TCS, the reaction was carried out at room temperature. The use of two equimolar amounts of TCS, resulted in the highest yield. The molar ratio of aldehyde, β-naphthol, acetonitrile and TCS was kept at 1:1:2:2, respectively (Scheme 1).

Thus we prepared a range of 1-amidoalkyl-2-naphthols under the optimized reaction conditions: stirring the β-naphthol (1 mmol), aryl aldehydes (1 mmol) and acetonitrile (2 mmol) in the presence of tetrachlorosilane (2 mmol) at room temperature. A series of 1-amidoalkyl-2-naphthols were prepared in high to excellent yields (Table 2).

In the case of aromatic aldehydes the three-component reaction proceeded smoothly to give the corresponding 1-amidoalkyl-2-naphthols in high yields. Due to the availability of a vast number of aromatic aldehydes, this three component reaction can be very useful to synthesis the desired products.

As Table 2 shows that, the reaction proved to be general and tolerated a variety of aromatic aldehydes with substituent carrying either electron-donating (Table 2, entries 2, 3, 4, 6, 7, and 10) or electron-withdrawing groups (Table 2, entries 5, 8, and 9). Although as can be seen from the results of table, this reaction is affected by electronic and steric factors.

The suggested mechanism for this reaction is depicted in Scheme 2. This mechanism involves 1,2-addition of tetrachlorosilane (TCS) to aldehyde to produce silyl ether intermediate [A]. On the other hand, 2-naphthol reacted with TCS and produce silyl enol ether [B] and HCl. As reported in literature [32] [33] the reaction of 2-naphthol with aromatic aldehydes in the presence of Lewis acid is known to give ortho-quinone methides (o-QMs). The same o-QMs, generated in-situ, intermediate [E] have been reacted with nitrilium salt [34] [35] [C] to form 1-amidoalkyl-2-naphthol after hydrolysis (Scheme 1). The reaction gave the desired products in the Ritter type reaction [36] [37] at room temperature. To our knowledge, there is no general protocol employing aromatic nitrile in such addition so far. Therefore, we examined the reaction of benzonitrile with aldehyde, β-naphthol and TCS as representative example to aryl nitrile. It is noteworthy that no reaction was observed under the above conditions to give the corresponding 1-amidoalkyl-2-naphthols typically.

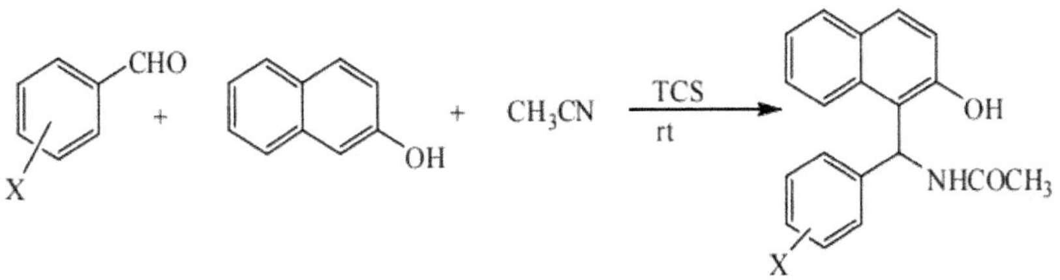

Scheme 1: Reaction of different aldehydes, acetonitrile and 2-naphthol.

Table 1: Effect of different molar ratio on the yield of product.

Molar ratio between benzaldehyde, 2-naphthol and acetonitrile	Reaction time (hr)	Yield (%)
1:1:1	15	73
1:1:2	15	83
2:1:1	15	60

Table 2: Reaction of aldehydes with β-naphthol and acetonitrile in the presence of tetrachlorosilane.

Entry	Reactant	Time (hr)	Product name	Yield (%)	Product structure
1	C_6H_5CHO	15	N-((2-hydroxynaphthalen-1-yl)(phenyl)methyl) acetamide	83	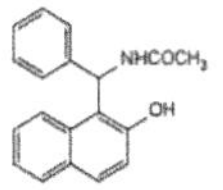
2	p-ClC$_6$H$_4$CHO	14	N-((4-chlorophenyl) (2-hydroxynaphthalen-1-yl)methyl) acetamide	78	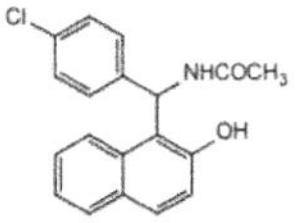
3	p-CH$_3$C$_6$H$_4$CHO	17	N-((2-hydroxynaphthalen-1-yl)(p-tolyl)methyl) acetamide	79	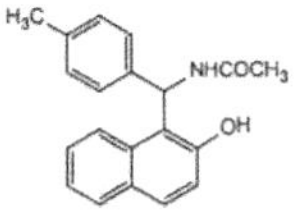
4	p-(Me)$_2$NC$_6$H$_4$CHO	18	N-((4-(dimethylamino)phenyl)(2-hydroxynaphthalen-1-yl)methyl)acetamide	65	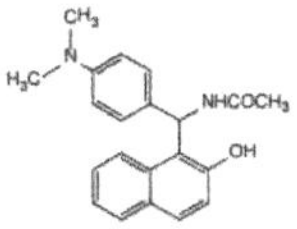
5	p-OHCC$_6$H$_4$CHO	13	N-((4-formylphenyl) (2-hydroxynaphthalen-1-yl)methyl) acetamide	70	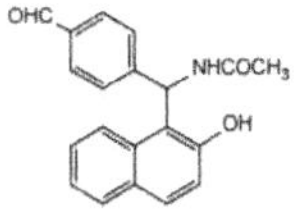
6	o-BrC$_6$H$_4$CHO	13	N-((2-bromophenyl) (2-hydroxynaphthalen-1-yl)methyl) acetamide	85	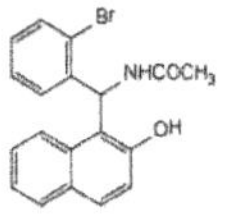

7	m-BrC₆H₄CHO	14	N-((3-bromophenyl) (2-hydroxy naphthalen-1-yl)methyl) acetamide	80	
8	o-O₂NC₆H₄CHO	18	N-((2-nitrophenyl) (2-hydroxy naphthalen-1-yl)methyl) acetamide	67	
9	m-O₂NC₆H₄CHO	18	N-((3-nitrophenyl) (2-hydroxy naphthalen-1-yl)methyl) acetamide	70	
10	p-CH₃OC₆H₄CHO	14	N-((4-methoxyphenyl) (2-hydroxy naphthalen-1-yl) methyl) acetamide	73	

The use of two equivalents of benzonitrile and tetrachlorosilane and one equivalent of zinc chloride as a binary reagent in the reaction resulted in the highest yield of N-((2-hydrox- ynaphthalen-1-yl) (phenyl)methyl) benzamide, which might be attributed to steric factors as well as to the low of nucleophilicity of benzonitrile (Scheme 3). As shown in Table 3, reaction of aldehydes with β-naphthol and benzonitrile in the presence of tetrachlorosilane and zinc chloride as a binary reagent under solvent- free conditions at room temperature produced highest yields from 1-benzamidomethyl-2-naphthol derivatives. Attempts to bring aliphatic aldehydes such as acetaldehyde into the reaction with 2-naphthol and acetonitrile under mild conditions were mostly unsuccessful. Therefore no other aliphatic Aldehydes were examined. To regard the purity of the compounds prepared, melting point and infrared spectra were matched with previously reported literature data. NMR and MS analysis to some compounds revealed the correct structure of the products obtained.

CONCLUSION

An efficient one-pot method has been developed for the synthesis of 1-amidoalkyl-2-naphthols from condensation of aldehyde, β-naphthol, acetonitrile and TCS as condensing reagent. The present methodology gives several advantages such as simple procedure, easy workup, high yields, and solvent free reaction conditions.

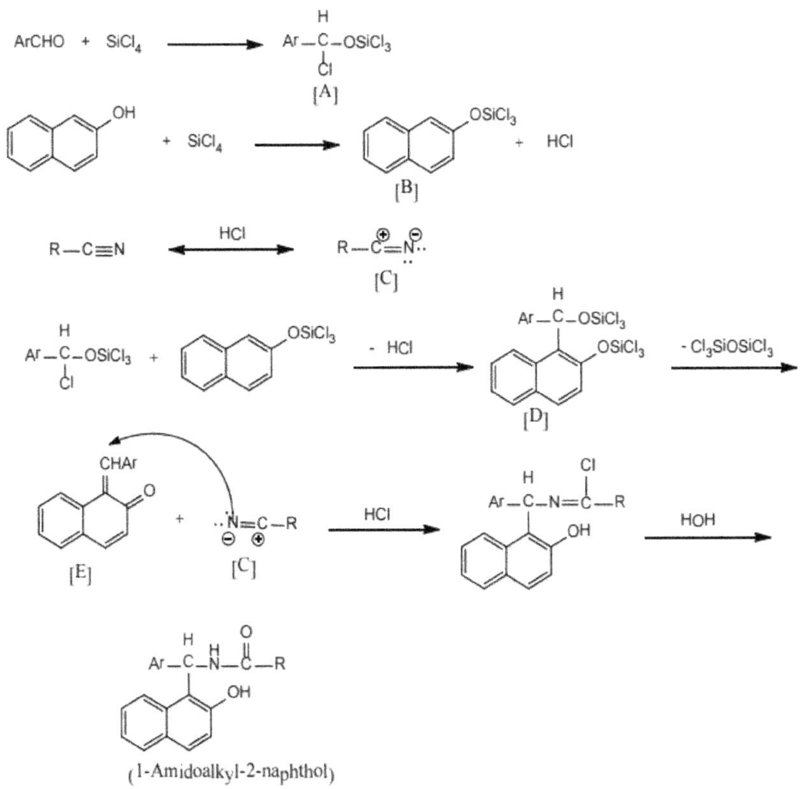

Scheme 2: Suggested mechanism for the formation of 1-amidoal-kyl-2-naph- thol.

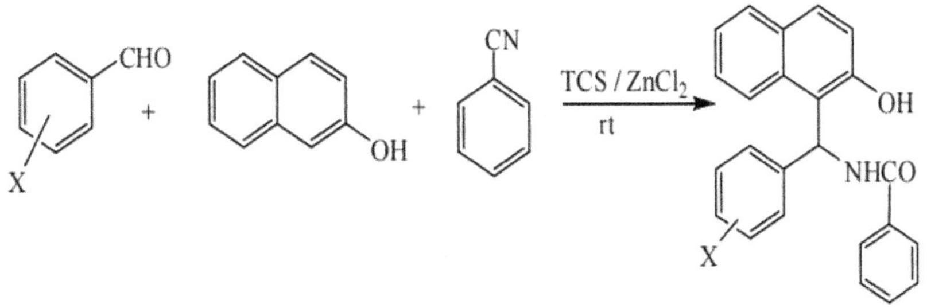

Scheme 3: Reaction of different aldehydes, benzonitrile and 2-naphthol.

Table 3: Synthesis 1-benzamidomethyl-2-naphthol derivatives in the presence of TCS/ ZnCl$_2$ at r.t. under solvent- free conditions.

entry	Reactant	Time (hr)	Product name	Yield (%)	Product structure
11	C$_6$H$_5$CHO	10	N-((2-hydroxynaphthalen-1-yl)(phenyl)methyl) benzamide	88	
12	p-ClC$_6$H$_4$CHO	8	N-((4-chlorophenyl) (2-hydroxy naphthalen-1-yl)methyl) benzamide	80	
13	p-CH$_3$C$_6$H$_4$CHO	8	N-((2-hydroxynaphthalen-1-yl)(p-tolyl)methyl) benzamide	76	
14	p- CH$_3$O C$_6$H$_4$CHO	8	N-((4-methoxyphenyl)(2-hydroxynaphthalen-1-yl) methyl) benzamide	77	
15	p-(Me)$_2$NC$_6$H$_4$CHO	11	N-((4-(dimethylamino)phenyl) (2-hydroxynaphthalen-1-yl) methyl)benzamide	73	
16	o-O$_2$NC$_6$H$_4$CHO	9	N-((2-nitrophenyl) (2-hydroxy naphthalen-1-yl)methyl) benzamide	82	
17	m-O$_2$NC$_6$H$_4$CHO	8	N-((3-nitrophenyl) (2-hydroxy naphthalen-1-yl)methyl) benzamide	85	
18	p-O$_2$NC$_6$H$_4$CHO	9	N-((4-nitrophenyl)(2-hydroxy naphthalen-1-yl)methyl) benzamide	79	

The salient feature of this methodology includes an easy purification, generality and in addition no cumbersome apparatus were needed.

CITE THIS PAPER

Samy B.Said,Mohammad M. A.Mashaly,Ahmed M.Sheta,Saad S.Elmorsy, (2015) New Method for Preparation of 1-Amidoalkyl-2-Naphthols via Multicomponent Condensation Reaction Utilizing Tetrachlorosilane under Solvent Free Conditions. *International Journal of Organic Chemistry*,**05**,191-199. doi: 10.4236/ijoc.2015.53019

REFERENCES

1. Azumaya, I., Kagechika, H., Yamaguchi, K. and Shudo, K. (1996) Facile Formation of Aromatic Cyclic N-Methylamides Based on cis Conformational Preference. Tetrahedron Letters, 37, 5003-5006. http://dx.doi.org/10.1016/0040-4039(96)00995-1

2. Elmorsy, S.S., Pelter, A. and Smith, K. (1991) The Direct Production of Tri- and Hexa-Substituted Benzenes from Ketones Under Mild Conditions. Tetrahedron Letters, 32, 4175-4176. http://dx.doi.org/10.1016/S0040-4039(00)79896-0

3. Elmorsy, S.S., Pelter, A., Smith, K., Hursthouse, M.B. and Ando, D. (1992) Investigations of the Tetrachlorosilane-Ethanol Induced Self Condensations of Ketones. Tetrahedron Letters, 33, 821-824. http://dx.doi.org/10.1016/S0040-4039(00)77724-0

4. Elmorsy, S.S., Khalil, A.M., Girges, M.M. and Salama, T.A. (1997) A New Approach to the Stereoselective Synthesis of β-Methylchalcones. Journal of Chemical Research, Synopses, No. 7, 232-233. http://dx.doi.org/10.1039/a607611d

5. Elmorsy, S.S., El-Ahl, A.S., Soliman, H.A. and Amer, F.A. (1995) Synthesis of Triazidochlorosilane (TACS). A Novel Silicon Mediated One Pot Conversion of Aldehydes to Nitriles. Tetrahedron Letters, 36, 2639-2640. http://dx.doi.org/10.1016/0040-4039(95)00302-S

6. Salama, T.A., El-Ahl, A.S., Khalil, A.M., Girges, M.M., Lackner, B., Steindi, C. and Elmorsy, S.S. (2003) A Convenient Regiospecific Synthesis of New Conjugated Tetrazole Derivatives via the Reaction of Dienones with the Tetrachlorosilane-Sodium Azide Reagent and their NMR Structural Assignment. Monatshefte für Chemie, 134, 1241-1252. http://dx.doi.org/10.1007/s00706-003-0045-x

7. Salama, T.A., Elmorsy, S.S., Khalil, A. M., Girges, M.M. and El-Ahl, A.S. (2007) Novel Uncatalyzed Hydrocyanation of Ketones utlizing Tetrachlorosilane-Potassium Cyanide Reagent. Synthetic Communications, 37, 1313-1319. http://dx.doi.org/10.1080/00397910701226897

8. Salama, T.A., Elmorsy, S.S., Khalil, A.M. and Ismail, M.A. (2007) A SiCl4-ZnCl2 Induced General, Mild and Efficient One-Pot, Three-Component Synthesis of β-Amido Ketone Libraries. Tetrahedron Letters, 48, 6199-6203. http://dx.doi.org/10.1016/j.tetlet.2007.06.128

9. Salama, A.T., Ismail, A.M., Khalil, M.A. and Elmorsy, S.S. (2012) Silicon-Assisted O-Heterocyclic Synthesis: Mild and Efficient One-Pot Syntheses of (E)-3-Benzylideneflavanones, Coumarin-3-Carbonitriles/

Carboxamides, and Benzannulated Spiropyran Derivatives. Archive for Organic Chemistry, 2012, 242-253.

10. Thomson, L.A. and Elman, J.A. (1996) Synthesis and Applications of Small Molecule Libraries. Chemical Reviews, 96, 555-600. http://dx.doi. org/10.1021/cr9402081

11. Hulme, C. and Gore, V. (2003) Multi-Component Reactions: Emerging Chemistry in Drug Discovery "From Xylocain to Crixivan". Current Medicinal Chemistry, 10, 51-80. http://dx.doi. org/10.2174/0929867033368600

12. Selvam, N.P. and Perumal, P.T. (2006) A New Synthesis of Acet-amido Phenols Promoted by Ce(SO4)2. Tetrahedron Letters, 47, 7481-7483. http://dx.doi.org/10.1016/j.tetlet.2006.08.038

13. Kantevari, S., Vuppalapati, S.V.N. and Nagarapu, L. (2007) Mont-morillonite K10 Catalyzed Efficient Synthesis of Amidoalkyl Naphthols under Solvent Free Conditions. Catalysis Commu-nications, 8, 1857-1862. http://dx.doi.org/10.1016/j.catcom.2007.02.022

14. Das, B., Laxminarayana, K., Ravikanth, B. and Rao, B.R. (2007) Iodine Catalyzed Preparation of Amidoalkyl Naphthols in Solution and under Solvent-Free Conditions. Journal of Molecular Catalysis A: Chemical, 261, 180-183. http://dx.doi.org/10.1016/j.molcata.2006.07.077

15. Patil, S.B., Singh, P.R., Surpur, M.P. and Samant, S.D. (2007) Cation-Exchanged Resins: Efficient Heterogeneous Catalysts for Facile Synthesis of 1-Amidoalkyl-2-naphthols from One-Pot, Three-Component Condensations of Amides/Ureas, Aldehydes, and 2-Naphthol. Synthetic Communications, 37, 1659-1664.http://dx.doi. org/10.1080/00397910701263858

16. Shaterian, H.R. and Yarahmadi, H. (2008) Sodium Hydrogen Sulfate as Effective and Reusable Heterogeneous Catalyst for the One-Pot Preparation of Amidoalkyl Naphthols. ARKIVOC, 2008, 105-114.http:// www.arkat-usa.org/get-file/23225/

17. Shaterian, H.R., Yarahmadi, H. and Ghashang, M. (2008) An Efficient, Simple and Expedition Synthesis of 1-Amidoalkyl-2-naphthols as "Drug Like" Molecules for Biological Screening. Bioorganic & Medicinal Chemistry Letters, 18, 788-792. http://dx.doi.org/10.1016/j.bmcl.2007.11.035

18. Patil, S.B., Singh, P.R., Surpur, M.P. and Samant, S.D. (2007) Ultrasound-Promoted Synthesis of 1-Amidoalkyl-2- naphthols via a Three-Component Condensation of 2-Naphthol, Ureas/Amides, and Aldehydes, Catalyzed

by Sulfamic Acid under Ambient Conditions. Ultrasonics Sonochemistry, 14, 515-518.http://dx.doi.org/10.1016/j.ultsonch.2006.09.006

19. Shaterian, H.R., Yarahmadi, H. and Ghashang, M. (2008) Silica Supported Perchloric Acid (HClO4-SiO2): An Efficient and Recyclable Heterogeneous Catalyst for the One-Pot Synthesis of Amidoalkyl Naphthols. Tetrahedron, 64, 1263-1269. http://dx.doi.org/10.1016/j. tet.2007.11.070

20. Mahdavinia, G.H. and Bigdeli, M.A. (2009) Wet Cyanuric Chloride Promoted Efficient Synthesis of Amidoalkyl Naphthols under Solvent-Free Conditions. Chinese Chemical Letters, 20, 383-386.http://dx.doi. org/10.1016/j.cclet.2008.12.018

21. Nagarapu, L., Baseeruddin, M., Apuri, S. and Kantevari, S. (2007) Potassium Dodecatungstocobaltate Trihydrate (K5CoW12O40?3H2O): A Mild and Efficient Reusable Catalyst for the Synthesis of Amidoalkyl Naphthols in Solution and under Solvent-Free Conditions. Catalysis Communications, 8, 1729-1734.http://dx.doi.org/10.1016/j. catcom.2007.02.008

22. Zare, A., Hasaninejad, A., Rostami, E., Moosavi-Zare, A.R., Roshankar, N., Khedri, F. and Khedri, M. (2010) An Efficient Solvent-Free Protocol for the Synthesis of 1-Amidoalkyl-2-naphthols Using Silica-Supported Molybdatophosphoric Acid. E-Journal of Chemistry, 7, 1162-1169. http:// dx.doi.org/10.1155/2010/512392

23. Zandi, M. and Sardarian, A.R. (2012) Eco-Friendly and Efficient Multi-Component Method for Preparation of 1- Amidoalkyl-2-naphthols under Solvent-Free Conditions by Dodecylphosphonic Acid (DPA). Comptes Rendus Chimie, 15, 365-369. http://dx.doi.org/10.1016/j. crci.2011.11.012

24. Shahrisa, A., Somayehesmati and Nazari, M.G. (2012) Boric Acid as a Mild and Efficient Catalyst for One-Pot Synthesis of 1-Amidoalkyl-2-naphthols under Solvent-Free Conditions. Journal of Chemical Sciences, 124, 927-931. http://dx.doi.org/10.1007/s12039-012-0285-6

25. Nandi, G.C, Samai, S., Kumar, R. and Singh, M.S (2009) Atom-Efficient and Environment-Friendly Multicomponent Synthesis of Amidoalkyl Naphthols Catalyzed by P2O5. Tetra-hedron Letters, 50, 7220-7222. http://dx.doi.org/10.1016/j.tetlet.2009.10.055

26. Nagawade, R.R. and Shinde, D.B. (2007) Zirconyl(IV) Chloride—Catalyzed Multicomponent Reaction of β-Naphthols: An Expeditious Synthesis of Amidoalkyl Naphthols. Acta Chimica Slovenica, 54, 642-646.

27. Sapanamalik, S. and Singh, R.K. (2012) Microwave Assisted Synthesis of 1-Amidoalkyl-2-naphthols Catalyzed by Anhydrous Zinc Chloride. Asian Journal of Chemistry, 24, 5669-5672.

28. Jiang, W.Q.L., An, T.J. and Zou, P. (2008) Molybdophosphoric Acid: An Efficient Keggin-Type Hetero-poloacid Catalyst for the One-Pot Three-Component Synthesis of 1-Amidoalkyl-2-naphthols. Chinese Journal of Chemistry, 26, 1697-1701.http://dx.doi.org/10.1002/cjoc.200890307

29. Dingermann, T., Steinhilber, D. and Folkers, G. (2004) Molecular Biology in Medicinal Chemistry. Wiley-VCH, Weinheim.

30. Shen, A.Y., Tsai, C.T. and Chen, C.L. (1999) Synthesis and Cardiovascular Evaluation of N-Substituted 1-Aminomethyl-2-naphthols. European Journal of Medicinal Chemistry, 34, 877-882.http://dx.doi.org/10.1016/S0223-5234(99)00204-4

31. Shen, A.Y., Chen, C.L. and Lin, C.I. (1992) Electrophysiological Basis for the Bradycardic Effects of 1-(1-Pyrrolidi-nylmethyl)-2-naphthol in Rodents. Chinese Journal of Physiology, 35, 45-54.

32. Khodaei, M.M., Khosropour, A.R. and Oghanian, H.M. (2006) A Simple and Efficient Procedure for the Synthesis of Amidoalkyl Naphthols by p-TSA in Solution or under Solvent-Free Conditions. Synlett, 2006, 916-920. http://dx.doi.org/10.1055/s-2006-939034

33. Haaf, W. (1963) Studies on Ritter Reaction. Chemische Berichte, 96, 3359-3369.http://dx.doi.org/10.1002/cber.19630961237

34. Ziegler, E., Kleineberg, G. and Meindle, H. (1963) Synthesen von Heterocyclen, 46. Mitt.: über Reaktionen mit Ketens?urechloriden (Kurze vorl?ufige Mitteilung). Monatshefte für Chemie, 94, 544-548. http://dx.doi.org/10.1007/BF00903495

35. Ritter, J.J. and Minieri, P.P. (1948) A New Reaction of Nitriles. I. Amides from Alkenes and Mononitriles. Journal of the American Chemical Society, 70, 4045-4048.http://dx.doi.org/10.1021/ja01192a022

36. Tillmanns, E.J. and Ritter, J.J. (1957) Notes-Nitriles in Nuclear Heterocyclic Syntheses. I. Dihydro-1,3 Oxazines. The Journal of Organic Chemistry, 22, 839-840. http://dx.doi.org/10.1021/jo01358a612

37. Smith, P.A.S. and Sullivan, J.M. (1961) The Cyclization of N-Alkenylthionamides to Thiazolines and Dihydrothiazines. The Journal of Organic Chemistry, 26, 1132-1136.http://dx.doi.org/10.1021/jo01063a038

Chapter 7

SYNTHESIS OF NEW FLUORINE/PHOSPHORUS SUBSTITUTED 6-(2'-AMINO PHENYL)-3-THIOXO- 1,2,4-TRIAZIN-5(2H, 4H)ONE AND THEIR RELATED ALKYLATED SYSTEMS AS MOLLUSCICIDAL AGENT AS AGAINST THE SNAILS RESPONSIBLE FOR BILHARZIASIS DISEASES

Abeer N. Al-Romaizan, Mohammed S. T. Makki, Reda M. Abdel-Rahman

Department of Chemistry, Faculty of Science, King Abdul Aziz University, Jeddah, KSA

ABSTRACT

New fluorine substituted 6-(5'-fluoro-2'-triphenylphosphiniminophenyl) 3-thioxo-1,2,4-triazin-5 (2H, 4H) one (2) was obtained via Wittig's reaction of the corresponding 6-(5'-fluoro-2'-amino- phenyl)-3-thioxo-1,2,4-triazinone (1). Behavior of compound 2 towards alkylating agents and/or oxidizing agents was studied were, N-hydroxyl (3), Mannich base (4,5), S-alkyl (6,7,8) and thiazolo [3,2-b][1,2,4] triazinones (10-14) and or 3-disulfide (18), 3-sulfonic acid 19 and 1,2,4-triazin-3,5- Dionne (20) derivatives obtained. Structures of the new products are established by elemental and spectral data. The new targets obtained screened as Molluscicidalagents against Biomophlaria Alexandrina snails responsible for Bilharziasis diseases, in compare with Baylucide as standard drug.

INTRODUCTION

The incorporation of fluorine atoms into a heterocyclic nitrogen molecule frequently provides properties of pharmacological interest as compared to their non-fluorinated analogs [1] -[5] . Also, bonded phosphorus atoms with S, O, N and C-atoms of heterocyclic systems enhance their important properties as herbicides, pesticides and insecticides [6] -[11] . On the other hand, 3-thioxo-1,2,4-triazin-5-one derivatives and their N- and S-alkyl derivatives have gained considerable attention due to their well as medicinal utility such as anti-HIV, anti AIDS and anticancer agents [12] -[16] . Literature reveals that no reports of a molecular scaffold containing these important cores. With this based upon these observations. The present work aims to synthesis and chemical reactivity of 1,2,4-triazinone bearing, fluorine, phosphorus and sulfur atoms through alkylation reactions and the new systems as Molluscicidal agents against Biomophalaria Alexandrina snails by removal from the wastewater (Clean water).

EXPERIMENTAL

Melting points were determined with an electro-thermal Bibbly Stuart Scientific Melting point SMPI (UK). A Perkin Elmer (Lambda EZ-210) double beam spectrophotometer (190 - 1100 nm) used for recording the elec- tronic spectra. A Perkin Elmer model RXI-FT-IR 55,529 cm^{-1} used for recording the IR spectra (EtOH as sol- vents). A Brucker advance DPX 400 MHz using TMS as an internal standard for recording the ^{1}H/^{13}C NMR spectra in deuterated DMSO (δ in ppm). AGC-MS-QP 1000 Ex model is used for recording the mass spectra. Hexafluorobenzene was used as external standard for ^{19}F NMR at 8425 MHz and ^{31}P (in CDCl$_3$, 101.25 MHz) [17] . Elemental analysis was performed on Micro Analytical Center of National Reaches Center-Dokki, Cairo, Egypt. Compound 1 was prepared according the reported method [14] and compound 15 as procedure published [18] .

6-(5'-Fluor phenyl)2'amino-3-thioxo-1,2,4-triazine-5(2H, 4H)one (1)

Equimolar mixture of 5-fluoroisatin (in 100 ml NaOH, 5%) and thiosemicarbazide (in 10 ml H$_2$O) reflux for 2 h, then cold and poured onto ice-HCl. The solid result was filtered off and crystallized from EtOH as yellow crystals to give 1. Yield (80%), m.p. 263°C - 265°C. Analytical data, Found C = 44.91, H = 2.90, F = 7.58, N = 23.40, S = 13.29%; Calculated for C$_9$H$_7$FN$_4$OS (238) C = 45.37, H = 2.94, F = 7.98 N = 23.52, S = 13.44%, M/Z (256, M + H$_2$O, 5%), base peak (68, 100%), 148 (21), 136 (18), 110 (30), 96 (50), 82 (58), 70 (78); UV: (λ_{max} EtOH) 280 nm. IR vcm^{-1} = 3424 (NH$_2$) 3258, 3169 (NH, NH), 1685 (C=O), 1618 (NH$_2$), 1545 (C=N), 1263 (C-F): 858, 818 (aryl CH) 685 (C-F) ^{1}H NMR (DMSO) = 14.66, 16.66, 10.90 (each 1H, s, 3NH), 8.68 - 8.06, 7.69 - 7.64, 7.39, 7.39 - 7.30 (3H, aryl protons) ^{13}C NMR = δ 179.47 (C=S), 162 (C=O), 159 - 157 (spin coupl- ing C-F), 138.54 (C=N), 131.82, 121.8, 121.51 (aromatic carbons), 78.14, 77.71 (C5-C6 1,2,4-triazine).

6-[5'-Fluoro-2'(triphenylphosphinimino)phenyl]-3-thioxo-1,2,4-triazine-5(2H, 4H)one (2)

A mixture of 1 (0.01 mol) and triphenyl phosphine (0.01 mol) in acetonitrile (20 ml), THF (20 ml) reflux for 2 h then cold. The solid produced and crystallized from EtOH to give 2 as deep yellowish crystals. Yield (70%) m.p. 249°C - 250°C. Analytical date Found C = 64.60, H = 3.96, F = 3.70, N = 11.01, S = 6.33%. Calculated for $C_{27}H_{20}FN_4OPS$ (498): C = 65.06, H = 4.01, F = 3.81, N = 11.24, S = 6.42%, M/Z (498.00) 370 (4), 290 (10), 171 (60), 159 (100), 128 (20), 118 (40), 102 (65), 92 (58), 76 (58), 65 (40); UV: (λ_{max} EtOH) 310 nm. IR vcm^{-1} = 3335 (NH), 1658 (C=O), 1380 (P=N), 1250 (C-F), 1087 (C-S) 1045, (P-N) 879 (aryl CH), 650 (C-F). ^{1}H NMR (DMSO): 14.56, 12.78 (each 1H, s, 2NH), 8.21, 7.76, 7.66, 7.65, 7.65, 7.64, 7.484, 7.840, 7.47, 7.464, 7.460, 7.45, 7.398, 7.391, 7.38, 7.37, 7.29, 7.28, 7.28, 7.27, 7.02, 7.01, 7.009, 7.005, 6.994, 6.990, 6.866, 6.859, 6.852, 6.845, (18H of aromatic protons). ^{13}C NMR (DMSO) = δ 179.74 (C=S), 163.0 (C=O), 138.62 (C-F), 131 (C=N), 118.94 - 107.94 (aromatic carbons), 7.76, 77.21 (C_5-C_6 of 1,2,4-triazine).

2,4-Di(hydroxymethyl)-6-(5'-fluoro-2'-triphenylphosphiniminophenyl)-3-thioxo-1,2,4-triazine-5-one (3)

A mixture of 2 (0.01 mol) and formaldehyde (0.02 mol), in methanol (50 ml) reflux for 2 h, cold. The solid obtained filtered off and crystallized from MeOH to give 2 as faint yellow crystals. Yield = (65%) m.p. 280°C - 282°C. Analytical date: Found C = 61.92, H = 4.21, F = 5.23, N = 9.93, S = 5.43% Calculated for $C_{29}H_{24}FN_4PSO_3$ (558): C = 62.36, H = 4.30, F = 3.40, N = 10.03, S = 5.73%. UV: (λ_{max} EtOH) = 363 nm. IR vcm^{-1} = 3346 (b, 2OH) 2974, 2889 (2CH$_2$), 1646 (C=O), 1382 (P=N), 1240 (C-F), 1086 (C-S), 1046 (P-N), 879 (aryl CH), 755 (C-F). ^{1}H NMR (DMSO) δ 8.34 - 6.84 (18 aromatic protons), 4.8, 4.4 (each s, 2H, alcoholic 2OH) 2.62, 2.58 (each s, 4H, 2CH$_2$). ^{13}C NMR (DMSO) = δ 179.86 - 179.68 (C=S), 163.07 (C=O), 159.62, 158.03 (C-F), (C-N), 138.60 (C=N), 132.121 - 107.94 (aromatic carbons), 77.75 77.32 (C5-C6 of 1,2,4-triazine), 40.57 - 40.46, 40.32 - 4.14 (2 CH$_2$).

2,4-Di(Piperidinomethyl)-6-(5'-fluoro-2'-triphenylphosphin-iminophenyl)-3-thioxo-1,2,-4-triazine-5-one (4)

A mixture of 2 (0.01 mol), piperidine (0.02 mol) and formaldehyde (0.02 mol) in methanol (50 ml) reflux for 2 h, cold. The solid produced filtered off and crystallized from MeOH to give 4 as yellow crystals. Yield = (60%) m.p. 179°C - 180°C. Analytical date found C = 67.41, H = 5.83, F = 2.55, N = 11.97, S = 4.33%. Calculated for $C_{39}H_{42}FN_6OPS$ (692): C = 67.63, H = 12.13, F = 2.74, N = 12.13, S = 4.62%. IR vcm^{-1} = 3062 (aromatic CH), 2936, 2840 (aliphatic CH$_2$), 1721 (C=O), 1538 (C=N), 1468 (deform CH$_2$), 1389) (P=N), 1248 (C-F), 1184 (C=S), 1049 (P-N), 885, 815, 754 (aryl CH), 709 (C-F). ^{1}H NMR (DMSO): δ 8.23 - 6.80 (18H, aromatic), 2.95, 2.92, 2.89 and 2.58 (CH$_2$ of piperdine, N-CH$_2$-N). ^{13}C NMR (DMSO) = δ 172 (C=S), 154 (C=O), 147.25 (C-F), 137.54 (C=N), 116.10 - 108 (aromatic carbons), 77.80, 77.39 (C_5-C_6 of

1,2,4-triazine), 40 - 58, 40.47, 40.33, 40.19, 40.05 (CH_2 of piperidine) 39.91 - 39.63 (N-CH_2-N).

2,4-Di(4'-arylaminomethyl)-6-(5'-fluoro-2'-triphenylphosphiniminophenyl)-3-thioxo-1,2,4-triazine-5- ones (5a & 5b)

A mixture of 2 (0.01 mol) and formaldehyde (0.02 mol), 4-fluoroaniline and/or 4-aminoantipyrine (0.02 mol) in methanol (50 ml) warm under reflux for 2 h, then cold. The solid thus obtained, filtered off and crystallized from EtOH to give 5a & 5b as yellow crystals.

Compound 5a, yield (75%) m.p. 210°C - 212°C. Analytical data: Found C = 65.80, H = 4.11, F = 7.47, N = 10.89, S = 4.01%. Calculated for $C_{41}H_{32}F_3N_6OPS$ (744), C = 66.12, H = 4.30; F = 7.66; N = 11.29, S = 4.30%. M/Z = 744 (M, 0, 0%), 370 (1), 367 (25), 290 (60), 272 (30), 248 (20), 218 (42), 169 (100), 128 (85), 102 (100), 65 (100). UV: (λ_{max} EtOH) 364 nm; IR vcm^{-1} = 3343 (aryl-NH), 2974, 2889 (CH_2), 1650 (C = O), 1382 (P = N), 1250 (C-F), 1086 (C-S), 1045 (P-N) 879 (aryl CH). ^{1}H NMR (DMSO) = δ 12.71 (s, 1H, NH), 12.55 (s, 1H, NH), 7.41, 7.39, 7.32, 7.06, 7.01, 6.99, 6.98, 6.97, 6.91, 6.90 & 6.89, 6.88, 6.878, 6.873, 6.86 & 6.85, 6.84, 6.83, 6.77 & 6.768, 6.761, 6.754, 6.742, 6.735 (aromatic CH), 5.18 - 5.15 & 5.14 - 5.13 (4H, CH_2 of N CH_2-NH). ^{13}C NMR = (DMSO) = δ 178.0 (C=S), 161.02 (C=O) 144.99 (C-F), 138.37 (C=N), 130.65 - 114.11 (aromatic car- bons) 77.57, 77.14 (C_5-C_6 of 1,2,4-triazine), 40.61 - 40.35, 39.94 - 39.66 (2N-CH_2-NH).

Compound 5b, yield (60%); m.p. 200°C - 202°C. Analytical data: Found C = 65.89, H = 4.91, F = 2.04, N = 14.83, S = 3.35%; Calculated for $C_{51}H_{46}FN_{10}O_3PS$ (928), C = 65.94, H = 9.95, F = 2.04, N = 15.08, S = 3.44%. IR vcm^{-1}: 3277 (b, NH), 2974 (aliphatic CH_3) 2840 (CH_2), 1697, 1660 (2C=O), 1604 (C=N), 1542 (C=N), 1482 (deform, CH_2), 1416 (P=N), 1370 (NCSN), 1274 (C-F), 1152 (C-S), 1045 (P-N), 878, 810, 750 (aromatic CH), 650 (C-F). ^{1}H NMR (DMSO) δ 7.48 - 7.26 (aromatic CH) 5.07 (s, 1H, NH), 3.1 - 3.0 (s, 1H, NH), 2.99 - 2.88, 2.84 - 2.82 (2 CH_2), 2.62 - 2.41, 2.34 - 2.23, 2.229 - 2.2221, 2.16 - 2.07 (4 Me). ^{13}C NMR = (DMSO) δ 179.81 (C=S) 159.11 (C=O), 142 (C-F), 138.0 (C=N), 129.38 - 123.31 (aromatic carbons), 77.54, 77.11 (C_5-C_6 of 1,2,4- triazin), 40.61 - 40.08 & 39.94 - 39.66 & 36.75 - 35.49 (N-CH_2), 18.42, 15.15 (N-Me, C-Me).

[6-(5'-Fluoro-2'-triphenylphosphiniminophenyl)-5-oxo-1,2,4-triazine-3-yl] thioacetic acid (6)

Equimolar mixture of 2 and monochloroacetic acid in DMF (20 ml) warm for 30/min, then poured onto ice. The solid yielded filtered off and crystallized form EtOH to give 6 as faint yellow crystals. Yield (80%), m.p. 187°C - 188°C. Analytical data: Found: C = 62.42, H = 3.81, F = 3.20, N = 9.85, S = 5.57%;

Calculated for $C_{29}H_{22}FN_4O_3PS$ (556). C = 62.58, H = 3.95, F = 3.41, N = 10.07, S = 5.75. IR vcm^{-1} = 3327 (b, OH, NH), 2973, 2884 (CH$_2$), 1659 (b, 2 C=O), 1440 (deform CH$_2$), 1380 (P=N), 1250 (C-F), 1087 (C-S), 1045 (P-N) 880 (Ar CH), 810 (Ar CH). ^{1}H NMR (DMSO) δ = 10.31 (s, 1H, NH), 8.06, 8.0, 7.98 - 7.97, 79.86 - 7.84, 7.73, & 7.726, 7.721, 7.14, 7.08, 7.67 & 7.66, 7.65, 7.63, 7.53 - 7.51, 7.48 - 7.35 & 7.34, 6.997, 6.992, 6.838 - 6.824 (18 CH, aromatic) & 4.74 (s, 1H, OH of COOH), 3.86 - 3.24 (2H, CH$_2$). ^{13}C NMR: (DMSO): δ 168.29 (C=S), 165.40 (C=O), 157.16 (C=O), 142.13 (C-F) 131.35 (C=N), 130.04 - 102.25 (aromatic carbons), 72.43, 72.22 (C$_5$-C$_6$ of 1,2,4-traizine), 34.95 - 34.81 (CH$_2$ carbon).

1,1-Di[6-(5'-Fluoro-2'-triphenylphosphiniminophenyl)-5-oxo-1,2,-4-triazine-3'yl]dimercaptoacetic acid (7)

A mixture of 2 (0.02 mol) and 1,1-dicholoracetic acid (0.01 mol) in DMF (20 ml) reflux for 30 min, cold then poured into ice. The resulted solid filtered off and crystallized from dioxin to give 7 as faint yellow crystals, yield (60%) m.p. 238°C - 240°C. Analytical data: Found C = 63.45, H = 3.49, F = 3.39, N = 10.39, S = 5.88%, Calculated for $C_{56}H_{40}F_2N_8O_4P_2S_2$ (1052) C = 63.87, H = 3.80, F = 3.61, N = 10.64, S = 6.08%. IR vcm^{-1} = 3425, 3259, 3170 (OH, NH, NH), 1865, 1680 (C=O), 1618 (C=N), 1476, 1452 (aliphatic CH), 1360 (P=N), 1252 (C-F), 1193 (C-S), 1045 (P-N), 903, 859, 818, 758 (aryl CH) 685 (C-F). ^{1}H NMR (DMSO): δ 12.79, 12.78 (each s, 2NH), 10.75 (s, 1H, OH), 8.21 - 6.84 (18 CH, aromatic), 2.82 - 2.59 (s, 1H, CH) ^{13}C NMR: δ 179.72 (C=S), 163 (C=O), 159 (C-S), 158.0 (C-S), 138.61 (C-F), 132.18 (C=N), 121.12 - 107.94 (aromatic carbons), 77.66, 77.45 (C$_5$-C$_6$) of 1,2,4-triazine), 40.57 - 40.46 (-CH-).

Tri[6-(5'-fluoro-2'-triphenxylphosphin-iminophenyl)-5-oxo-1,2,-4-triazine-3'yl]trimercaptoacetic acid (8)

A mixture of 2 (0.03 mol) and 1,1,1-trichloroacetic acid (0.01 mol) in DMF (20 ml) warm for 30 min then cold and poured on to ice. The produced solid filtered off and crystallized from Et OH to give 8 as reddish crys- tals. Yield (60%); m.p. 189°C - 190°C. Analytical data: Found C = 63.89, H = 3.45, F = 3.55 N, 10.67, S = 5.83%. Calculated for $C_{83}H_{58}F_3N_{12}O_5P_3S_3$ (1548); C = 64.34, H = 3.74, F = 3.68, N = 10.85, S = 6.2%. UV (λ_{max} EtOH) 359 nm. IR vcm^{-1} = 3500 - 3100 (b, 3NH, OH) 1716 (C=O), 1624 (NH = OH of 1,2,4-triazinone) 1537 (C=N) 1471 (aliphatic CH). 1390 (P=N), 1300 (NCSN), 1260 (C-F), 1200 (C-S), 1645 (P-N), 920, 850, 780 (aryl CH), 650 (C-F). ^{1}H NMR (DMSO) = δ 12.95, 12.72, 12.33 (each s, 3H, NH), 10.84 (s, 1H, OH), 8.51, 8.23, 8.01, 7.92, 7.89, 7.71, 7.7, 7.69, 7.65, 7.63, 7.59, 7.58, 7.57 - 7.54, 7.50 - 7.47, 7.40 - 7.37, 7.33 - 7.31, 7.02 - 6.98, 6.86 - 6.83 (aromatic CH). ^{13}C NMR = (DMSO) δ 179.61 (C=S), 163 (C=O), 159.5 (C-S), 157 (C-F) 138.58 (C=N), 137.79

(C=N), 132.91 (C=N), 131.99 - 107.93 (aromatic CH), 77.92, 77.49 (C_5-C_6 of 1,2,4-tria- zine).

6(5'-Fluoro-2'-triphenylphosphin-iminophenyl)2,-4-dihydro-thiazolo[3,2-b][1,2,4]triazine-3,7-dione (9)

Equimolar mixture of 2 and monochloroactic acid in DMF (20 ml) reflux for 2 h then cold and poured onto ice. The solid obtained filtered off and crystallized from dioxan to give 9 as brown ppt, Yield (60%) m.p. 224°C. Compound 6 (0.50 mg) heat above its melting point (60°C higher) for 10 min, cold then treat with MeOH. The solid produced filtered off and crystallized from dioxan to give 9 as brown ppt. Yield (58%), m.p. 225°C - 227°C. Analytical data Found C = 64.40, H = 3.51, F = 3.35, N = 9.93, S = 5.48% Calculated for $C_{29}H_{20}FN_4O_2PS$ (538); C = 64.68, H = 3.71 F = 3.53, N = 10.40. S = 5.94%. UV: (λ_{max} EtOH) 352 nm. IR vcm^{-1} = 3204 (b-OH), 1694 (C=O), 1623 (C=N), 15,636, 1475 (CH$_2$), 1380 (P=N), 12,999 (C-F), 1148 (C-S) 816, (aromatic CH), 711 (C-F). ^{1}H NMR (DMSO) = δ 10.79 (s, 1H, Phenolic OH), 8.23 (s, 1H, CH of thazole), 7.95 - 7.47, 7.35, 7.35, 6.98, 6.80 (aromatic CH). ^{13}C NMR = (DMSO): δ 167.21 (C=S), 147.47, (C=O), 136.64 (C-F), 132.97 (C=N), 131.92 - 128.54, 118.80 - 118.14, 113.79 - 113.73 (aromatic carbons), 111.04 - 110.99 (-CH=), 77.80, 77.38 (C_5-C_6 of 1,2,4-triazine).

6(5'-Fluoro-2'-triphenylphosphiniminophenyl)-5-oxo-3-(cyanomethylthia)-2H-1,2,4-triazine (10)

A mixture of 2 (0.01 mol) and chloroacetinitrile (0.01 mol) in DMF (20 ml) warm (10 min) then cold and poured onto ice. The result solid filtered off and crystallized from dioxan to give 10 as faint Yellow crystals. Yield (70%); m.p. 214°C - 215°C. Analytical data: Found C = 64.39, H = 3.58, F = 3.11, N, 12.85, S = 5.75%. Calculated for $C_{29}H_{21}FN_5OPS$ (537); C = 64.80, H = 3.91, F = 3.53, N = 13.03, S = 5.95%. M/Z = 537 (5%) 281 (20), 207 (60), 149 (20), 113 (30), 85 (100), 58 (100). UV: (λ_{max} EtOH) 321 nm. IR vcm^{-1} = 3424, 3167 (NH, S-CH=C=NH) 2100 - 2085 (C≡N), 1646 (C=O), 1595 (C=N), 1481 (CH$_2$), 1370 (P-N), 967, 839, 762 (aryl CH), 700 (C-F). ^{1}H NMR (DMSO) = δ 13.90, (s, 1H, NH), 12.76 (s, 1H, HC=NH), 8.22 - 6.81 (aromatic CH), 4.69 (1H, HC=NH) 2.59 (2H, CH$_2$). ^{13}C NMR (DMSO): δ 158.11 (C=O), 147.0 (C-F) 132 (C=N), 131.86 - 128.44 (aromatic carbons), 112.21 (C≡N), 77.96, 77.53 (C_5-C_6 of 1,2,4-triazine), 40.133 (-CH=NH), 33.63 (CH$_2$).

3-Amino-6(5'-fluoro-2'-triphenylphosphiniminophenyl)-thiazolo[3,2-b][1,2,4]triazine-7-one (11)

Compound 10 (0.5 gm) in DMF (20 ml) warm for 2 h then cooled and poured onto ice. The solid produced filtered off and crystallized from EtOH to give 11 as broom ppt, Yield (66%); m.p. 223°C - 225°C. Analytical data: Found C =

64.51 H = 3.38, F = 3.21 N = 12.55, S = 5.62%. Calculated for $C_{29}H_{21}FN_5OPS$ (537), C = 64.80, H = 3.91, F = 3.53, N = 13.03, S = 5.95%. M/Z, 537 (2%), 370 (2), 226 (2), 168 (100), 140 (60), 114 (30), 62 (18), 70 (18). IR vcm^{-1} = 3348, (b-NH$_2$), 16430 (C=O), 1383 (P=N), 1250 (C-F), 1086 (C-S), 1045 (P-N), 878 (aryl CH). ^{1}H NMR (DMSO) = δ 8.11 (s, 1H, = CH thiazole), 7.72 - 7.011, 6.98 - 6.80 (aromatic CH), 3.99 - 3.84 (2H-NH$_2$). ^{13}C NMR (DMSO) = 162.54 (C=O), 132.16 (C-F), 132.00 (C=N), 131.99 (C-S), 131.66 - 131.64 (=CH-), 128.61 - 120.55 (aromatic carbons), 77.59, 77.38 (C$_5$-C$_6$ of 1,2,4-triazine), 40.51 (-N-C=N).

3-(4'-Fluoro benzoyl)amino-6-(5'-fluoro-2'-triphenylphosphiniminophenyl)-thiazolo[3,2-b][1,2,4] triazine-7-one (12)

Equimolar mixture of 11 and 4-fluorobenzoyl chloride in DMF (20 ml) warm for 10 min then cold and poured onto ice. The resulted solid filtered off and crystallized from EtOH to give 12 as deep-Yellowish crystals. Yield (75%). m.p. 205°C - 207°C. Analytical data: Found C = 65.19 H = 3.41, F = 5.49 N = 10.51, S = 4.59%. Calcu- lated for $C_{36}H_{25}F_2N_5O_2PS$ (660), C = 65.55, H = 3.80, F = 5.76, N = 10.60, S = 4.80. IR vcm^{-1} = 3342, (b-NH), 1651 (b, 2C=O), 1381 (P=N), 1326 (NCSN) 1230 (C-F), 1086 (C-S), 1045 (P-N), 879 (aryl CH). ^{1}H NMR (DMSO) = δ 13.70 (s, 1H, NH), 9.89 (s, = CH of thiazole) 8.411, 8.17, 8.07 - 8.05, 8.01, 7.99, 7.95, 7.79, 7.66 - 7.64, 7.44 - 7.42, 7.28, 7.27 (aromatic CH). ^{13}C NMR = (DMSO): δ 167.53 (C=O), 162.54 (C=O) 138.59 (C-F) 132.25 (C-N), 132.19 (C=N), 129.33 - 127.27, 117.78 - 115.17, 112.15, 12.09, 110.52, 108.12, 107.95 (aromatic carbons), 77.64. 77.43 (C$_5$-C$_6$ of 1,2,4-triazine). Schiff base (13)

Equimolar amounts of 11 and 4-fluorobenzaldehyde in absolute ethanol (20 ml) reflux for 30 min then cooled. The solid thus obtain filtered off and crystallized from EtOH to give 13 as Yellowish ppt. Yield (70%); m.p. 248°C - 250°C. Analytical data: Found C = 66.85, H = 3.61, F = 5.75 N = 10.59, S = 4.71%. Calculated for $C_{36}H_{24}F_2N_5OPS$ (643), C = 67.18, H = 3.73, F = 5.90, N = 10.88, S = 4.97%. IR vcm^{-1} = 3100, 2880 (aromatic & aliphatic CH), 1700 (C=O), 1600 (C=N), 1483 (C-P), 1370 (P=N), 1230 (C-F), 1200 (C-S), 1045 (P-N), 880, 840, 810 (aryl CH), 650 (C-F)^{1}H NMR (DMSO) = δ 9.97 (s, 1H, -CH=N-), 8.62 (s, 1H, -CH = thiazole) 8.23, 8.22, 8.09 - 8.00, 7.94 - 7.92, 7.71 - 7.63, 7.56 - 7.53, 7.49 - 7.45, 7.26 - 7.23, 7.12 - 7.10, 7.0 - 6.96, 6.89 - 6.84, 6.81 - 6.79 (aromatic CH).

6-(5'-Fluoro-2'-triphenyl   phosphiniminophenyl)-3-oxo-3phenyl-thiazolo[3,2-b][1,2,4]triazine-7-one (14)

A mixture of 2 (0.01 mol) and phenacylbromide (0.01 mol) in ethanolic KOH, (20 ml, 5%) reflux for 2 h, cold then poured onto ice-HCl. The solid produced filtered off and crystallize from dioxan to give 14 as brown ppt. Yield (60%);

m.p. > 300°C. Analytical data: Found C = 69.88, H = 3.59, F = 3.01 N, 9.00, S = 5.13%, Calcu- lated for $C_{35}H_{24}FN_4OPS$ (598); C = 70.23, H = 4.01, F = 3.17, N = 9.36 S = 5.35%. IR vcm^{-1} = 3080, 3030 (aromatic CH), 1680 (C=O), 1380 (P=N), 1240 (C-F), 1180 (C-S), 1045 (P-N), 880, 850 (aryl CH).

Diaylthioether (16)

A mixture of 2 (0.01 mol) and Schiff base 15 (0.01 mol) in dry C_6H_6 (100/ml) reflux 8 h, cold and used peter- either 100°C - 120°C to complete precipitation. The solid obtained filtered off and crystallized dioxan to give 16 as Yellowish crystals. Yield (80%); m.p. 204°C - 205°C. Analytical data: Found C = 66.53, H = 4.31, F = 4.44, N = 11.88, S = 3.66%, Calculated for $C_{45}H_{36}F_2N_7O_2PS$ (807); C = 66.91, H = 4.46, F = 4.70, N = 12.14, S = 3.96%. M/Z = (807.0.0), 580 (5), 515 (4), 462 (8), 423 (10), 370 (5), 339 (20), 282 (20), 225 (20), 207 (60), 176 (100), 149 (56), 119 (38), 85 (90), 58 (100). UV: (λ_{max} EtOH) 323 nm. IR vcm^{-1} = 3332 (NH), 2973, 2886 (ali- phatic CH), 1636 (C = O), 1488 (CH$_3$), 1381 (P=N), 1324 (NCSN), 1250 (C-F), 1086 (C-S), 1045, (P-N), 880, 755 (aryl CH). ^{1}H NMR (DMSO) = δ 12.77, (s, 1H, NH), 10.75 (s, 1H, NH), 9.68 (s, 1H, S-CH-Ar), 8.23, 7.83 - 7.28, 7.27 - 7.00, 6.99 - 6.84 (aromatic CH), 3.69 (s, CH$_3$-N) 2.79 (s, CH$_3$-C).^{13}C NMR (DMSO): δ 164.73 (C= O), 163.08 (C=O), 160.63, 155.18 (C-F), 151.9 (C-S), 134.64, 134.25, 134.23 (C=N), 129.487 - 115.492 (aro- matic carbons), 77.72, 77.50 (C$_5$-C$_6$ of 1,2,4-triazine), 67.00 (S-CH=NH), 39.95 - 39.81, 39.67 - 39.55 (2 CH$_3$).

2,3-Diaryl?2,3-dihydro-4-thioxo-7-(5'-fluoro-2'-triphenylphosphiniminophenyl)-1,3,5-thiazolo[3,2-b]

[1,2,4] triazine-8-one (17)

A mixture of 16 (0.01 mol) and CS$_2$ (5 ml) in DMF (20 ml) reflux for 4 h, cold then powered onto ice. The resultant solid filtered off and crystallized from dioxan to give 17 as yellowish crystals. Yield (75%), m.p. 254°C - 255°C. Analytical data: Found C = 64.88, H = 3.85, F = 4.38, N = 11.40, S = 7.45%, Calculated for $C_{46}H_{34}F_2N_7O_2PS_2$ (849); C = 65.01, H = 4.00, F = 4.47, N = 11.54, S = 7.53%; M/Z (849, 0.0%), 370 (2), 329 (40), 290 (100), 159 (100), 128 (100), 102 (100), 96 (100), 65 (100). UV: (λ_{max} EtOH) 34.7 nm. IR vcm^{-1} = 2873 (aliphatic CH), 1684 (C=O), 1614. 1593 (C=N) 1475, 1425 (CH$_3$), 1318 (P=N), 1264 (C-F), 1199 (C=S), 1130 (C-S), 1052 (P-N), 985, 899, 854, 814, 732 (aryl CH). ^{13}C NMR (DMSO) = δ 179.70 (C=S), 163.07 (C=O), 155.08 (C-F), 138.62 (C=N, 1,2,4-triazine), 132.23, 132.13 (C=C pyrazole), 129.43 - 115.44, 112.09 - 107.96 (aromatic carbons), 77.71, 77.28 (C$_5$-C$_6$ of 1,2,4-triazine) 66.94 (S-CH-NH), 40.57, 39.76 (2 CH$_3$).

Di-Heteroaryldisulfide (18)

Compound 2 (0.05 gm) and $FeCl_3$ (0.5 gm) in MeOH (20 ml) reflux for 3 h, then filtered. The solid produced filtered off and crystallized from dioxan to give 18 as deep-yellowish crystals. Yield (80%), m.p. 238°C - 240°C. Analytical data: Found C = 64.85, H = 10.89, F = 3.55, N = 10.89, S = 6.22%. Calculated for $C_{54}H_{38}F_2N_8O_2P_2S_2$ (994); C = 65.19, H = 11.26, F = 3.82, N = 11.26, S = 6.43%. IR vcm^{-1} = 3300, 3200 (NH, NH), 1680 (C=O), 1600 (C=N), 1350 (P=N), 1100 (C-S), 1040 (P-N), 900, 850, 800 (aryl CH), 650 (C-F). ^{1}H NMR (DMSO) = δ 14.55, 12.78 (each s, 2H, NH, NH), 8.20 - 6.85 (aromatic CH). ^{13}C CNMR (DMSO): δ 179.90, 179.74 (2C-S), 159 - 66, 158.66 (2C=O), 138.63 (C-F), 132.18 (C=N), 121.12 - 107.95 (aromatic carbons), 77.65, 77.43 (C_5-C_6 of 1,2,4-triazine).

6-(5'-Fluoro-2'-triphenylphosphiniminophenyl)-5-oxo-2H-1,2,4-triazine-3-sulfonic acid (19)

Compound 2 (0.05 gm) in ethanol (10 ml) and H_2O_2, (0.5 ml) add with stirring for 2 h. The solid obtained fil- tered off and crystallized from EtOH to give 19 as yellowish crystals yield (75%); m.p. 258°C - 260°C. Analyti- cal data: Found C = 59.00, H = 3.44, F = 3.25, N = 9.87, S = 5.45%. Calculated for $C_{27}H_{20}FN_4O_4PS$ (546), C = 59.34, H = 3.66, F = 3.47, N = 10.25, S = 5.86%. IR vcm^{-1} = 3300 (NH), 1696 (C=O), 1390 (NCSN), 1360 (P=N), 879, 820, 780, (aryl CH). 670 (C-F). ^{1}H NMR (DMSO) = δ 12.79 (s, 1H, NH), 10.72 (s, 1H, SO_2-OH), 8.13 - 6.76 (aromatic CH). ^{13}C NMR (DMSO): δ 179.75 (-S=O), 163.09 (C=O), 159.67 (C-F), 138.62 (C=N), 132.23 (C-S), 121.08 - 107.95 (aromatic carbons), 77.60, 77.39 (C_5-C_6 of 1,2,4-triazine).

6-(5'-Fluoro-2'-triphenylphosphiniminophenyl)-1,2,4-triazine-3,5-(2H, 4H) dione (20)

Compound 2 (0.05 gm) in ethanol (10 ml) and $KMnO_4$ solution (ethanolic 1%, 1 ml) add drop wise then stirr- ing for 2 h. The produced solid filtered off and crystallized from Et OH to give 20 as yellowish crystals yield (50%); m.p. 273°C - 275°C. Analytical data: Found C = 66.89, H = 4.01, N = 11.35, F = 3.55. Calculated for $C_{27}H_{20}FN_4O_2P$ (482), C = 67.21, H = 4.14, N = 11.61, F = 3.94. M/Z: (482, 0.0%), 370 (10), 206 (101, 148, (16), 128 (24), 110 (35), 96 (55), 83 (78), 68 (100). IR vcm^{-1} = 3426, 3259, 3170 (OH, NH, NH), 1766, 1681 (2C=O), 1619 (C=N) 1452 (C-P) 1301 (P=N), 1252, (C-F) 1048 (P-N), 905, 861, 820, 802, 784, 761 (CH), 687 (C-F), ^{1}H NMR (DMSO) = δ 12.73, 10.82 (each s, NH, OH), 7.88 - 6.84 (aromatic CH). ^{13}C NMR = δ 153.58 (C=O), 152.78 (C=O), 147.84 (C-F). 132.58 (C=N) 130.51 - 111. (Aromatic carbons), 77.85, 77.60 (C_5-C_6 of 1,2,4-tri- azine).

RESULTS AND DISCUSSION

Chemistry

A recent work on the synthesis and chemistry of bioactive sulfur bearing 1,2,4-triazinone moiety was reported [16] [19] . In continuation of this attitude the present investigation reports the synthesis of fluorine and phosphorus-substituted 6-amino-phenyl-3-thioxo-1,2,4-triazin-5-(2H, 4H) one (1) and study that behavior towards vari- ous alkylating agents. Treatment of 5-fluoroisatin with thiosemicarbozide in alkaline medium [14] [15] pro- duced 6-(2'-amino-5'-fluorophenyl-3-thioxo-1,2,4-triazin-5-(2H'4H) one (1). Warm compound 1 with triphe- nylphosphine in acetonitrile produced the yield 2 (Scheme 1).

In the imino [yield, 2] a negative charge of nitrogen is bonded to positive charge of phosphorus stabilized by partial overlap of the filled N-P orbital. This stabilization increase due to the charge on the α-carbon atom is spread by 1,2,4-triazine resonance. Abdel-Rahman [14] [15] reported that N-alkyl of 3-thioxo-1,2,4-triazinones exhibited a wide biological spectrum anti HIV and anticancer properties. Similarly, hydroxyl methylation of compound 2 by boil with formaldehyde-methanolproduced 2,4-di(hydroxylmethyl)-6-(5'fluoro-2'-triphenylphos- phiniminophenyl)-3-thioxo-1,2,4-triazin-5-one (3). Also, reflux of compound 2 with secondary and primary amines such as piper dine, 4-fluoroaniline and 4-amino-antipyrine in the presence of formaldehyde methanol, furnished the Mannich bases 4 and 5 (Scheme 2).

Formation of 3 and 4 was may be as (Figure 1).

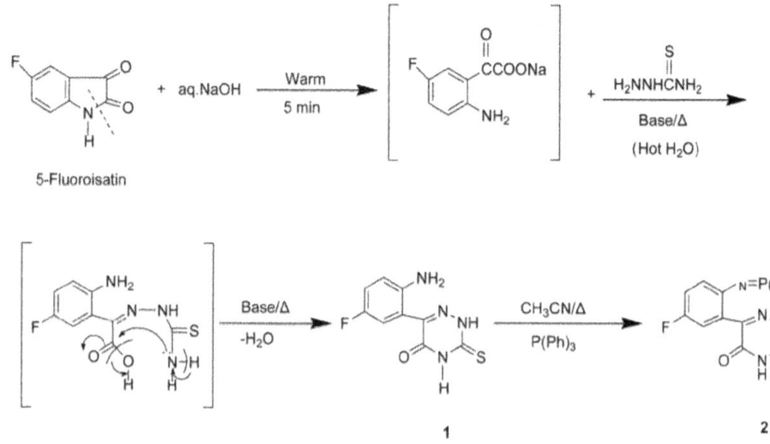

Scheme 1: Formation of compounds 1 & 2.

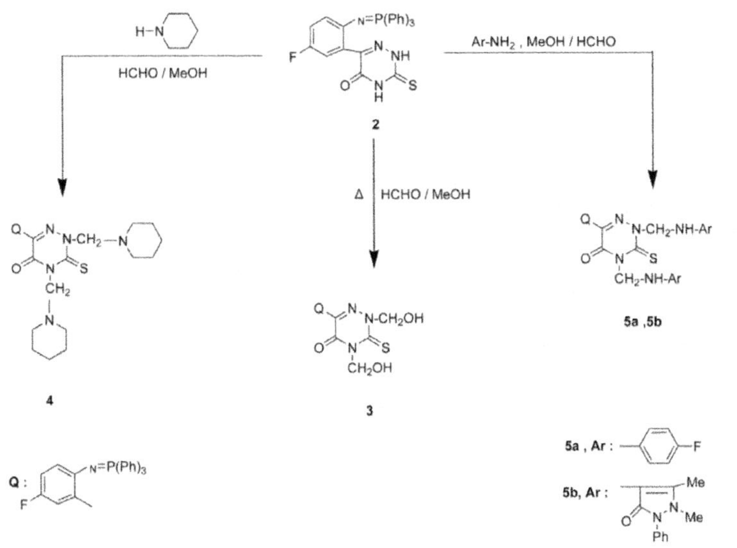

Scheme 2: Formation of compounds 3 - 5.

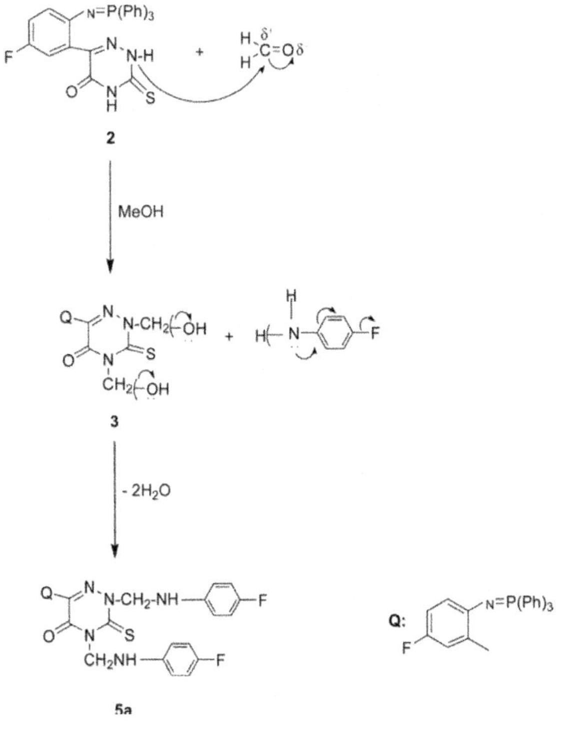

Figure 1: Formation of compounds 3 & 5a.

Due to a higher nucleophilicity of sulfur atoms, the direct displacement of an acidic proton of mercapto group by a simple electrophile can be easily occur via treatment of compound 2 with haloacetic acids. Thus treatment of compound 2 with halo aliphatic acids such as mono/di/trichloroacetic acids in DMF afforded the substituted thiaacetic acids 6-8 (Scheme 3).

The multicomponent reaction (MCR) was considered as powerful synthetic tool for preparing target mole- cules of biological relevance in an efficient manner. Thus, treatment of compound 2 with active methylene reagents as chloroacetonitrile in warm DMF [20] produced 3-cyanomethyl thai-6-iminophosphorane-1,2,4-trinazin- 5-(2H)one (10). The latter compound 10 use for the synthesis of thiazolo [3,2-b][1,2,4]triazinones (11-13) sys- tems (Scheme 4). Acidic hydrolysis of 10 by warm with diluted HCl for short time (10 min) yielded the com- pound 6. Boil compound 6 with DMF along time afforded 6-iminophosphorane-2,3-dihydoro-thiazolo [3,2-b] [1,2,4] triazine-3,7-dione (9) (Scheme 4).

Heat compound 10 on heating with DMF a long time (2 hours), produced 3-aminothiazolo-1,2,4-triazine 11. Presence of an amino group in structure 11 was deduced from treat with 4-fluorobenzoylchloride (DMF) and/or with 4-fluorobenzaldehyde (EtOH) yield the anilido 12 and/or Schiff's base 13 (Scheme 4). Treatment of com- pound 2 with α, β-bifunctional oxygen-halogen reagents as phenacyl bromide in ethanolic KOH, yielded 3- phenyl-6-iminophosphorane-thiazolo [3,2-b][1,2,4]triazin-7-one (14) (Scheme 5). The nitrogen-sulfur contain- ing fused heterobicyclic structures have demonstrated a high degree of binding affinity when they serve as Ligands for various biological receptors [12] [13] . Thus addition of Mercator group (as nucleophilic) of compound 2 to an Schiff's base 15 in boil dry dioxan yielded the thioether16, which upon ring closure reaction by reflux with CS_2 in DMF furnished 2,3-diaryl-2,3-dihydro-7-iminophosphorane-4-thioxo-1,3,5-thiadiazino[3,2-b]

[1,2,4]-triazin-8-one (17) (Scheme 5).

Abdel-Rahman et al. [21] -[25] reported that thioethers, sulfide and sulfonic acid bearing a 1,2,4-triazine moieties. Exhibited a very interesting medicinal activity as anti-HIV and anticancer agents. Recently, Slawinski et al. [25] synthesized 2-mercaptobenzene sulfonamide bearing a 1,2,4-trinzines exhibited a significant activity against cell lines of colon cancer, renal cancer, and melanoma, as well as good selectivity toward non-small cell lung cancer. Similarly, oxidation of compound 2 via treatment with $FeCl_3$ in boiling methanol and/or with H_2O_2 in ethanol by stirred at room temperature furnished the disulfide 18 and/or 3-sulfonic-1,2,4-triazinone 19. Final- ly, treatment of 2 with ethanolic $KMnO_4$ at room temperature [21] led to the direct formation of 6-(5'-fluoro- 2-triphenylphosphiniminophenyl)-1,2,4-triazin-3,5(2H, 4H) dione (20) (Scheme 5).

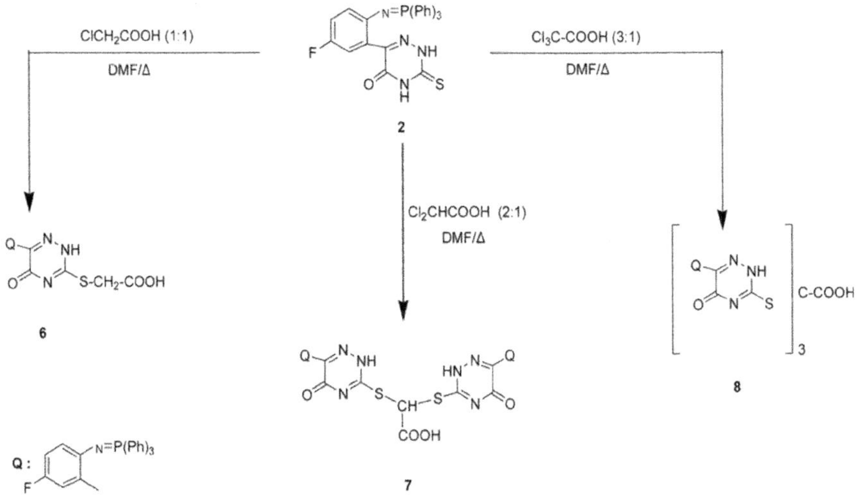

Scheme 3: Formation of compounds 6 - 8.

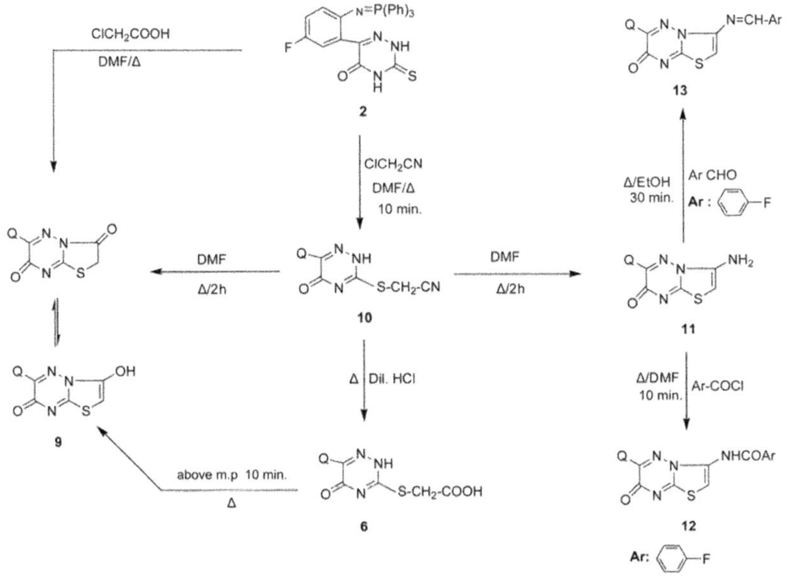

Scheme 4: Formation of compounds 9 - 13.

Elucidation the Former Structures

UV Spectra

The electronic conjugated molecule of compound 2 exhibited λ_{max} at 310 nm while that of compounds 3 (363), 5a (364), 8 (359) and 16 (323) nm. A higher absorption bands of new acyclic systems than that of 2 confirm that N- and S-substitution were formed. On the other hand, the absorption bands of fused heterobicycle compounds 9 (352), 17 (347) and 10 (321) nm is higher than the start 2 (310) nm. This is attributing to extension of hetero- conjugation of heterobicyclic systems through a type of cylization.

IR Spectra

The new compounds obtained recorded the absorption bands at 1380 - 1390, 1250 - 1230 cm^{-1}due to presence of both P=N and C-F functional groups. Compounds 3-5 showed a lack of band at 3200 - 3100 cm^{-1} for NH=OH of 1,2,4-triazinones, while that of compounds 6-8 and 10 recorded the absorption band at 3343 and 1643 cm^{-1} attributed to presence of ^{4}NH & ^{5}C=O of 1,2,4-triazinone. Only compounds 9-14 showed a lack of the absorption bands at 1200 - 1100 cm^{-1} for C=S, which confirm that heterocyclization. In addition to the compounds 6-9 & 18, 20 exhibited a two absorption bands at v 1700 and 1665 cm^{-1} due to the presence of two carbonyl groups. Also, IR absorption spectra of compounds 3-8, 9-10 and 16 recorded the absorption bands at v 2975 and 2885 cm^{-1} attributed to aliphatic functional groups [1] [14] [15] [26] .

NMR Spectral Study

1) ^{1}H NMR spectrum of 1 showed a resonated signals at δ 14.6, 12.6 and 10.9 ppm for 3NH with δ 8.6 - 0.80, 7.69 - 7.64, 7.41 - 7.31 ppm for three aromatic protons, while that of 3 exhibited a signals at δ 5.24 and 4.98 ppm attributed to two OH with δ 2.92 - 2.88, 2.62 - 2.58 ppm for two CH$_2$ protons. Compounds 3, 4 and 5 showed a lack's of ^{4}NH and ^{2}NH of 1,2,4-triazine moiety, while that of 5 recorded additional signals at δ 1.9 and 1.75 ppm of two methyl groups of antipyrine moiety. ^{1}H NMR spectra of 6-8 recorded δ at 12.7, 4.7 ppm for NH and OH protons, while that of 9 showed a signal at δ 10.5 and 8.5 ppm, attributed to OH and CH = of thiazole moiety. In addition to compound 10 recorded a signals at δ 13.90, 2.59 ppm for NH, CH$_2$ protons, while that of 11 exhibited only signals at δ 8.01 and 3.99 ppm for = CH thiazole and amino-protons. Moreover ^{1}H NMR spectrum of 16 showed a signals at δ 12.76 and 10.75 ppm for two NH of 1,2,4 trinazine while a lacks of these (2NH) protons of 17, with presence of CH proton of thiadiazine moiety at δ 9.68 ppm. ^{1}H

NMR spectra of compounds 18 recorded the presence of δ at 14.55 and 12.79 ppm attributed to 2NH of 1,2,4-triazine protons, while that of 19 exhibited a signals at δ 12.8 and 10.7 p pm for NH and CH. (SO$_2$-OH) protons, with signals of aromatic protons. Finally, compound 20 exhibited δ at 12.73 and 10.82 ppm attributed to NH and OH protons [14] [15] [19] [26] .

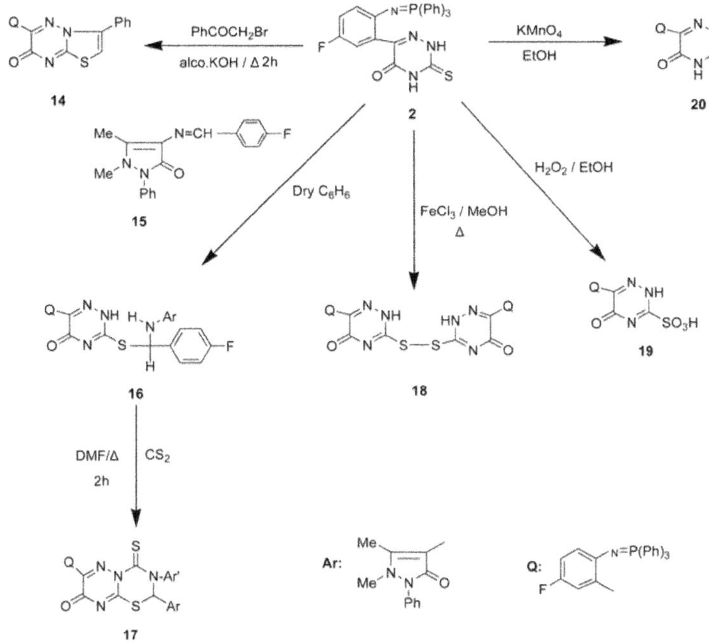

Scheme 5: Formation of compounds 14 - 20.

2) ^{13}C NMR spectra of all the synthesized compounds showed a resonated signals at δ 180, 165 - 163, 140 - 138, 135 - 121 and 112 p pm attributed to C=S, C=O, C=N, aromatic and C-F carbons. Also, ^{13}C NMR spectra of compounds 3-6, 9 and 10 recorded signals at δ 39 - 33 ppm for CH$_2$carbons. Only the compound 10 showed an additional signal at δ 112 p pm for C≡N carbon. Finally,^{13}C NMR spectra of the entire compound exhibited a resonated signals at 77 - 75 ppm for C5-C6 of 1,2,4-triazine [27] (Figure 2).

3) ^{19}F NMR spectral study recorded a signal at δ −126 to −125 ppm.

4) ^{31}P NMR spectral study exhibited a signal at δ 30 - 29 p pm attributed to P=N [17] .

Mass Fragmentation Study

Mass fragmentation pattern study of some selective synthesized compounds indicated that fused heterobiycyclic systems 11 have a more base peak, while that of acyclic structures 1and 16 have only base peak which indicate that their less stability. A higher stability of fused heterobicyclic systems is due to the delocalization of net charge over all the active centers (Figure 3 to Figure 5).

MOLLUSCICIDAL ACTIVITY

Based upon the earlier work by Abdel-Rahman et al. [7] [16] on the synthesis of phosphono substituted-1,2,4- triazine derivative and their molluscicdal activities against Biomophalaria Alexandrina Snails responsible for Bilharziasis diseases, the prepared compounds were tested as killing of that snails (shell in diameter 5 - 8). The intermediate host of sohistosomamausoni in Giza Govern state that was not treated with molluscicides.

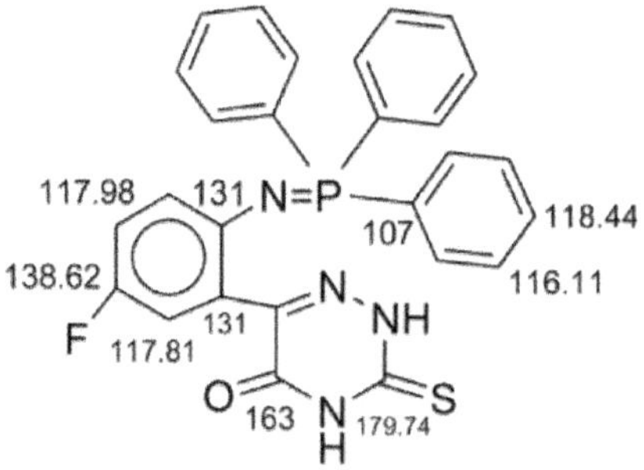

Figure 2. ^{13}C NMR data of compound 2.

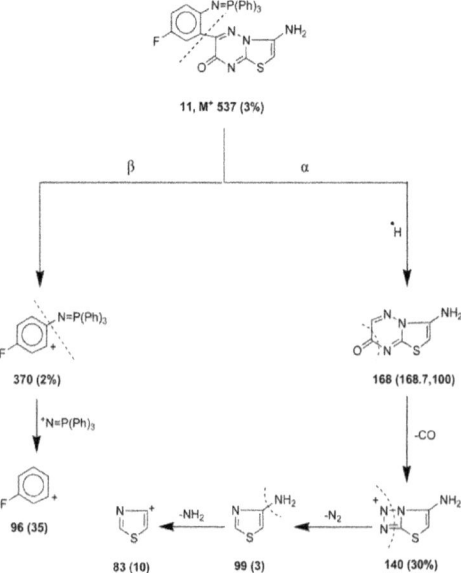

Figure 3: Mass fragmentation pattern of compound 11.

The snails were adapted to laboratory conditions for two weeks before being used in toxicity tests to be sure that the snails are strong and healthy. Snails were kept in plastic aguaria filled with de chlorinated tap water at room temperature (25°C - 27°C). Stock solution (500 μg·ml⁻¹) of the tested compounds were synthesized in the least volume of ethanol and completed of the least volume of ethanol and completed to the required volume with de chlorinated tap water on the basis of weight volume. A series of more diluted solutions were then prepared fol- lowing the instructions given by WHO organization [28] [29] . The result given in (Table 1) revealed that the high activity towards snails in the following sequences: 18 > 2 > 20 > 3 > 8 and 9 > 10 > 6 > 17 >> 5a and 5b > 7 > 14 at 100 ppm in compared with Baylucide as standard reference.

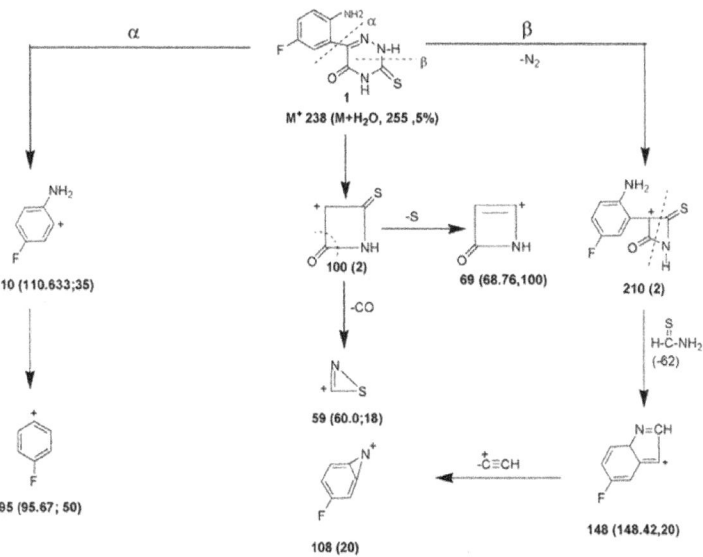

Figure 4: Mass fragmentation pattern of compound 1.

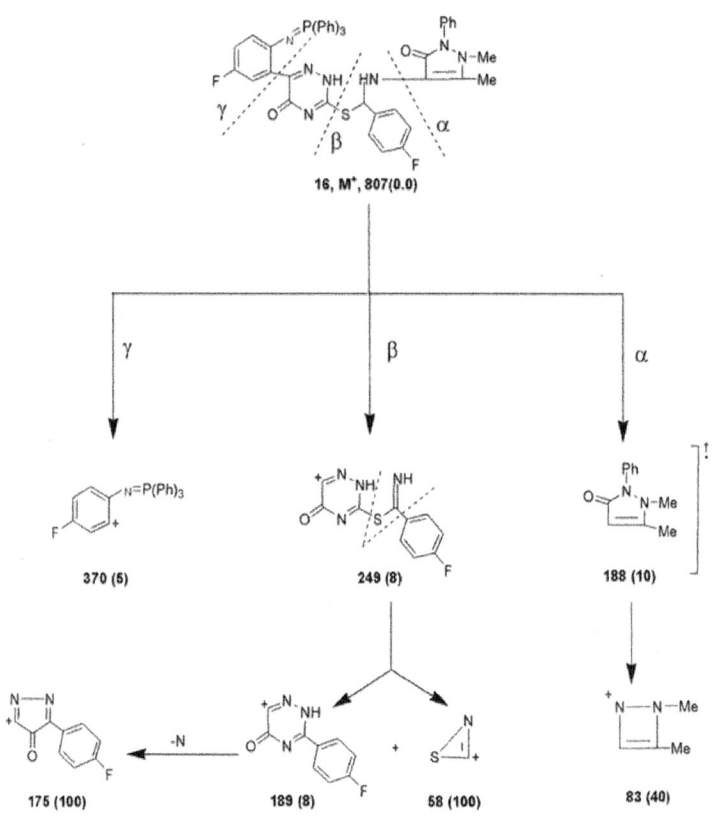

Figure 5. Mass fragmentation pattern of compound 16.

In general, the strong effect of the compounds 2, 3, 8, 18 and 20 is due to presence both the S-S, S-H and O-H functional groups which agree with bio-oxidation-reduction processes. The moderate effect of the compounds 5a, 6, 9, 10 and 17 is attributed to thioether and cyclic sulfur nitrogen systems. Finally, the lethal effect of the compounds 4, 5b, 7, 11 and 14 may be to absence of SH and/or OH of Mannich base and for thiazolotriazine systems which led to the inhibition of delocalization electron-density over all the center of systems. Also, presence of hetero-elements (F, P, S, O) and N elements in corporated with 1,2,4-trinazines led to increases of electro-negativity, over all the molecular structure and enhance the electrostatic force and hydrophobic properties [17] [18] [31] -[33] . Thus, total electron-barrier of molecular distribution of the evaluated systems synthesized led to highly inhibition of the enzymatic effect on the living processes for the tested snails by causing break of a vital cyclic of that snails, and enhance the possibility killing of these snails. QSAR study of the obtained resulted from (Table 1), and based on the introduction of P, S and F in the synthesized 1,2,4-triazines, in compared with the mortality of tested snails, indicated that, increases of P and S percent % led to increase of mortality, while, increase of F percentage % led to decrease of mortality of snails. Also, very high electronegative of fluorine atom can modify the electronic distribution in the molecule affecting its absorption distribution and metabolism. In conclusion, 3-thioxo-1,2,4-triazine-5-ones bearing an P, S and F elements and their related S-alkyl derivatives, enhance the mortality of snails, which cause Bilharziasis Diseases than that their non-fluorinated and non-phosphinated systems. Also, increases of P and S percentage % led to higher mortality of the tested snails, in hope to obtain more clean water from waste water.

Table 1: The molluscicidal activity of the synthesized systems (2 - 20) mortality of snails various concentration (ppm).

Comp. No.	25 ppm	50 ppm	100 ppm
2	30	60	80
3	30	50	80
4	20	30	60
5a	20	40	70
5b	10	20	50
6	20	40	70
7	20	30	50
8	30	50	80
9	30	50	70
10	30	50	70
12	10	20	30
13a	10	20	30
13b	10	20	30
14	10	30	40
16	10	20	30
17	20	40	70
18	40	60	90
19	30	40	50
20	30	60	80
Reference standard, Baylucide		100	100

CONCLUSION

New fluorine substituted 6-(5'-fluoro-2'-triphenylphosphiniminophenyl) 3-thioxo-1,2,4-triazin-5 (2H, 4H) one (2) was obtained via Wittig's reaction of the corresponding 6-(5'-fluoro-2'-aminophenyl)-3-thioxo-1,2,4-triazinone (1). 3-thioxo-1,2,4-triazine-5-ones bearing an P, S and F elements and their related S-alkyl derivatives, enhance the mortality of snails, which cause Bilharziasis Diseases than that their non-fluorinated and non-phosphinated systems. Also, increases of P and S percentage % led to higher mortality of the tested snails, in hope to obtain more clean water from waste water.

ACKNOWLEDGEMENTS

The authors are thankful to Prof. M. M. El-Sayed for helping in testing the molluscicidal activity in Theodor Bilharz Research institute, Giza, Egypt.

REFERENCES

1. Abdel-Rahman, R.M. and Ali, T.E. (2013) Synthesis and Biological Evaluation of Some New Polyfluorinated 4- Thiazolidinone and α-Aminophosphonic Acid Derivatives. Monatshefte fur Chemie, 144, 1243-1252. http://dx.doi.org/10.1007/s00706-013-0934-6

2. Makki, M.S.T., Bakhotmah, D.A. and Abdel-Rahman, R.M. (2012) Highly Efficient Synthesis of Novel Fluorine Bearing Quinoline-4-Carboxylic Acid and the Related Compounds as Amylolytic Agents. International Journal of Organic Chemistry, 2, 49-55.http://dx.doi.org/10.4236/ijoc.2012.21009

3. Makki, M.S.T., Bakhotmah, D.A., Abdel-Rahman, R.M. and El-Shahawy, M.S. (2012) Designing and Synthesis of New Fluorine Substituted Pyrimidine-Thion-5-Carbonitrles and the Related Derivatives as Photochemical Probe Agent for Inhibition of Vitiligo Disease. International Journal of Organic Chemistry, 2, 311-320.http://dx.doi.org/10.4236/ijoc.2012.223043

4. Abdel-Rahman, R.M., Makki, M.S.T. and Bawazir, W.A. (2011) Synthesis of Some New Fluorine Heterocyclic Nitrogen Systems Derived from Sulfa Drugs as Photochemical Probe Agents for Inhibition of Vitiligo Disease Part I. E-Journal of Chemistry, 8, 405-414.

5. Abdel-Rahman, R.M., Makki, M.S.T. and Bawazir, W.A. (2010) Syntheisis of Fluorine Heterocyclic Nitrogen Systems Derived from Sulfa Drugs as Photochemical Probe Agents for Inhibition of Vitiligo Disease Part II. E-Journal of Chemistry, 7, 593-5102.

6. Abdel-Rahman, R.M. (2003) Synthesis of New Phosphaheterobicyclic Systems Containing 1,2,4-Triazine Moiety— Part IX: Straight forward Synthesis of New Fluorine Bearing 5-Phospha-1,2,4 Triazine/1,2,4-Triazepine-3-Thiones; Part X: Synthesis of New Phosphaheterobicyclic Systems, Containing a 1,2,4-Triazine Moiety. Trends in Heterocyclic Chemistry, 8, 187-195.

7. Ali, T.E., Abdel-Rahman, R.M., Hanafy, F.J. and El Edfawy, S.M. (2008) Synthesis and Molluscicidal Activity of Phosphorus—Containing Heterocyclic Compounds Derived from 5,6-Bis (4-brome phenyl)-3-hydrazino-1,2,4-triazine. Phosphorus, Sulfur, and Silicon, 183, 2565-2577.

8. Abdel-Rahman, R.M., Ibrahim, M.A. and Ali, T.E. (2010) 1,2,4-Triazine Chemistry Part II, Synthetic Approaches for Phosphorus Containing 1,2,4-Triazine Derivatives. European Journal of Chemistry, 1, 388-396. http://dx.doi.org/10.5155/eurjchem.1.4.388-396.154

9. Blakley, B., Brousseau, P., Fournier, M. and Voccia, I. (1999) Immunotoxicity of Pesticides. Toxicology Industrial Health, 15, 119-132.http://dx.doi.org/10.1177/074823379901500110

10. Sengupta, A.K., Bajaj, O.P. and Agarwal, K.C. (1980) Synthesis and Insecticide Activity of N^4-Aryl-N^1-(0.0-Dial- kylthiophosphoryl) Piperazines. Journal of the Indian Chemical Society, 57, 1170-1171.

11. Du, Y.M., Tian, J., Liao, H., Bai, C.J., Yan, X.L. and Liu, G.D. (2009) Aluminum Tolerance and High Phosphorus Ef- ficiency Helps Stylosanthes Better Adapt to Low-P Acid Soils. Annals of Botany, 103, 1239-1247. http://dx.doi.org/10.1093/aob/mcp074

12. Abdel-Rahman, R.M. (2001) Role of Uncondensed 1,2,4-Triazine Derivatives as Biological Plant Protection Agents. Pharmazie, 56, 195-212.

13. Abdel-Rahman, R.M. (2001) Role of Uncondensed 1,2,4-Triazine Compounds and Related Heterobicyclic Systems as Therapeutic Agents. Pharmazie, 56, 18-30.

14. Abdel-Rahman, R.M. (1991) Synthesis and Anti Human Immune Virus Activity of Some New Fluorine Containing Substituted 3-Thixo-1,2-4-Triazin-5-Ones. Farmaco, 46, 379-389.

15. Abdel-Rahman, R.M. (1992) Synthesis of Some New Fluorine Bearing Tri-Substituted 3-Thioxo-1,2,4-Triazine-5- Ones as Potential Anti Cencer Agent. Farmaco, 47, 319-326.

16. Makki, M.S.T., Abdel-Rahman, R.M. and El-Shahawi, M.S. (2012) Synthesis and Voltammetric Study of Some New Macrocylis of Arsenic

(III) in Wastewater and as Molluscicidal Agents against Biomophalaria Alexandrina Snails. Comptes Rendus Chimie, 15, 617-626.

17. Basavaiah, D., Chandrashekar, V., Das, U. and Reddy, G.J. (2005) A Study toward Understanding the Role of a Phos- phorus Stereogenic Center in (5S)-1,3-Diaza-2-Phospha-2-oxo-3-Phenylbicyclo(3.3.0) Octane Derivatives as Catalysts in the Borane-Mediated Asymmetric Reduction of Prochiral Ketones. Tetrahedron: Asymmetry, 16, 3955-3962. http://dx.doi.org/10.1016/j.tetasy.2005.10.038

18. Ali, P., Pamakanth, P. and Meshram, J. (2010) Exploring Microwave Synthesis for Co-Ordination: Synthesis, Spectral Characterization and Comparative Study of Fluorine Substituted Transition Metal Complexes with Binuclear Core De- rived from 4-Amino-2,3-Dimethyl-1-Phenyl-3-Pyrazolin-5-One. Journal of Coordination Chemistry, 63, 323-329.http://dx.doi.org/10.1080/00958970903305437

19. Abdel-Rahman, R.M. (2000) Chemistry of Uncondensed 1,2,4-Triazines: Part II-Sulfur Containing 5-oxo-1,2,4-Triazin 3-yl Moiety. Phosphorus, Sulfur and Silicon, 166, 315-357.http://dx.doi.org/10.1080/10426500008076552

20. Abdel-Rahman, R.M. and Islam, I.E. (1993) Synthesis and Reactions of Acetonitrile Derivatives Bearing a 5,6-Diphe- ny-1,2,4-Triazin-3-yl Moiety. Indian Journal of Chemistry, Section B, 32, 526-529.

21. Abdel-Rahman, R.M., Seada, M., Fawzy, M.M. and El-Baz, I. (1993) Synthesis and Anti-Canceranti Human Immune Virus Activities of Some New Thioether Bearing a 1,2,4-Triazine-3-Hydrazones. Farmaco, 48, 397-406.

22. EL-Gendy, Z., Morsy, J., Allimony, H.A., Abdel-Monem, W.R. and Abdel-Rahman, R.M. (2003) Synthesis of Heter- obicyclic Nitrogen System Bearing the 1,2,4-Triazine Moiety as Anti-HIV and Anti-Cancer Drugs, Part II. Phosphorus, Sulfur and Silicon, 178, 2055-2071.http://dx.doi.org/10.1080/10426500390228738

23. Abdel-Rahman, R.M. and El-Mahdy, K. (2012) Biological Evaluation of Pyramidpyrimidines as Multi-Targeted Small Molecule Inhibitors and Resistance Modifying Agents. Heterocycles, 85, 2391-2414. http://dx.doi.org/10.3987/REV-12-745

24. Zaki, M.T., Abdel-Rahman, R.M. and El Sayed, A.Y. (1995) Use of Arylidenrhodanines for the Determination of Cu(II) Hg(II) and CN⁻. Analytica Chimica Acta, 307, 127-138.http://dx.doi.org/10.1016/0003-2670(95)00048-5

25. Slawinski, J. and Gdaniec, M. (2005) Synthesis, Molecular Structure, and in Vitro Antitumor Activity of New 4- Chlore-2- Mercaptobenzenesulfonamide Derivatives. European Journal of Medicinal Chemistry, 40, 377-389.http://dx.doi.org/10.1016/j. ejmech.2004.11.014

26. Abdel-Rahman, R.M. and Abdel-Malik, N.S. (1990) Synthesis of Some New 3,6-Diheteroarryl-1,2,4-Triazine-5-Ones and Their Effect on Amylolytic Activity of Some Fungi. Pakistan Journal of Science and Industrial Research, 33, 142-147.

27. Ebraheem, M.A., Abdel-Rahman, R.M., Abdel-Haleem, A.M., Ibrahim, S.S. and Allimony, H.A. (2008) Synthesis and Antifungal Activity of Novel Polyheterocyclic Compound Containing Fused 1,2,4-Triazine Moiety. Arkivoc, 21, 202- 213.

28. WHO (1953) Expert Committee on Bilharziasis, 65, 33.

29. WHO (1965) Snail Control Information of Bilharziasis Monograph Series. 50, 124.

30. Smarti, B.E. (2001) Fluorine Substituent Effect (on Bioactivity). Journal of Fluorine Chemistry, 109, 3-11. http://dx.doi.org/10.1016/S0022-1139(01)00375-X

31. Billard, T., Gille, S., Ferry, A., Bartelemy, A., Christophe, C. and Langlois, B.R. (2005) From Fluoral to Heterocycles: A Survey of Polyfluorinated Iminiums Chemistry. Journal of Fluorine Chemistry, 126, 189-196. http://dx.doi.org/10.1016/j.jfluchem.2004.08.007

32. Isanbor, C. and O'Hagan, D. (2006) Fluorine in Medicinal Chemistry: A Review of Anti-Cancer Agents. Journal of Fluorine Chemistry, 127, 303-319.http://dx.doi.org/10.1016/j.jfluchem.2006.01.011

33. Sandford, G. (2007) Elemental Fluorine in Organic Chemistry (1997-2006). Journal of Fluorine Chemistry, 128, 90- 104. http://dx.doi.org/10.1016/j.jfluchem.2006.10.019

Chapter 8

SYNTHESIS OF NOVEL FLUORINE SUBSTITUTED ISOLATED AND FUSED HETEROBICYCLIC NITROGEN SYSTEMS BEARING 6-(2'-PHOSPHORYLANILIDO)-1,2,4-TRIAZIN-5-ONE MOIETY AS POTENTIAL INHIBITOR TOWARDS HIV-1 ACTIVITY

Reda M. Abdel-Rahman, Mohammed S. T. Makki, Abeer N. Al-Romaizan*

Department of Chemistry, Faculty of Science, King Abdul Aziz University, Jeddah, KSA

ABSTRACT

Novel 6-(5'-fluoro-2'-diphenylphosphorylanilido)-3-hydrazino-1,2,4-trizin-5 (2H) one (3) is achieved from hydrozinolysis of the corresponding 3-thioxo-analoges 2. Compound 2 is also obtained from phosphorylation of 6-(5'-fluoro-2'-aminophenyl)-3-thioxo-1,2,4-triazin-5(2H) one (1). Novel fluorine substituted isolated and/or fused heterobicyclic nitrogen systems bearing and/or containing, 6-phosphoryl anilido-1,2,4-trizin-5 (2H) one moiety (4 - 22) have been synthesized from ring closure reactions of compound 3 with π-acceptors activated carbon compounds in different medium and conditions. Structures of the products are characterized by MS, IR, UV-VIS, CH, N, and ^{1}H/^{13}CNMR spectral data. The new products have been evaluated as potential inhibitors towards HIV-1 activity.

INTRODUCTION

Organophosphorus systems are ubiquitous in nature and exhibit many applications in the field of agriculture medicine and industry [1] [2] . Many multi-ring phosphorus heterocycles are used as pesticide [3] , bactericide [4] -[6] , antibiotics [4] , and acts as HIV protease inhibitors [7] . Thus, synthesis of new phosphorus bearing a heterocycles has attracted the attention of researchers. Phosphorylation of organic compounds often improves their biological activity, especially through a vital energy, because the P-O is the store of energy for metabolism process. For example, phophorylated-N in the nucleocapsid affects the interaction between the N-atom and the genomic RNA. The charge repulsion between the negatively charged phosphoserine and the negatively charged RNA may weaken the interaction between N-atom and RNA, thus enabling the viral polymerase to gain access to genomic RNA and to initiate viral RNA transcription and replication [8] . On the other hand, chemistry of N-phosphorylheterocycles showed that these compounds from two dimensional polymeric chains via intermolecular P-O^{-+}H-N hydrogen bonds [9] . Also, phosphodiester compound had a type of action, especially enzymetic of DNA replication [10] on DNA ligase as (Figure 1(.

It is interesting that fluorine containing aheterocycles bearing functional groups exhibits highly effective in biological process, pharmaceuticals, agrochemicals, polymers and a wide range of consumer products [11] . It reflects its resistance to metabolic change due to the strength of the C-F bond providing biological stability and the application of its nonstick-interfacial physical characteristics [12] -[16] . Abdel-Rahman et al. [17] -[21] reported the synthesis and chemical reactivity of 3-thioxo-1,2,4-triazin-5 one derivatives as a bioactive molecules, especially as anti-cancer, anti-AIDS, Amyllolytic, cellobiase and antimicrobial agents. Based upon these observations, the aim of this work is to study the formation of 6-(5'-fluoro-2'-phosphoryl anilido-3-hydrazino-1,2,4- triazin-5 (2H) one then study their behaviour as electron donor towards different electron acceptors such as carbo-sulfur, oxygen, halogen and nitrogen compounds; finally, a type of isolated and/or fused heterobicyclic systems obtained and evaluation as potential inhibitors towards HIV-1 virus.

EXPERIMENTAL

Melting points were determined with an electro-thermal Bibbly Stuart Scientific Melting point SMPI (UK). A perkin Elmer (Lambda EZ-2101) double beam spectrophotometer (190 - 1100 nm) used for recording the electronic spectra. A Perkin Elmer model RXI-FT-IR 55,529 cm^{-1} used for recording the IR spectra (EtOH as solvents). A Brucker advance DPX 400 MHz using TMS as an internal standard for recording the ^{1}H/^{13}C NMR spectra in deuterated DMSO (δ in pp m). AGC- MS-QP 1000 Ex model is used for recording the mass spectra. Hexafluorobenzene was used as external standard for ^{19}FNMR at 84.25 MHz and ^{31}P (in CDCl$_3$, 101.25 MHZ).

Elemental analysis was performed on Micro analytical Center of National Reaches Center-Dokki, Cairo, Egypt. Compound 1 prepared according to the reported method [17] .

6-(2'-aminos-5'-fluorophenyl-3-thioxo-1,2,4-triazin-5(2H)one (1)

A mixture of 5-fluoroisatin (0.01 mol) in sodium hydroxide solution (5%, 50 ml) warm for 10 min, then thiosemicarbazide (0.01 mol, in hot water, 10 ml) add and complete the refluxing for 2 h. The reaction mixture cooled then poured onto ice and neutralize with diluted HCl. The solid thus obtained filtered off and crystallized from ethanol to give 1 as yellow crystals, yield 80%; m.p. 265°C. Analytical data; found: C, 44.91; H, 2.90; F, 7.58: N, 23.40; S, 13.29%.Calculated for $C_9H_7FN_4OS$ (238); C, 45.37; H, 2.94; F, 7.98; N, 23.52; S, 13.44%.UV: λ_{max} (EtOH) = 280 nm. IR vcm^{-1}: 3424(NH$_2$), 3258, 3169 (NH, NH), 1685 (C=O), 1618 (deform. NH$_2$), 1545 (C=N), 1263 (C-F), 858,818 (aryl CH), 685 (C-F). ^{1}H NMR (DMSO) δ = 14.66, 12.66, 10.90 (each s, 1H, 3NH), 8.68 - 8.06, 7.69 - 7.64, 7.39 - 7.30, (s, 3H, aryl protons). ^{13}C NMR (DMSO): δ 179.47(C=S), 162 (C=O), 159 - 157 (C-F), 138.54 (C=N), 131.82, 121.8, 121.51 (aromatic carbons, 78.14, 77.71 (C$_5$ - C$_6$, 1,2,4-trinzine). M/Z (Int. %); 256 (M + H$_2$O, 5%), 68 (100), 148(21), 138(18), 110 (30), 96 (50), 82(58.0), 70(78).

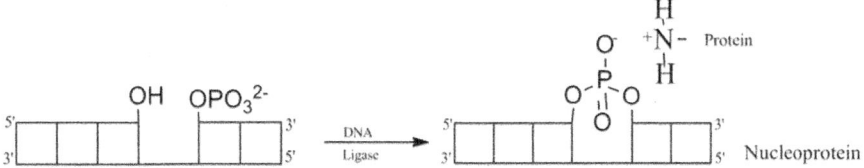

Figure 1: The reaction of DNA ligase phospho diester link.

6-(5'-Fluoro-2'-diphenylphosphorylanilido)-3-thioxo-1,2,4-triazin-5(2H) one (2)

Equimolar mixture of compound 1 and diphenylphosphoryl chloride in DMF (20 ml) warm for 1h, cooled then poured onto ice. The produce solid filter off and crystallized from methanol to give 2 as deep yellow crystals. Yield 70%; m.p. 218-220°C. Analytical data; found: C, 53.39; H, 3.51; F, 3.88: N, 11.69%.Calculated for $C_{21}H_{16}FN_4PSO_4$(470); C, 53.61; H, 3.70; F, 4.04; N, 11.91%. IR vcm^{-1}: 3133.8 (NH), 1688.4 (C=O), 1574 (C=N), 1370 (Cyclic NCSN), 1262 (C-F), 1200 (P=O), 1150 (C-S), 10.96 (Ph-O-P), 858, 801 (substituted phenyls) M/Z: (Int.%); 473 (M + 3, 5.11), 110 (100).

3-Hydrazino-6-(5'-fluoro-2'-diphenylphosphorylanilido)-1,2,4-triazin-5(2H)one (3)

A mixture of compound 2 (1 gm) and hydrazine hydrate (2ml) in ethanol (50ml) reflux for 2h. cooled. The solid thus produce filter off and crystallized from ethanol to give 3 as orange crystals, yield 85%; m.p. 178°C - 180°C. Analytical data; found: C, 53.45; H, 3.59; F, 3.80%.Calculated for $C_{21}H_{18}FN_6PO_4$(468), C, 53.84; H, 3.84; F, 4.02; N, 17.14%. IR v cm^{-1} = 3381,

3340, 3220 - 3190 (NH, NH, NH$_2$), 1682 (C=O), 1558 (C=N), 1266.9 (C-F), 1210-1156 (P=O). 1050 (Ph-O-P), 860,805 (substituted phenyls).^{1}H NMR(DMSO) δ = 10.7, 10.2, 8.7 (each s, 3H, 3NH, 1,2,4-triazino NH-P=O), 7.511, 7.27-2.24, 7.23-7.22 (each s, 3H, aryl protons); 7.143 - 7.002; 6.860 - 6.796 (each m, 10H, phenyl protons), 2.88 (s, 2H, NH$_2$ of hydrazine), ^{13}C NMR (DMSO) δ = 163.78, 159.44, 157.86, 134.97, 129.07, 127.65, 123.66, 122.87, 120.12, 120.09, 113.47, 113.31, 110.77, 110.71, 107.93, 105.27, 105.10, 77.76, 77.33.

3-(3'-Amino-4'-carboxy-5-(4-chlorophenyl)-4',5'-dihydro-pyrazolin-1'-yl)-6-(5'-fluoro-2'-diphenylphosph- orylanilido)-1,2,4-triazin-5(2H)one (4)

Equimolar amounts of 3 and α-(4-chlorophenylidene) cyano acetic acid in ethanol (100 ml) and a few drops piperidine (0.5 ml) reflux for 8h, cooled, then added off and crystallized from dioxan to give 4 as yellowish crystals, yield 65% m.p., 158°C - 160°C. Analytical data; found: C, 54.81; H, 3.11; F, 2.55; Cl, 5.09; N, 14.29%. Calculated for C$_{31}$H$_{24}$FClN$_7$PO$_6$(675). C, 55.11; H, 3.55; F, 2.81; Cl, 5.18; N, 14.51%; M/S: 675 (1.11), 95(100). IR vcm^{-1}: 3420 (OH), 3381, 3155 (NH, NH$_2$), 1680 (C=O) 1558(C=N), 1478 (deform aliphatic CH), 1269 (C-F), 1200 (P=O), 1043 (Ph-O-P), 993.861, 804 (Aryl CH).^{1}H NMR (DMSO) δ = 10.67, 10.14, 8.72 (each δ, 3H, 3NH, 1,2,4-triazinare NH-P=O), 9.77 (δ, 1H, OH), 8.15 - 7.52, 7.36 - 7.29, 7.13 - 7.00, 6.991 - 6.811 (each m, 19H, 7H aryl, 10H phenyls pyrazole protons); 3.46 (δ, 2H, NH$_2$).^{13}C NMR (DMSO) δ: 179.64; 163.76; 163.05; 159.42; 157.84; 138.58; 134.94; 128.65; 127.59; 123.72; 123.66; 121.13; 121.07; 117.67; 117.51; 113.45; 113.28; 112.06; 112.01; 110.75; 110.70; 109.66; 108.09; 107.92; 105.25; 105.08; 77.78; 77.35; 23.64.

3-(3'-Methyl-5'-arylamino-pyrazolin-1'-yl)-6(5'-fluoro-2'-diphenylphosphorylanilido)-1,2,4-trinzin-5(2H) one (5)

A mixture of 3 (0.01 mol) and acetyl acetanilide derivative (0.01 mol) in DMF (50 ml) reflux for 2h, cooled then poured onto ice. The solid thus yield filtere off and crystallized from dioxan to give 5 as deep-yellowish crystals, yield 70%; m.p. 100°C - 101°C. Analytical data; found C, 56.03; H, 3.41; F, 2.23; N, 16.17; S, 3.91%. Calculated for C$_{36}$H$_{29}$FN$_9$PSO$_6$(765). C, 56.47; H, 3.79; F, 2.48; N, 16.47; S, 4.18%. IR vcm^{-1}: 3424, 3220 - 31.70 (NH, NH), 2932 (CH$_3$), 1694 (C=O), 1620 (C=N), 1358 (SO$_2$-NH), 1268 (C-F), 1220 (P=O), 1054 (Ph-O-P), 952, 909, 810, 758 (aryl CH).^{1}H NMR (DMSO) δ = 12.28, 12.27, 10.36, 10.21 (each s, 4H, 4NH) 9.7 (s, 1H, C$_4$- of pyrazole), 7.979, 7.971, 7.556, 7.545 (m, 4-H of pyridine); 7.34, 7.336, 7.322, 7.091(m, 4H of aryl-P-SO$_2$NH) 6.95 - 6.83; 6.66 - 6.52; 6.41 - 6.16 (each m, 13H of FC$_6$H$_3$, 2C$_6$H$_5$). 0.46 - 0.45 (s, 3H, CH$_3$). ^{13}C NMR(DMSO) δ = 179.64, 163.75, 163.03, 162.41, 158.08, 157.99, 157.47; 151.97; 138.58, 134.94, 130.23,

129.10, 123.72, 123.66, 117.66, 117.50, 115.11, 113.41, 113.25, 112.97, 112.07, 112.02, 110.76, 110.71, 108.09, 107.92, 105.21, 105.04, 77.83-77.40, 36.40, 31.29.

3-(3'-Amino-4',5'-dihydro-5'-oxo-pyrazolin-1'-yl)-6(5'-fluoro-2'-diphenylphosphorylanilido)-1,2,4-triazin- 5(2H)one (6)

Equiomolar of 3 and ethyl cyanoacetate in THF (100ml) reflux for 4H, cooled. The solid thus produce filtered off and crystallized from THF to give 6 as faint yellow crystals, yield 68%; m.p. 200-202°C. Analytical data; found: C, 53.53; H, 3.31; F, 3.29; N, 18.01%. Calculated for $C_{24}H_{19}FN_7PO_5(535)$; C, 53.83; H, 3.55; F, 3.55; N, 18.31%. IR vcm^{-1}: 3425 (NH$_2$), 3220-3190 (NH

$\rightleftarrows$ OH), 1694 (C=O), 1471 (deformation. CH$_2$), 1310 (N-N), 1250 (C-F), 1210 (P=O), 1058 (Ph-O-P), 952, 915, 808 (aryl CH).^{1}H NMR (DMSO) δ: 12.7, 10.67, 8.8 (each s, 3H, 3NH), 10.24 (s, 1H, OH of pyrazole), 7.66 - 7.400, 7.39-7.00, 6.99-6.79 (each m, 14H, aryl & phenyl protons). ^{13}C NMR(DMSO) δ:179.54, 163.70, 159.31, 157.74, 134.91, 128.98, 127.31, 123.68, 117.57, 113.29, 113.13, 111.95, 110.69, 110.64, 108.23, 105.08, 104.91, 78.0-77.54.

3-(3',5'-Diaminopyrazolin-1'-yl)-6-(5'-fluoro-2'-diphenylphosphorylanilido)-1,2,4-triazin-5(2H)one (7)

A mixture of 3 (0.01mol) and malononitrile (0.01 mol) in ethanol (50ml) and piperidine (0.5 ml) reflux for 8h, cooled. The solid obtained filtered off and crystallized from ethanol to give 7 as deep orange crystals. Yield 70%; m.p. 205°C - 207°C. Analytical data; found: C, 53.55, H, 3.12; F, 3.31; N, 20.71%. Calculated for $C_{24}H_{21}FN_8PO_4$ (535); C, 53.83; H, 3.55; F, 3.55; N, 20.93%. IR: vcm^{-1} 3430 - 3380 (NH$_2$), 1700 (C=O), 1620 (deform. NH$_2$), 1580 (C=N) 1265 (C-F), 1200 (Ph-P=O), 1050 (Ph-O-P), 930, 910, 850, 800(aryl CH).^{1}H NMR (DMSO) δ: 12.7, 10.7, 10.2 (each s, 3H, 3NH), 8.8 (s, 1H, NH-P=O), 7.55 - 7.311, 7.131 - 6.991, 6.87 - 6.76 (each m, 14 H, aryl & phenyl protons), 3.89 (s, 4H, 2NH$_2$). ^{13}C NMR (DMSO)δ: 163.73, 159.38, 157.80, 134.93, 127.49, 123.67, 113.38, 113.22, 110.66, 105.18; 105.01, 77.84-77.41, M/S (Int.%): 534 (536, M + 2, 1.55%), 248 (1.11), 97 (3.18) 96 (5.55), 95 (100), 93 (18.11), 68 (42.00), 67 (3.11), 62 (37.15).

3-(5'-Phenyl-3'-oxo-2,3,4,5-tetrahydro-pyrazolin-1'-yl)-6-(5'-fluoro-2'-diphenyl phosphorylanilido)-1,2, 4-triazin-5-(2H)one(8)

Equimolar amounts of 3 and cinnamoyl chloride in DMF (20ml) reflux for 4 h, cooled then poured onto ice. The produce solid filtered off and crystallized from dioxan to give 8 as deep yellowish crystals, yield 60%, m.p. 160°C - 162°C Analytical data; found: C, 59.89; H, 3.55; F, 3.01; N, 13.75%. Calculated for $C_{30}H_{23}FN_6PO_5$ (597); C, 60.30; H, 3.85; F, 3.18; N, 14.07%. IR vcm^{-1}: 3428 (OH), 3180 (NH), 2937 (CH$_2$), 1694 (C=0), 1610 (C=N), 1482 (deform. CH$_2$), 1314 (N-N), 1230 (C-F), 1169 (P=O), 1059 (Ph-O-P), 954, 909, 809 (aryl CH).^{1}H NMR (DMSO) δ: 14.7, 13.5, 12.6 (each s, 3H, 3NH), 10.80 (s, 1H, OH of pyrazole), 8.8 (s, 1H, NH-P=O), 7.99 - 7.41, 7.04 - 7.32, 7.10 - 6.93, 6.88 - 6.40 (each m, 20H, aryl and phenyl protons), 3.23 (s, 2H, NH$_2$). ^{13}C NMR (DMSO) δ: 163.70, 163.58, 162.90, 162.24, 159.34, 157.83, 157.76, 144.07, 134.29, 130.01, 128.77, 127.84, 121.16, 118.94, 118.84, 117.43, 114.23, 113.13, 112.97, 111.94, 111.89, 111.48, 111.43, 110.64, 110.59, 108.30, 108.12, 78.13 - 77.91, 36.26, 31.14.

3-(3'-(4''-Nitrophenyl)-5'-(4''-fluorophenyl)-4'-5'-dihydro-pyrazolin-1'-yl)-6-(5'-fluoro-2'-diphenyl-p- hos-phorylanilido)-1,2,4-triazin-5(2H)one (9)

A mixture of 3 (0.01 mol) and a chalcone (0.01 mol) in ethanol (50 ml), and piperidine (0.5 ml) reflux for 8 h, cooled then poured on ice-HCl. The solid produce filtered off and crystallized from THF to give 9 as yellow crystals, yield 82%, m.p. 148°C - 150°C. Analytical data; found: C, 59.49; H, 3.51; F, 5.21; N, 13.29%. Calculated for C$_{36}$H$_{26}$F$_2$N$_7$PO$_6$ (721); C, 59.91; H, 3.60; F, 5.27; N, 13.59%. IR vcm^{-1}: 3180 (NH), 2827 (CH$_2$), 1675 (C=O), 1595 (C=N), 1547 (C=N), 1500, 1423 (deform. CH$_2$), 1291 (C-F), 1225 (Ph-P=O), 1097 (Ph-O-P), 921, 847, 762 (aryl CH), 684 (C-Cl).^{1}H NMR (DMSO) δ = 10.57, 10.115 (each s, 2H, 2NH), 8.160 - 8.157, (s, 1H, NH-P=O), 8.118 - 8.095, 7.838 - 7.49, 7.23 - 7.010, 6.92 - 6.90, 6.73 - 6.70 (each m, 23H, aryl & phenyl protons) 2.49 - 2.48 (δ, 2H CH$_2$). ^{13}C NMR (DMSO) δ = 163.72, 159.35, 157.78, 134.92, 130.85, 130.80, 129.46, 127.41, 123.86, 123.73, 123.67, 121.10, 113.34, 113.18, 111.98, 110.72, 110.67, 108.09, 107.92, 105.13, 104.97, 77.92 - 77.50, 39.56.

3-(3',5'-Dioxo-2',3',4',5'-tetrahydro-pyrazolin-1'-yl)-6-(5'-fluoro-2'-diphenyl phosphorylanilido)-1,2,4- triazin-5'(2H)one (10)

Equimolar mixture of 3 and diethyl malonate in THF (100 ml) reflux for 8 h, cooled. The solid thus obtain filtered off and crystallized from dioxan to give 10 as faint yellow crystals, yield 66%, m.p. 189-190°C. Analytical data; found: C, 53.35; H, 3.18; F, 3.35; N, 15.38%. Calculated for C$_{24}$H$_{18}$FN$_6$PO$_6$(536); C, 53.73; H, 3.35; F, 3.54; N, 15.67%. IR vcm^{-1}: 3300 (NH), 3169 (NH), 1694 (C=O), 1626 (C=N), 1579 (C=N), 1471 (deform. CH$_2$), 1304 (N-N), 1265 (C-F), 1196 (P=O), 1040 (Ph-O-P), 903, 863, 809 (aryl CH).^{1}H NMR (DMSO) δ = 12.7, 10.69, 10.35 (each s, 3H, NH, OH, OH) 9.21 (s, 1H, NH-P=O), 8.15 (s, 1H, C$_4$ of pyrazole), 7.79 - 7.38, 7.10 - 7.003, 6.99 - 6.79 (each m, 15H, aryl & phenyl protons), 4.18 - 4.17, (δ, CH$_2$ of pyrazole). ^{13}CNMR (DMSO) δ = 179.39, 163.58, 162.91, 159.18, 157.61, 134.86, 126.95, 123.74, 123.68, 117.44, 117.28, 113.13, 112.97, 111.94, 110.64, 110.58, 108.12, 107.95, 104.89, 104.72, 78.29 - 77.85, 61.09, 13.94. M/S (Int.%): 536 (538, M + 2, 2.28), 99 (5.5), 96 (13.11), 95 (100), 93 (36.11), 69 (21.85), 68(42.35), 62(5.18).

3-(5'-Aryl-3'-thioxo-2',3'-dihydro-1',2',4'-triazol-1'-yl)-6-(5'-fluoro-2'-diphenylphosphorylanilido)-1,2,4-triazin-5(2H) one (11)

A mixture of 3 (0.01 mol) and P-methoxybenzoylisothiocyanate (0.01 mol) in dioxan (20ml) reflux for 4 h, cooled. The solid produce filtered and crystallized from dioxan to give 11 as orange yellowish crystals. Yield 65%, m.p. 141°C - 192°C. Analytical data, found: C, 55.62; H, 3.41; F, 2.74; N, 15.02; S, 4.72%. Calculated for $C_{30}H_{23}FNPSO_5$ (643); C, 55.98; H, 3.57; F, 2.95; N, 15.24; S, 4.97%. IR vcm^{-1} = 3220-3180 (b, NH), 1693 (C=O), 1624, 1537 (C=N), 1477 (deform. MeO), 1305(N-N), 1266(C-F), 1155 (C-S), 1099 (P=O), 1039(P-O), 980, 840, 809 (aryl CH). ^{1}H NMR(DMSO) δ = 12.74, 10.80 - 1066, 8.73 (each s, 3H, 3NH), 8.40 - 7.52, 7.38 - 7.30, 7.29 - 7.00, 6.99 - 6.800 (each m, 17H) aryl & phenyl protons), 3.68 - 3.67 (s, 3H, OMe). ^{13}C NMR(DMSO) δ: 179.64, 163.77, 163.04, 159.40, 158.02, 157.85, 138.58, 134.94, 132.10, 127.60, 123.72, 123.66, 121.13, 121.07, 117.69, 117.53, 113.46, 113.28, 112.02, 110.76, 110.71, 108.10, 107.93, 105.25, 105.07, 77.81 - 77.35, 66.91.

3-(3'-(4'-Methoxyphenyl)-5'-thioxo-4',5'-dihydro-1'2',4'-triazol-1'-yl)-6-(5'-fluoro-2'-diphenylphos p-

horylanilido)-1,2,4-triazin-5(2H)one (12)

A mixture of 3 (0.01 mol) and 4-methoxybenzoyl isothiocynate (0.01 mol) is DMF (20 ml) reflux for 4 h, cooled then poured onto ice. The produce solid filtered off and crystallized from ethanol (to give 12 as orange crystals, yield 70%, m.p. 186°C - 188°C. Analytical data: found: C, 55.69; H, 3.55; F, 2.70; N, 14.89; S, 4.79%. Calculated for $C_{30}H_{23}FN_7PSO_5$ (643); C, 55.98; H, 3.57; F, 2.95; N, 15.24; S, 4.97%. IR vcm^{-1}: 3382 (NH), 3318 (NH), 3204 (NH), 1684 (C=O), 1562 (C=N), 1484 (deform. MeO), 1311 (N-N), 1270 (C-F), 1169 (C=S), 1041 (Ph-O-P), 995, 881, 807 (aryl CH).

3-Methyl-1H-7-(5'-fluoro-2'-diphenyl   phosphorylanilido)-1,2,4-triazino[4,3-b][1,2,4] triazin-4,8-dione (13)

A mixture of 3 (0.01 mol) and sodium pyruvate (0.01 mol) in sodium hydroxide solution (5%, 100 ml) warm under reflux for 2 h, cooled them poured onto ice HCl. The solid thus obtained filtered off and crystallized 70%, m.p. 220°C - 222°C. Analytical data; found: C, 55.19; H, 3.31; F, 3.38; N, 15.88%. Calculated for $C_{24}H_{18}FN_6PO_5$ (520); C, 55.38; H, 3.46; F, 3.65; N, 15%. IR vcm^{-1}: 3380, 3162 (NH, NH) 1683 (C=O), 1589, 1553 (C=N), 1458 (deform. CH$_3$), 1308 (N-N), 1250 (C-F), 1204 (P=O), 1040 (Ph-O-P), 909, 861, 804 (aryl CH). ^{1}H NMR (DMSO) δ: 14.55, 12.8 (each s, 2H, 2NH), 10.06 (s, 1H, NH) 7.66 - 7.40, 7.35 - 7.26, 6.93 - 6.76 (each m, 13H, aryl & phenyl protons), 5.58 (s, 1H, OH), 1.256 (s, 3H, CH$_3$). ^{13}C NMR(DMSO)δ:177.40, 176.13, 158.14, 139.40, 128.24, 119.11, 118.95, 118.82, 117.83, 114.59,

114.43, 114.35, 114.29, 112.36, 112.19, 112.09, 112.04, 111.84, 111.77, 111.60, 110.87, 109.95, 109.90, 108.90, 108.37, 77.67 - 77.24, 55.38, 17.21.

Indolo[3,4-e][1,2,4] triazino [4,3-b]-1,2,4-triazin-11-one (14)

Equimolar amounts of 3 and isatin in DMF (20 ml) reflux for 2h, cooled then poured onto ice. The yielded solid filtered off and crystallized from dioxan to give 14 as deep-brown crystals, yield 80%, m.p. 198-200°C. Analytical data found: C, 59.89; H, 3.01; F, 2.98; N, 16.59%. Calculated for $C_{29}H_{19}FN_7PO_4$ (579); C, 60.10; H, 3.28; F, 3.28, N, 16.92%. IR vcm⁻¹: 3139 (NH), 1694 (C=O), 1626, 1540 (C=N), 1305 (N-N), 1250 (C-F), 1157 (P=O), 1080 (Ph-O-P), 970, 900, 850, 811 (aryl CH).¹H NMR(DMSO) δ=10.76-10.61 (s, 1H, NH), 8.011 (s, 1H, NH-P=O), 7.67 - 7.28, 7.27 - 7.12, 7.05 - 6.94, 6.87 - 6.816 (each m, 17H, aryl and phenyl protons). ¹³CNMR (DMSO) δ: 163.77, 163.68, 162.45, 159.48, 157.32, 141.29, 136.82, 134.95, 95.133.65, 130.07, 122.27, 120.80, 120.30, 117.73, 117.57, 113.54, 113.50, 113.38, 112.10, 112.04, 110.79, 110.73, 108.09, 107.92, 105.34, 105.17, 77.67 - 77.24. M/Z (Int. %) 579 (581, M + 2; 13.15%) 331 (31.88), 275 (75.19), 248 (1.15); 141 (18.20), 95 (100), 93 (20.51), 63 (5.13).

3-Phenyl-1,2,3,4-tetrahydro-7-(5'-fluoro-2'-diphenyl phosphorylanilido)-1,2,4-triazino [4,3-b] [1,2,4] triazin- 8-one (15)

A mixture of 3 (0.01 mol) and phenacyl bromide (0.01 mol) in ethanolic KOH (5%, 20 ml) reflux for 2 h, then poured onto ice-HCl. The solid produced filtered off and crystallized from THF to give 15 as brownish crystals. Yield 65% m.p., 98°C - 100°C. Analytical data; Found: C, 59.98; H, 3.55; F, 3.20; N, 14.51%. Calculated for $C_{29}H_{22}FN_6PO_4$ (568); C, 61.26; H, 3.87; F, 3.34; N, 14.78%. IR vcm⁻¹: 3158 (NH), 1655 (C=O), 1580 (C=N), 1474 (deform. CH₂), 1308 (N-N), 1230 (C-F), 1180 (P=O), 1092 (Ph-O-P), 980, 950, 900, 850, 801 (aryl CH).¹H NMR(DMSO) δ = 12.75 (s, 1H, NH), 10.78 (s, 1H, NH-P=O), 7.84 - 7.55, 7.49 - 7.30, 7.29 - 7.00, 6.99 - 6.79 (each m, 18H, aryl & phenyl protons). 3.4 (s, 2H, CH₂). ¹³C NMR (DMSO) δ:179.66, 163.04, 159.62, 158.02, 138.60, 129.51, 129.14, 128.67, 128.45, 128.18, 127.18, 126.89, 126.67, 125.86, 125.44, 123.31, 121.05, 120.20, 117.72, 117.56, 113.90, 113.74, 112.21, 112.13, 112.07, 110.47, 110.04, 108.10, 107.92, 77.72-77.29, 36.62.

7-(5'-Fluoro-2'-diphenylphophorylanilido)-1,2,3,4-tetrahydro-1,2,4-triazino[4,3-b][1,2,4]triazin-4,8-dione (16)

A mixture of 3 (0.01 mol) and monochloroacetic acid (0.01 mol) in DMF (20 ml) reflux for 2 h, cooled then poured onto ice. The solid thus obtained filtered off and crystallized from THF to give 16 as yellowish crystals. Yield 70%, m.p. 165°C - 167°C. Analytical data; found: C, 54.01; H, 3.25; F, 3.58;

N, 16.31%. Calculated for $C_{23}H_{18}FN_6PO_5$ (508); C, 54.33; H, 3.54; F, 3.74; N, 16.53%. IR vcm^{-1}: 3382 (NH), 3224 (NH), 1685 (C=O), 1590 (C=N), 1477 (deform. CH_2), 1250 (C-F), 1180 (P=O), 1041 (Ph-O-P), 807, 760 (aryl CH). ^{1}H NMR (DMSO) δ: 12.97, 10.70, 10.42 (each s, 3H, 2NH, OH), 9.19 (s, 1H, NH-P=O), 7.99 - 7.00, 6.99 - 6.86, 6.85 - 6.79 (each m, 13H, aryl & phenyl protons), 4.33 (s, 2H, CH_2). ^{13}C NMR (DMSO) δ = 173.41, 163.77, 163.67, 159.42, 157.84, 157.70, 136.69, 136.04, 121.23, 121.06, 113.87, 113.81, 113.24, 113.08, 111.48, 111.43, 110.966, 110.699, 111.644, 109.28, 107.00, 106.83, 105.02, 104.02, 78.08 - 77.65, 31.22.

7-(5'-Fluoro-2'-diphenylphosphorylanilido)-1,2-dihydro-1,2,4-triazino [4,3-b][1,2,4] triazin-3,8-dione (17)

Equimolar mixture of 3 and 1,1-dichloroacetic acid in DMF (20ml) reflux for 2h, cooled then poured onto ice. The yield obtained filtered and crystallized from dioxan to 17 as yellowish crystals. Yield 78%, m.p. 180°C - 182°C. Analytical data: found C, 54.03; H, 2.89; F, 3.55; N, 16.31%. Calculated for $C_{23}H_{16}FN_6PO_5$ (506); C, 54.54; H, 3.16; F, 3.75; N, 16.60%. IR vcm^{-1}: 3100 (NH), 1791 (C=O), 1655 (C=O), 1590 (C=N), 1547 (C=N), 1295 (N-N), 1235 (C-F), 1158 (P=O), 1094 (Ph-O-P), 984, 865, 830 (aryl CH). ^{1}H NMR (DMSO) δ: 12.7 (s, 1H, NH), 10.53 (s, 1H, NH-P=O), 9.07 (s, 1H, CH=N), 7.91 - 7.16, 6.92 - 6.81, 6.77 - 6.71 (each m, 13H, aryl & aryl protons). ^{13}C NMR(DMSO) δ:179.58, 163.82, 163.01, 162.39, 159.50, 157.92, 142.32, 138.56, 136.67, 136.03, 121.28, 121.12, 117.45, 115.59, 115.43, 113.90, 112.13, 111.47, 110.67, 109.50, 109.34, 108.08, 107.91, 107.12, 106.95, 77.88 - 77.45, 66.36.

7-(5'-Fluoro-2'-diphenylphosphorylanilido)-1,2,3,4-tetrahydro-1,2,4-triazino[4,3-b][4,34]triazin-3,4,8- trione (18)

A mixture of 3 (0.01 mol) in dry C_6H_6 treat with oxalyl chloride (0.01 mol) added dropwise then, TEA added drop wise (few drops) then heated under reflux for 2h, cooled. The solid produce filtered off and crystallized from C_6H_6 to give 18 as deep-yellowish crystals, yields 60%, m.p. 168°C - 170°C. Analytical data: found: C, 52.49; H, 2.85; F, 3.51; N, 15.88%. Calculated for $C_{23}H_{16}FN_6PO_6$ (522); C, 52.87; H, 3.06; F, 3.63; N, 16.09%. IR vcm^{-1}: 3382, 3319, 3156 (3NH), 1680 (C=O), 1557 (C=N), 1270 (C-F), 1143 (P=O), 1041 (Ph-O-P), 994, 860, 804 (aryl CH). ^{1}H NMR (DMSO) δ: 14 (s, 1H, NH), 12.9 (s, 1H, NH), 10.8 (s, 1H, NH-P=O) 9.3, 8.50, 8.0 (each s, aryl protons), 7.45, 7.35, 7.039, 7.035, 7.031, 7.02, 6.886, 6.878, 6.870, 6.86 (m, 10H phenyl protons). ^{13}C NMR (DMSO) δ: 167.31, 163.07, 159.53, 158.95, 129.48, 128.26, 120.21, 118.50, 118.26, 112.38, 112.33, 112.19, 112.14, 77.62 - 77.202. M/Z (Int.%) 522 (525, M + 3, 18.55), 248 (3.15), 96 (23.11), 95 (100), 93 (8.9), 68 (15.0).

The acid hydrazido derivative 20

A mixture of 3 (0.01 mol) and oxazolone19 (0.01 mol) in ethanol (50 ml) with H_2O (20 ml) reflux for 2 h, cooled then poured onto ice. The produced solid filtered off and crystallized from dioxan to give 20 as yellow crystals. Yield 58%, m.p. 155-157°C. Analytical data; found: C, 59.89, H, 3.55, F, 5.00; N, 13.05%. Calculated for $C_{37}H_{28}F_2N_7PO_6$ (735); C, 60.40; H, 3.80; F, 5.17; N, 13.13%. IR vcm⁻¹: 3600 - 3200 (b, OH, NH, NH), 1592 (CONH), 1480 (deform. CH=C), 1240 (C-F), 1170 (P=O), 1050 (Ph-O-P), 850, 752 (aryl CH).

3-[5'-(4"-Fluorophenylidene)-3'-phenyl-1'H-6'-oxo-1',2',4'-triazin-2'-yl]-6-(5'-fluoro-2'-diphenylphospho- rylanilido)-1,2,4-triazin-5-(2H)one (21)

Compound 20 (1 mg) and glacial acetic acid (10ml) reflux for 2h, cooled then poured onto ice. The solid thus obtained filtered off and crystallized from ethanol to give 21 as deep yellowish crystals. Yield 80%, m.p. 233°C - 235°C, Analytical data; found: C, 61.81; H, 3.49; F, 5.09; N, 13.55%. Calculated for $C_{37}H_{26}F_2N_7PO_5$ (717); C, 61.92; H, 3.62; F, 5.29; N, 13.66%. IR vcm⁻¹: 3192 (NH), 2880, 2810 (aliphatic CH), 1699 (C=O), 1632 (CONH), 1479 (deform. CH=), 1303 (N-N), 1240 (C-F), 1210 (P=O), 1056 (Ph-O-P), 900, 870, 810, 801 (aryl CH).

6-[5'-fluoro-2'-(diphenylphosphato)aminophenyl]-3,3,3-triphenyl-3-λ⁵-1,2,4,3-triazaphopholino[4,5-b] [1,2,4]triazine-7(8H) one (22)

A mixture of compound 2 (0.01 mol) and triphenyl phosphine (0.01 mol) in acetonitryl (20ml) warm for 30 min, cooled. The solid obtained filtered off and crystallized from dioxan to give 22 as deep yellow crystals, yield 70%, m.p. 283°C - 285°C. Analytical data; found: C, 64.01; H, 4.15; F, 2.39, N, 11.35%. Calculated for $C_{39}H_{31}FN_6P_2O_4$ (728); C, 64.28; H, 4.25; F, 2.60; N, 11.53%. IR vcm⁻¹: 3300 - 3100 (b, NH, NH), 1694 (C=O), 1620 (C=N), 1307(N-N), 1264 (C-F), 1179 (P=O), 1027 (Ph-O-P), 910, 880, 793, 745 (aryl CH).

RESULTS AND DISCUSSION

α-Aminophosphonic acids continue to elicit study due to interest in their biological properties as herbicides [22] , plant growth regulators [23] and most notably those species heaving a direct P-N bond are investigated as transition state analogues of the tetrahedral transition-state involved in peptide hydropysis [24] . Ali et al. [25] [26] studied the reactivity of α-amino phosphonates as dipolar ion structure and have type of tautomeric formula due to the higher e-withdrawing of two phenoxy and P=O groups (Figure 2). Thus, α-aminophosphonate group had a higher degree of stability towards any attack of reagents, which attribute to presence of differ factors of stability [27].

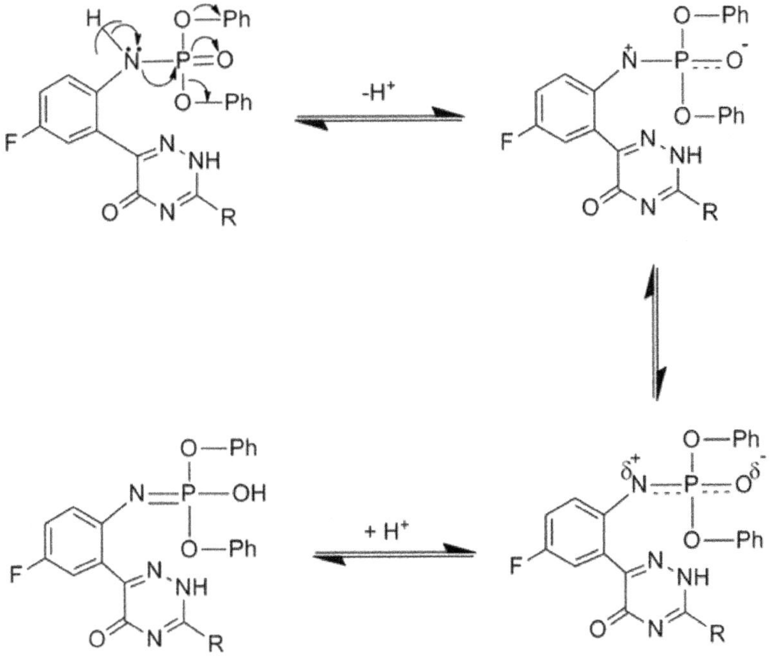

Figure 2: A possible present formula of the new synthesized isolated systems.

Phosphorus elements in these systems, was determine by using the spectrophotomeric. The method is based on the development of floated complex of molybdophosphonic acid (MPA) and methylene blue (MB) with N, N'-diphenylbenzamide (DPBA) in toluene and its subsequent dissolution in acetone [28] .

A series of some new fluorine substituted phosphoryl-amino-1,2,4-trinzines bearing a functionally pyrazole ring have been obtained via cycloaddition and/or cyclocodensation of 3-hydrazino-1,2,4-trinzinone3 with π-ac- ceptor activated carbon atom reagents. In addition, a type of 1,2,4-trinzino [4,3-b] [1,2,4] triazindiones have been also obtained from cyclocondensation of compound 3 with α, β-bifunctional oxygen and halogen compounds in different conditions. The former structures of the new products confirmed from correct elemental analysis and their spectral measurements. Keeping in view the diverse medicinal activities associated with organo-heterocyclilc systems substituted fluorine, phosphorus and 3-thioxo-1,2,4-trinzinone, which intend to construct novel fluorinated substituted phosphoryl amino bearing 1,2,4-triazinone moiety hoping to active additive effects towards their HIV-1 activity.

Chemistry

The starting material 3-hydrazino-6-(5'-fluoro-2'-diphenylphosphorylanilido)-1,2,4-trizino-5(2H) one (3) obtained from treatment of 6-(5'-fluoro-2'-aminophenyl)-3-thioxo-1,2,4-triazin-5(2H)one (1) with diphenylphosphoryl chloride in DMF to give 6-(5'-fluoro-2'-diphenylphosphorylanilido)-3-thioxo-1,2,4-triazin-5-one (2) followed by hydrozinolysis in boiling ethanol (Scheme 1).

Recently, the most reactions of activated nitrites take place in basic medium leading to novel heterocyclic systems [29] [30] . Similarly, amino-pyrazolyl-1,2,4-triazinano derivatives 4-7 were obtained from the interaction between compound 3 and arylidenecyanoacetic acid (EtOH-piperidine), acetyl acetanilide (DMF), ethyl cyanoacetate (TFH) and/or malono nitrile (EtOH-piperidine) (Scheme 2).

These reactions are carried out via cycloaddition and/or cyclocondensation reactions [31] (Figure 3).

It is interested that 3-perhydropyrazo-1-yl-1,2,4-trizinones 8-10 also obtained from the ring closure reaction of compound 3 with cinnmoyl chloride (DMF), chalcone (EtOH-piperidine) and or diethylmalonate (THF) (Scheme 3).

Formation of 8 may be take place via aroylation then cycloadditon reaction [32] (Figure 4).

The greater reactivity of the polyfunctional compound as aroylisothiocynate towards the hydrazino group as bi-nucleophile is presumably due to its favourable location between both carbonyl and thiocarbonyl functional groups [33] .

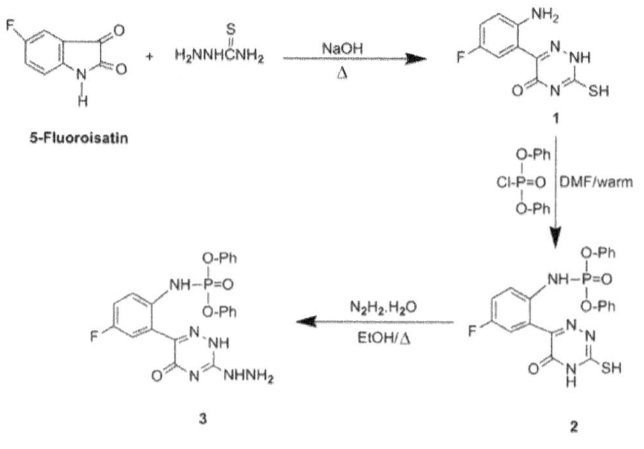

Scheme 1: Formation compounds 1-3.

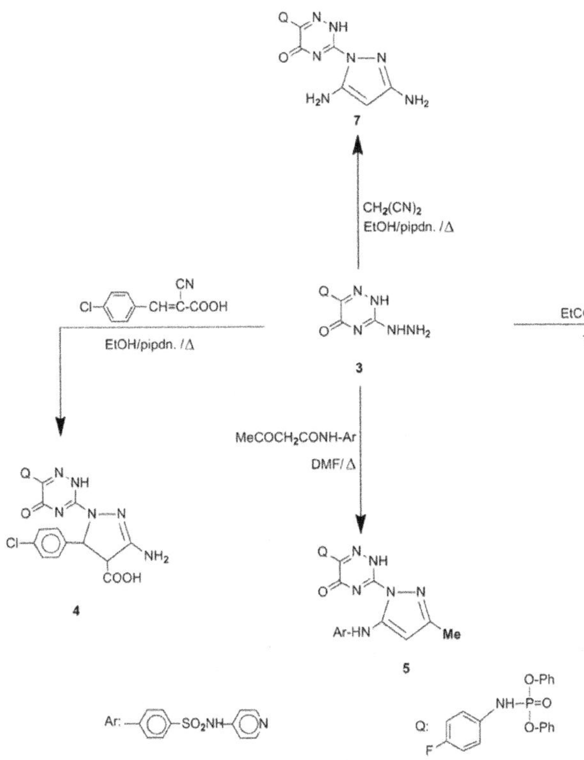

Scheme 2: Formation compounds 4-6.

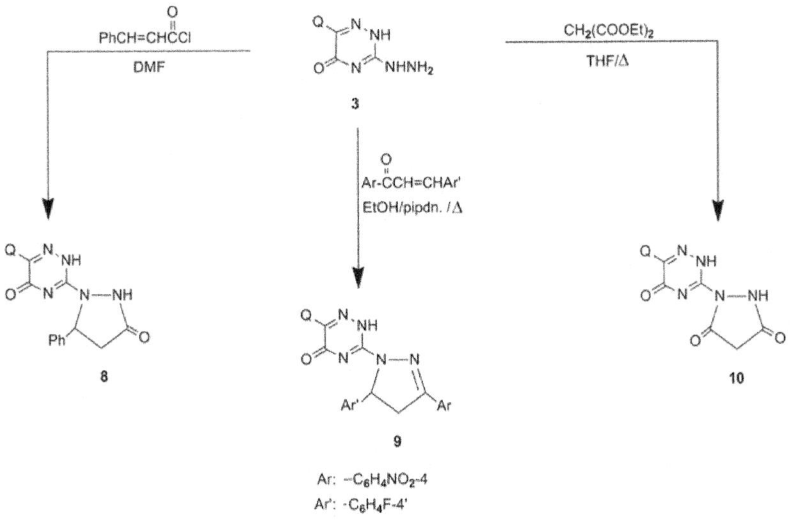

Scheme 3: Formation compounds 8-10.

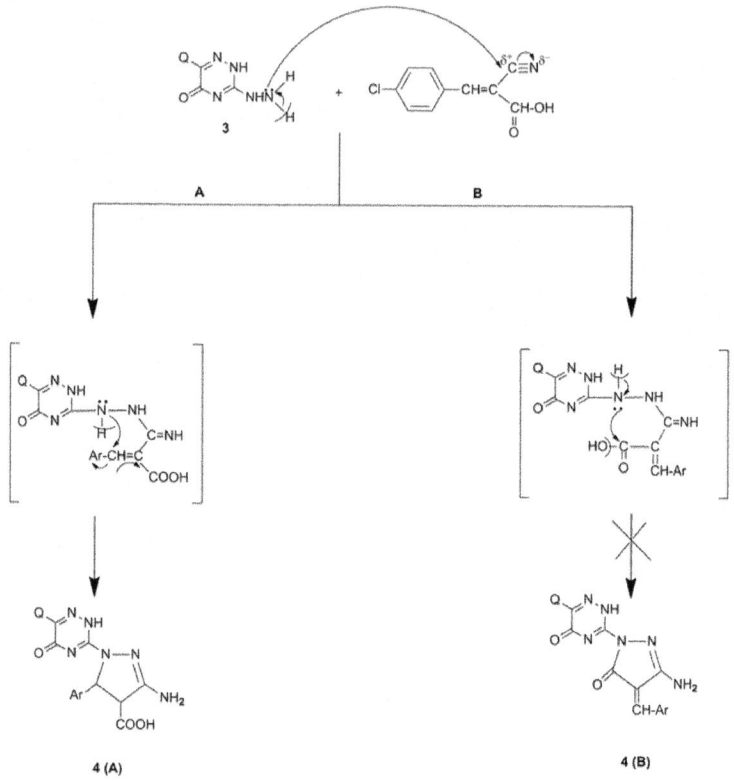

Figure 3: Formation of compound 4 from compound 3.

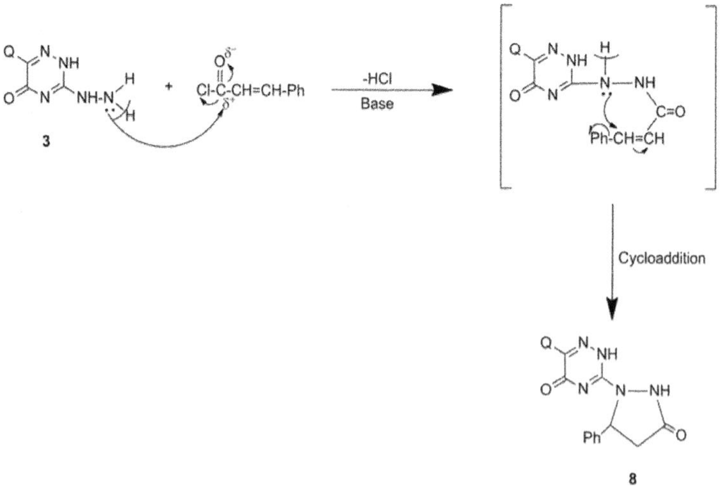

Figure 4: Formation of compound 8 from compound 3.

Thus, treatment of 3-hydrazino-1,2,4-trizinone 3 with aroylisothiocyanate in boiling non-polar solvent as dioxan yield 3-(5'-aryl-3'-mercapto-1',2',4'-triazol-1'-yl)-6-(5'-fluoro-2'-diphenly phosphorylanilido)-1, 2,4-triazin-5(2H) one (11), while that reaction when carried out in DMF, 3-(3'-aryl-5'-mercapto-1', 2',4'-tri- azol-1'-yl)-6-(5'-fluoro-2'-di-phenylphosphorylanilido)-1,2,4-trizin-5-(2H) one (12) isolated (Scheme 4). Formation of compounds 11 and 12 starting from compound 3 were outlined in)Scheme 4(.

Reactivity of α,β-bifunctional carbonyl compounds towards an hydrazino groups arrived us to synthesize new fused heterobicyclic nitrogen systems. Thus, the interaction between compound 3 and sodium pyruvate in warming sodium hydroxide solution afforded 1H-3-methyl-7-aryl-1,2,4-triazino[4,3-b][1,2,4] traizin-4,8-dione (13), while cyclo-condensation of compound 3 with isatinas 1,2-bicarbonyl compound in boiling DMF yield indolo [2,3-e] [1,2,4] trinzino [4,3-b][1,2,4] triazin one (14). Refluxing of 3 with phenacyl bromide in ethanolic KOH furnish the tetrahydro-1,2,4-triazino [4,3-b][1,2,4] triazin-8-one (15) (Scheme 5).

A large degree of the biological activity is attributed of the nature of substituent's and a degree of electronic distribution over the active center of the 1,2,4-triazines [34] [35] . Thus, direct nucleophilic displacement of chlorine atoms by nitrogen or other nucleophilic can easily occur if present α-carbonyl groups. Based on these facts, treatment of compound 3 with activated halogen as chloroacetic acid (DMF), dichloroacetic acid (DMF) and or oxalyl chloride (C$_6$H$_6$/TEA), produce perhydro 1,2,3,4-tetrahydro-1,2,4-triazino [4,3-b] [1,2,4] triazin- 4,8-dione (16); 1,2-dihydro-1,2,4-triazino [4,3-b] [1,2,4] triazin-4,8-dione (17) and 1,2,3,4-tetrahydro-1,2, 4- triazino [4,3-b] [1,2,4] triazin-3,4,8-trione (18) derivatives (Scheme 6).

In view of interesting results obtained from the reaction of 1,3-oxazolium salts and of 1,3-oxazol-2-one derivatives with hydrazine derivatives[36] , [37] , it was worthwhile to investigate the behavior of oxazolone19 towards hydrazine-derivative. Similarly, 3-hydrazino-6-aryl-1,2,4-triazin-5 (2H) one (3) when react with oxazole derivation 19 in boiling aqueous ethanol, the acid hydrazide derivative 20 isolated. Ring closure reaction of 20 by refluxing with glacial acetic acid afforded 3-(1'H-3-phenyl-5'-arylidene-6'-oxo-1,2,4-triazin-2'-yl-1-6-(5'-fluoro-   2'-diphenylphosphorylanilido)-1,2,4-triazin-5(2H)one(21)Finally,6-[5'-fluoro-2'-(diphenylphosphato)aminoph-enyl]-3,3,3-triphenyl-3-λ^5-1,2,4,3-triazaphopholino[4,5-b][1,2,4]triazine-7(8H) one (22) isolated from treat compound 3 with triphenylphosphine in THF (Scheme 7).

Elucidation the Former Structures

UV Spectra Study

UV absorption study of the new compounds, synthesize give us a good indication about electronic distribution and molecular configuration as possible.

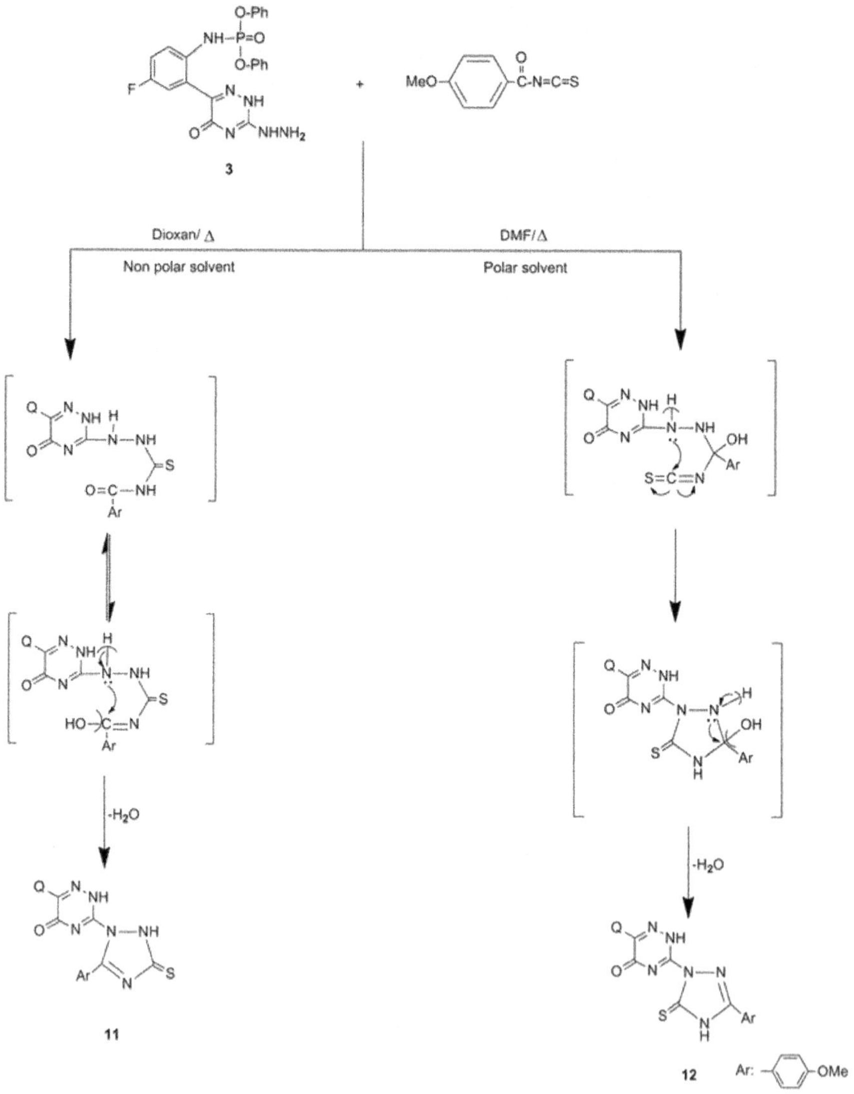

Scheme 4: Formation compounds 11 & 12.

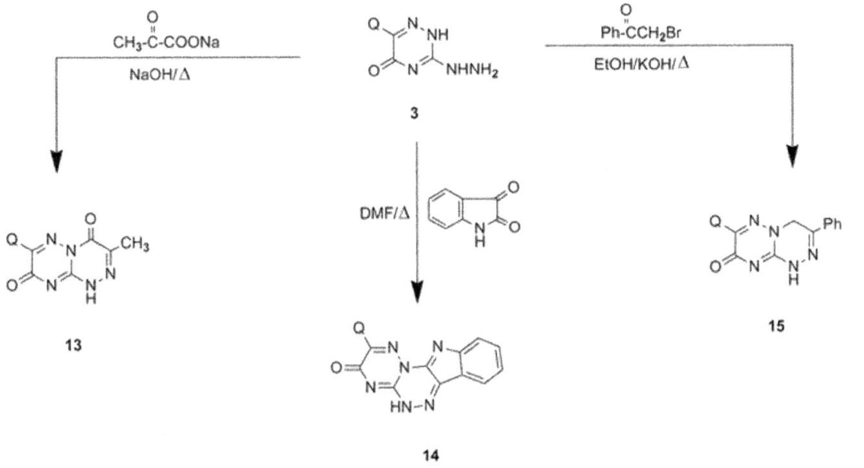

Scheme 5: Formation compounds 13-15.

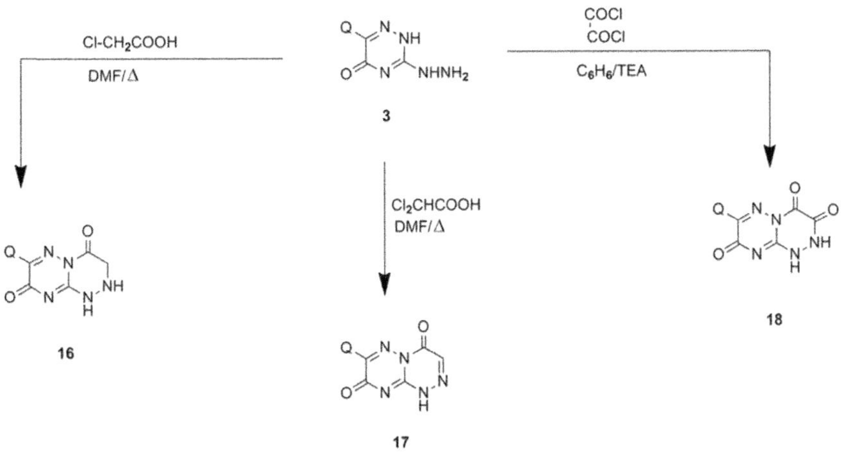

Scheme 6: Formation compounds 16-18.

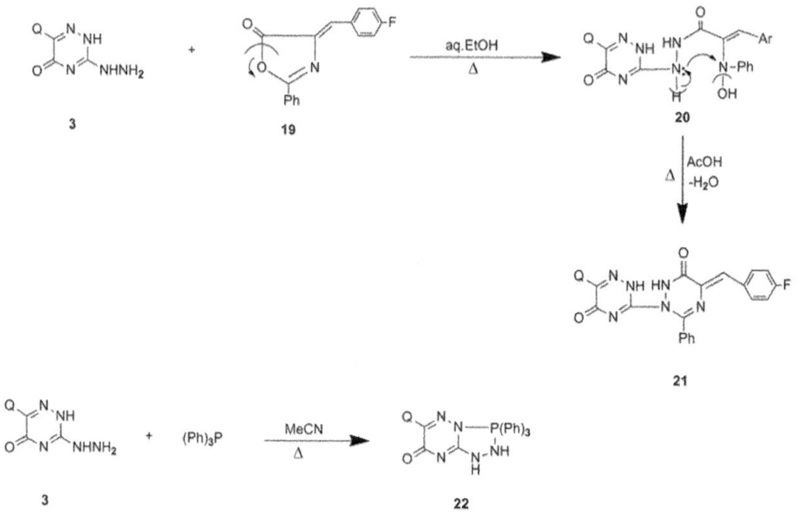

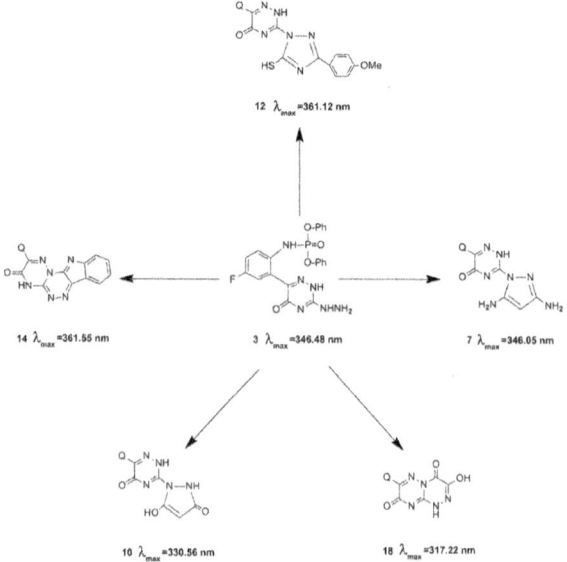

Scheme 7: Formation compounds 19-22.

In general, all the new compounds record the presence of n-π*, n-σ*, π-π* and σ-σ* electronic transition. UV-absorption spectrum of compound 3 as state recorded λ_{max} (EtOH) at 346.48 nm, while that of compounds 14 (361.55) and 12 (361.12). A higher value of λ_{max} for 14 and 12 than 3 is may be a lack's of -OH groups (which generate of H-bending (Figure 5).

Figure 5: UV-absorbtion data of compound 3 and some prepared compounds.

On the other hand, UV absorption spectra of selected compound record a lower λ_{max} than compound 3. λ_{max} of 10 (330.56) and 18 (317.22) nm. A lower λ_{max} of these compounds than the start is may be due to the presence of O-H group, which generate a type of H-bonding. The Keto-enol forms of 10 and 18 led to the inhibition of heteroconjugative, in addition a type of H-bonding which is closed to 1,2,4-triazine moiety (Figure 6).

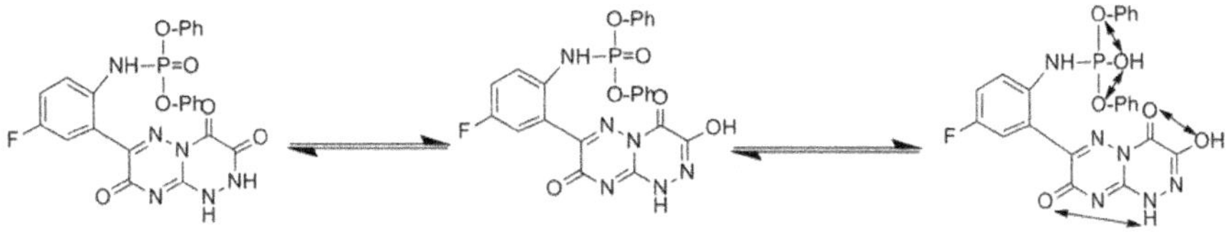

Figure 6: A possible conformer structure of 18 (fused system).

IR-Absorption Study

The IR absorption spectral data show that most of the new compounds lack's a band of NH-P=O, which is due to a type of H-bonding present (Figure 6), while that of these compounds record only NH bond of new 1,2, 4-triazinones and/ or pyrazoles moiety. In addition, IR absorption spectra of all the synthesized compounds exhibit an absorption bands at 3300 - 3190 (NH), 1690 - 1650 (C=O). Moreover presence of a characteristic bands at v 1390 - 1370 for cyclic NCN, 1250 and 1220 - 1200 cm^{-1} for C-F and P=O functional. Also, all the new compounds record v at 1100 - 1050 cm^{-1} attribute to Ph-O-P group. Only, the compounds 3, 4, 6, and 7 record a v for NH$_2$ at 3390 - 3400 cm^{-1}, while that of the compounds 4, 6, 8, 10, 13, 16-21 showed v for C=O in addition the original C=O of 1,2,4-trinzine. Finally, IR spectra of the compounds 4, 5, 6, 8-13 and 15, 16 record a type of bands characteristic for aliphatic groups (deformation 1480 - 1440 cm^{-1}).

NMR Spectral Study

1) ^{1}H NMR Spectral Study

The NH proton signals in all the new compounds appear as doublets at δ 8.8 - 8.2 ppm (JP-N-H=6 - 5.5 H$_z$) is due to its coupling with phosphorus. Also, H-bonding with oxygen of P=O and the deshielding effect of phenoxyphosphoryl group (Ph-O-P=O) are obviously the contributing factors for the downfield shift of the NH proton. On the other hand, phenoxy protons resonated at δ 7.50 - 7.1 ppm and their integration corresponds to five protons with no splitting of the signals. This shows that all the protons as magnetically equivalent. Normally, exo and endo NH protons of 1,2,4-triazinone reveal at $\delta 12$ - 11 ppm, while what of the 1,2,4-tiazinone addujent of CH_2 or NH protons show as enolic protons at $\delta 11$ - 10 ppm (5,7,17,18,9).In addition, all the new compounds 4 & 6, exhibited a resonated signals at $\delta 5$ - 4 ppm as NH_2 protons and NH proton at $\delta 11$ - 10 ppm. Finally, the perhydro pyrazolyl-1,2,4-triazinones and the 1,2,4-trinzino-1,2,4-trinzinones which containing an aliphatic protons show a resonated signals at $\delta 4$ - 3 and 1 - 0.5 ppm for CH_2 and CH_3 protons.

2) ^{13}CNMR Spectral Study

The ^{13}C Chemical shifts of phenoxy moiety are agreeing well with the reported values [38] . But, the coupling constants for ^{2}Jare concurring with those of equationally oriented P-O-Ar groups [39] showing the 1,2,4-trinzine ring has probably half chair conformation with phosphorus atom projecting upwards and the O-Ar group orienting equationally. The carbons C, which are connected to phosphorous through NH, resonated at δ 112.97 with ^{2}J P-N-C (d, J 8.1) and the difference in their chemical shifts may be attributed to the variation of shielding effect of NH [40] . In addition, all the new compounds record the resonated attribute to C=O (170 - 160), C-F (150 - 140), C=N(140 - 130), C-N (111 - 110) of 1,2,4-triazinone, with a differ type of carbons of pyrazolyl as well as carbons of other 1,2,4-triazine formed (Figure 7 and Figure 8).

Mass Spectrometric Measurements

The mass spectral investigations of the isolated heterobicyclic system is differ than the fused heterobicyclic systems (14 & 18) for example, M/Z of 7 and/or 10 recorded the molecular ion Peak's at m/z 534 and 336 respectively with a lower abounds percent's, which indicating the fragile nature of these systems. While M/Z of compound 14 showed the highest value peak at 579 with a base peak at m/z 95, which give us a high degree of stability for this Skelton. Moreover, M/Z of 18 exhibits a molecular ion peak at m/z 522 with moderate abounds percent. From these data, we can conclude that fused heterobicyclic nitrogen systems are more stabilized [41] than other isolated heterobicyclic systems.

It's interest that in all mass fragmentation pattern, 4-fluorophenyl ion is a base peak followed by N-phosphorus oxide ions, while that heterocycle supported that a large fragmentation bath way (Figures 9-12).

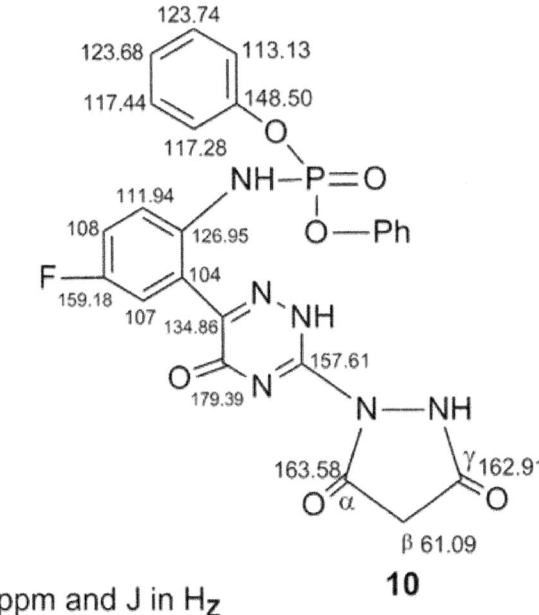

δ in ppm and J in Hz

Figure 7: ¹³C NMR data of compound 10.

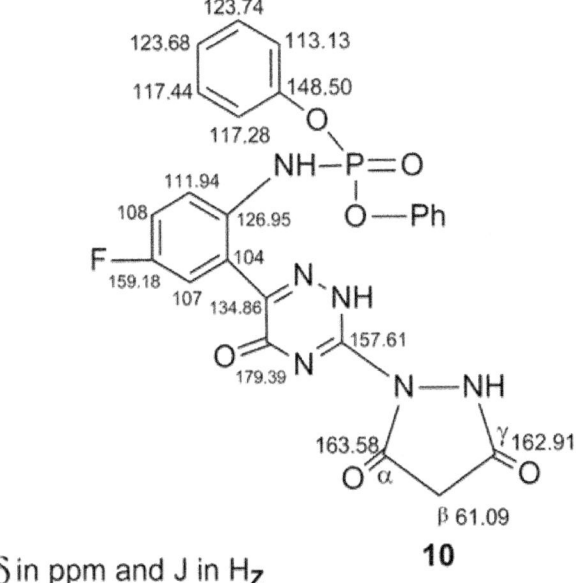

δ in ppm and J in Hz

Figure 8: ¹³C NMR data of compound 18.

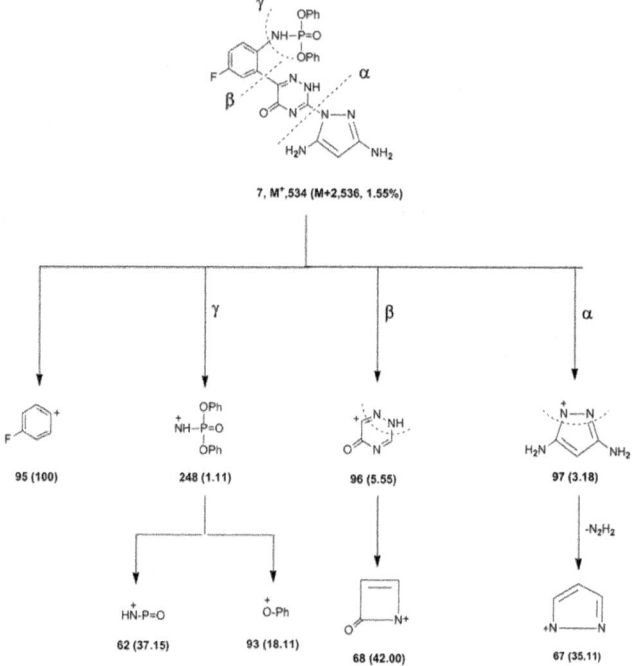

Figure 9: Mass Fragmentation pattern of compound 7.

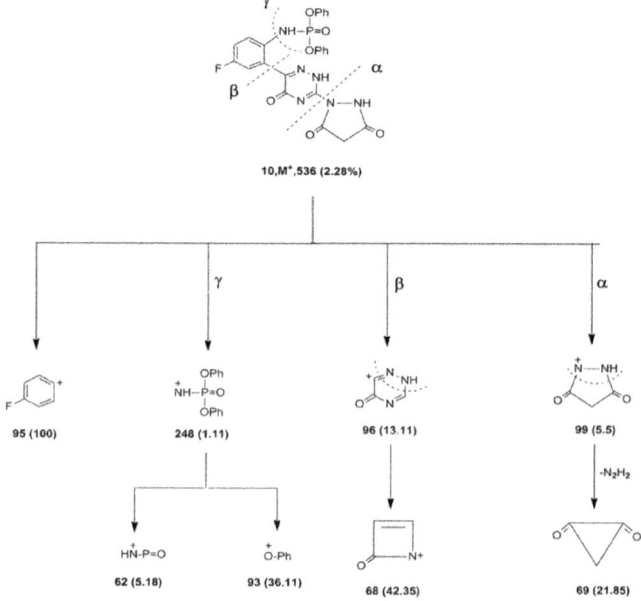

Figure 10: Mass Fragmentation pattern of compound 10.

HIV-1 INHIBITION (ENZYME INHIBITION)

Human immunodeficiency virus type-1 is the causative agent of acquired immunodeficiency syndrome (AIDS) which is one of the most serious health problems [42] . Since reverse transcriptase (RT) is an essential enzyme for the replication [43] of HIV, it is the most favoured target for the antiviral chemotherapy against HIV infection [44] . 3'-Azido-2',3'-dideoxythymidine (AZT) [45] and 2',3'-dideoxyinosine (DDI) [46] , 2',3'-dideoxycyty- dine (DDC) [47] and 2',3'-didehydro-3-deoxythymidine (DT4) [48] are the well-known potent nucleoside reverse transcriptase inhibitors clinical use, but unfortunately they produce serious side effects such as bone marrow suppression. The search for a more effective and less toxic agent has brought into focus potent yet structurally different non-nucleoside HIV-1 reverse transcriptase inhibitors (NNRTIs) [49] . Shakil et al., [50] reported that increase or decrease of electro-negativity and hydrophobicity of the bioactive drugs, cytotoxicity will also increase or decrease accordingly. So less electronegative and less hydrophobic substituents would be preferred to design the less cytotoxic drugs. The large number of research papers published every year indicate that the development of an effective drug for the inhibition of HIV-1 via enzymes inhibitors [51] [52] . HIV PIs for example, prevent the cleavage of the gag and gag-pol precursor polyproteins to the structural proteins and functional proteins, thus arresting maturation and thereby blocking infectivity of the nascent virions. e.g. Tipranavir showed loss cross-resistance to HIV strains that were resistant to the established (peptidomimetic) inhibitors of HIV protease. Also, tipranavir retained marked activity against HIV-1 isolates derived from patients with multidrug resistance to other PIs (Figure 13).

In search for new poly substituted 1,2,4-triazine bearing a phosphoryl group. The present work is synthesize novel fluorine substituted phosphorylanilido-1,2,4-triazin-ones and evaluate as potential inhibitors for HIV-1. The procedure used in the National Cancer Institute's Test for agents active against HIV is designed to detect agents acting at any stage of the virus reproductive cycle [53] . The assay basically involves the killing of T_4 lymphocytes by HIV. Small amounts of HIV are added to cells and two cycles of virus reproduction are necessary to obtain the required cell killing. Agents that interact with virions, cells or virus gene-products to inter- fere with viral activities will protect cells from cytolysis. The tetrazolium salt XTT is added to all wells and cultures are incubated to allow formazan color development by viable cells used analyzed spectrophotometrically (Figure 14).

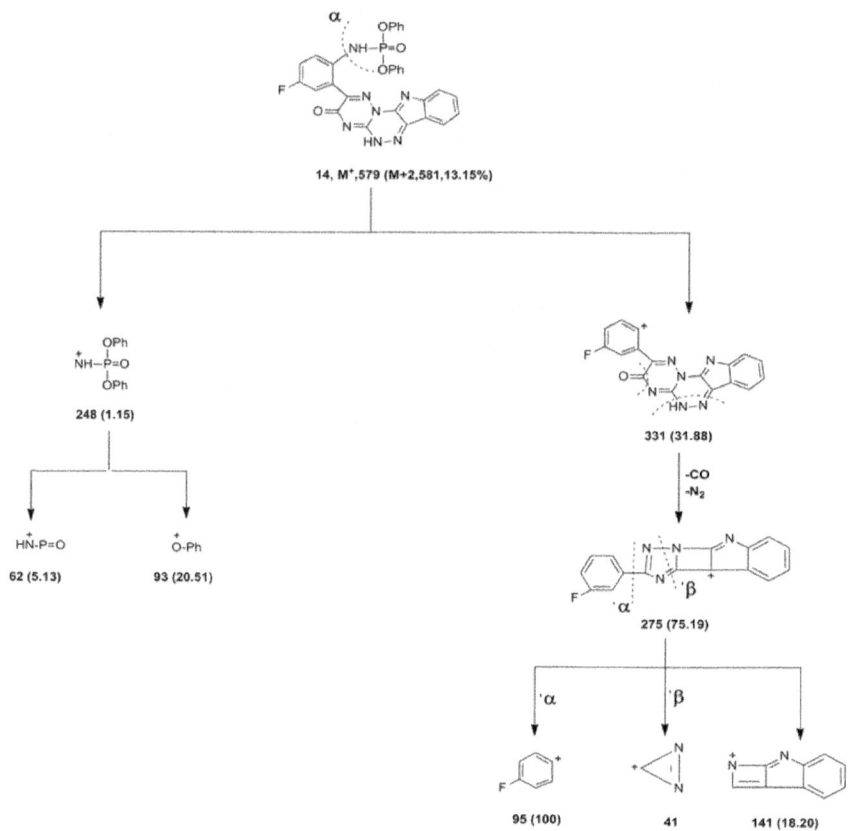

Figure 11: Mass Fragmentation pattern of compound 14.

Anti-HIV-1 screening results of some of compounds (Table 1) shows that these systems found inactive, probably due to their inability to exist in butterfly like conformation as explained in a similar case [54] . Only the compounds 7, 12, 18 and 22 exhibited a higher protection% (concentration required in protect MT-4 cells against the cytopathogenicity of HIV by 50%). The order reactivity increases as 18 > 10 > 7 > 12 > 22. A higher effect's of compound 18 (Table 2) is may be due to a higher possibility to form a type of H-bonds with proteins of virus, which led to a moderate degree of inhibition of HIV-1 activity. Also, compound 18 had a differ type of phenolic bonds, which give a variety of differ effects towards HIV-1 activity. Moreover, compound 18 had a less possibility to form a type of intra-molecular H-bond, thus their hydroxyl group become a higher degree of free action with a active center of proteins of HIV-1 virus.

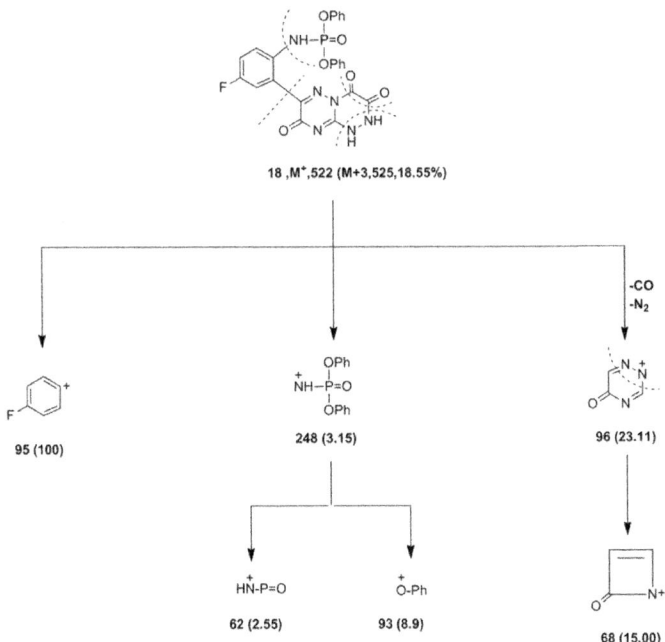

Figure 12: Mass Fragmentation pattern of compound 18.

Table 1: The in vitro anti-HIV-1 screening results of new compounds 3-22 in MT-4 cells.

Compd. No.	Dose (Molar)	Anti-HIV-1 Activity			
		IC_{50} (μg/mL)	Max Production %	Percent of Control	
	2.00×10^{-6}			Infected	Uninfected
3		>55.13	13.5	10.11	101.11
6		>20.20	19	17.38	106.32
7		>27.20	26.5	63.51	97.93
10		>19.40	30.5	80.29	96.11
12		>28.57	25.0	35.40	105.77
14		>88.30	9.5	0.04	0.51
16		>3.99	16.00	14.16	101.91
18		>42.60	35.00	83.47	104.97
22		>94.40	23.5	18.19	104.12

Table 2. The in vitro anti-HIV-1 screening results of compound 18.

Index	Concentration	Dose	Percent of Protection	Percent of Control	
				Infected	Uninfected
IC_{50} (Molar)	1.12×10^{-5}	6.33×10^{-9}	7.55	13.99	101.90
EC_{50} (Molar)	4.26×10^{-7}	2.0×10^{-8}	10.10	17.11	106.20
TIC_{50} (IC/E_C)	$2.63 \times 10^{+1}$	6.35×10^{-8}	12.00	18.15	104.21
		2.00×10^{-7}	30.01	32.31	105.30
		6.33×10^{-7}	60.66	62.15	98.10
		2.00×10^{-6}	82.23	83.47	104.99
		6.34×10^{-6}	75.66	80.13	99.10
		2.00×10^{-5}	-7.55	0.040	0.55

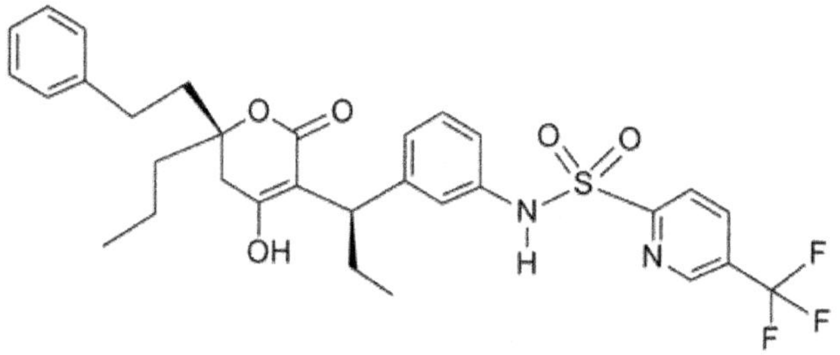

Figure 13: Tipranavir (PNU-140690).

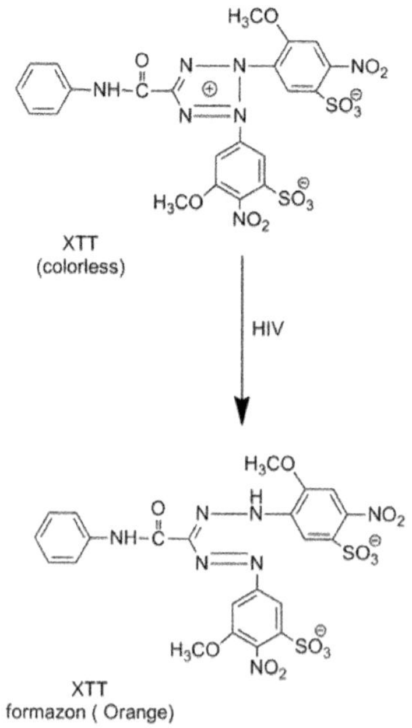

Figure 14: The standard indicator for HIV present (XXT).

CONCLUSION

In search for new anti-HIV-1, novel fluorine substituted isolated and fused heterobicyclic nitrogen systems bearing 6-(2'-phosphorylanilido)-1,2,4-triazin moiety have been obtained from a ring closure reactions of the corresponding 3-hydrazino-1,2,4-triazinone with π-acceptable reagents. Some compounds exhibited a mark's inhibitors as anti-HIV-1 activity in hope to a possible control on the HIV-1 activity.

REFERENCES

1. Hartley, F.R. and Dahl, O. (1996) The Preparation and Properties of Tervalent Phosphorus Acid Derivatives. In: Brecuer, E. and Heartly, F.R., Eds., The Chemistry of Organophosphorus Compounds: Ter- and Quinque-Valent Phosphorus Acids and Their Derivatives, John Wiley & Sons, New York, 653.

2. Prakasha, T.K., Day, R.O. and Holmes, R.R. (1994) Pentacoordinated Molecules. 101. New Class of Bicyclic Oxy- phosphoranes with an Oxaphosphorinane Ring: Molecular Structures and Activation Energies for Ligand Exchange. Journal of the American Chemical Society, 116, 8095-8104. http://dx.doi.org/10.1021/ja00097a016

3. Fest, C. and Schmidt, K.J. (1982) The Chemistry of Organophosphorus Pesticides. Springer-Verlag, Berlin, 12. http://dx.doi.org/10.1007/978-3-642-68441-8

4. Bhatia, M.S. and Jit, P. (1976) Phosphorus-Containing Heterocycles as Fungicides: Synthesis of 2,2'-Diphenylene Chlorophosphonate and 2,2'-diphenylene Chlorothiophosphonate. Experienia, 32, 1111. http://dx.doi.org/10.1007/BF01927572

5. Manne, P.N., Deshmukh, S.D., Rao, N.G.N., Dodale, H.G., Tikar, S.N. and Nimbalkar, S.A. (2000) Efficacy of Some Insecticides against Helicoverpa armigera (HUB). Pestology, 34, 65.

6. Hendlin, D., Stapley, E.O., Jackson, M., Wallick, H., Miller, A.K., Woll, F.J., Miller, T.W., Chaiet, L., Kahan, F.M., Foltz, E.L., Woodruff, H.B., Mata, J.M., Hernandez, S. and Mochales, S. (1969) Phosphonomycin a New Antibiotic Produced by Strains of Streptomyces. Science, 166, 122-123. http://dx.doi.org/10.1126/science.166.3901.122

7. Polozov, A.M. and Cremer, S.E. (2002) Synthesis of 2H-1,2-Oxaphosphorin 2-oxides. Journal of Organometallic Chemistry, 646, 153-160. http://dx.doi.org/10.1016/S0022-328X(01)01207-4

8. Wu, X., Lei, X. and Fu, Z.F. (2003) Rabies Virus Nucleoprotein Is Phosphorylated by Cellular Casein Kinase II. Biochemical and Biophysical Research Communication, 304, 333-338.

9. Gholivand, K., Shariatinia, Z., Mahzouni, H.R. and Amiri, S. (2007) Phosphorus Heterocycles: Synthesis, Spectroscopic Study and X-Ray Crystallography of Some New Diazaphosphorinanes. Structural Chemistry, 18, 653-660.http://dx.doi.org/10.1007/s11224-007-9197-3

10. Page, D.S. (1981) Principles of Biological Chemistry. 2nd Edition, Willard Grant Press, Boston.

11. Hugcl, H.M. and Jackson, N. (2012) Special Feature Organo-Fluorine Chemical Science. Applied Sciences, 2, 558-565. http://dx.doi.org/10.3390/app2020558

12. Ismail, F.M.D. (2002) Important Fluorinated Drugs in Experimental and Clinical Use. Journal of Fluorine Chemistry, 118, 27-33. http://dx.doi.org/10.1016/S0022-1139(02)00201-4

13. Elliott, A.J. (1995) Chemistry of Organic Fluorine Compounds. In: Hudlicky, M. and Pavlath, A.E., Eds., Chemistry of Organic Fluorine Compounds II: A Critical Review, American Chemical Society, Washington DC, 1119-1125.

14. Dolbier Jr., W.R. (2005) Fluorine Chemistry at the Millennium. Journal of Fluorine Chemistry, 126, 157-163. http://dx.doi.org/10.1016/j.jfluchem.2004.09.033

15. Smart, B.E. (2001) Fluorine Substituent Effects (on Bioactivity). Journal of Fluorine Chemistry, 109, 3-11. http://dx.doi.org/10.1016/S0022-1139(01)00375-X

16. Ojima, I., McCarthy, J.R. and Welch, J.T., Eds. (1996) Biomedical Frontiers of Fluorine Chemistry. American Chemical Society, Washington DC. http://dx.doi.org/10.1021/bk-1996-0639

17. Al-Romaizan, A.N., Abdel-Rahman, R.M. and Makki, M.S.T. (2014) Synthesis of New Fluorine/Phosphorus Substituted 6-(2'-Amino phenyl)-3-thioxo-1,2,4-triazin-5 (2H, 4H) One and Their Related Alkylated Systems as Molluscicidal Agent as against the Snails Responsible for Bilharziasis Diseases. International Journal of Organic Chemistry, 4, 154-168.

18. Abdel-Rahman, R.M. (2001) Role of Uncondenced1,2,4-triazinecompounds and Related Heterobicyclic Systems as Therapeutic Agents. pharmazie, 56, 18-30.

19. Abdel-Rahman, R.M. (2001) Role of Uncondensed

1,2,4-Triazinederivatives as Biological Plant Protection Agents. Pharmazie, 56, 195-212.

20. Abdel-Rahman, R.M. (1999) Synthesis and Chemistry of Fluorine Containing 1,2,4-triazines. Pharmazie, 54, 791-804.

21. Abdel-Rahman, R.M. (2002) Synthesis of New Phosphaheterobicyclic Systems Containing 1,2,4-Triazine Moiety Part IX; Straightforward Synthesis of New Fluorine Bearing 5-Phospha-1,2,4-triazin/1,2,4-triazepine-3-thions-part X. Trends in Heterocyclic Chemistry, 8, 187-195.

22. Treov, K.D. (2006) Chemistry and Application of H-Phosphonates. El-Services, Amsterdam, 256.

23. Cherkasov, R.A. and Galki, V.I. (1998) The Kabachnik-Fields Reaction: Synthetic Potential and the Problem of the Mechanism. Russian Chemical Reviews, 67, 857-940.http://dx.doi.org/10.1070/RC1998v067n10ABEH000421

24. Annie Bligh, S.W., Mc Grath, C.M., Failla, S. and Finocchiaro, P. (1996) α-Aminophosphate Monoester in One Step. Phosphorus, Sulfured Silicon, 118, 189-194.

25. El-Sayed Ali, T. (2008) Synthesis and Characterization of Novel Bis-(α-Aminophosphonates) with Terminal Chromone Moieties. ARKIVOC, 2, 71-79.

26. El-Sayed Ali, T., Abdul-Ghaffar, S.A., El-Mahdy, K.M. and Abdel-Karim, S.M. (2013) Synthesis, Characterization, and Antimicrobial Activity of Some New Phosphorus Macrocyclic Compounds Containing Pyrazole Rings. Turkish Journal of Chemistry, 37, 160-169.

27. Abdel-Rahman, R.M. and Ali, T.E. (2013) Synthesis and Biological Evaluation of Some New Polyfluorinated-4- Thiazolidinone and α-Amino Phosphonic Acid Derivatives. Monatshefte fur Chemie, 144, 1243-1252.

28. Chakravarty, S. and Mishra, R.K. (1994) Spectrophotometric Determination of Phosphorus with N,N'-Diphenylbenzami- dine and Methylen-Blue. Journal of Indian Chemical Society, 71, 717-719.

29. Bukowski, L. (2001) Some Reactions of 2-cyanomethyl-3-methyl-3H-imidazo[4,5-b]pyridine with Isothiocyanates. Antituberculotic Activity of the Obtained Compounds. Pharmazie, 56, 23-27.

30. Abdel-Rahman, R.M. and Abdel-Monem, W.R. (2007) Chemical Reactivity of 3-Hydrazino-5,6-diphenyl-1,2,4-tri- azine towards π-Acceptors Activated Carbonitriles. Indian Journal of Chemistry, 46B, 838-846.

31. Abdel-Rahman, R.M. (1988) Reactions of 3-Hydrazino-5,6-diphenyl-1,2,4-triazine with Unsymmetrical 1,3-bicarbonyl Compounds: Synthesis of Some New 3-(3',5'-Disubstituted pyrazol-1'-yl(-5-6-diphenyl-1,2,4-triazines of Their Antimicrobial Activity. Indian Journal of Chemistry, 27B, 548-553.

32. Zaher, H.A., Abdel-Rahman, R.M. and Abdel-Halim, A.M. (1987) Reactions of 3-Hydrazino-5,6-diphenyl-1,2,4-tri- azine with α,β-Bifunctional Compounds. Indian Journal of Chemistry, 26B, 110-115.

33. Gamal, A.A. (1997) Studies on 3-Amino-1,2,4-triazole. Journal of the Indian Chemical Society, 74, 624-625.

34. El-Gendy, Z., Morsy, J.M., Allimony, H.A., Abdel-Monem, W.R. and Abdel-Rahman, R.M. (2001) Synthesis of Heterobicyclic Nitrogen Systems Bearing the 1,2,4-triazine Moiety as Anti-HIV and Anticancer Drugs, Part III. Pharmazie, 56, 376-382.

35. El-Gendy, Z., Morsy, J.M., Allimony, H.A., Abdel-Monem, W.R. and Abdel-Rahman, R.M. (2003) Synthesis of Heterobicyclic Nitrogen Systems Bearing a 1,2,4-Triazine Moiety as Anticancer Drugs, Part IV. Phosphorus, Sulfur and Silicon, 179, 2055-2071.http://dx.doi.org/10.1080/10426500390228738

36. Haddadin, M.J. and Hassner, A. (1973) Cycloaddition Reactions. XIV. Thermal and Photochemical Reactions of Some Bicyclic Aziridine Enol Ethers. Journal of Organic Chemistry, 38, 3466-3471. http://dx.doi.org/10.1021/jo00960a005

37. Abdel-Rahman, R.M. and El-Gendy, Z. (1989) Synthesis of Some New 1,2,4-Benzotriazine Derivatives from 2-Methyl Benzoxazole. Indian Journal of Chemistry, 28B, 1072-1076.

38. Buchannan, G.W., Whitman, R.H. and Malaiyandi, M. (1982) A Carbon-13 Nuclear Magnetic Resonance Spectral Investigation of Substituted Triphenyl Phosphates. Organic Magnetic Resonance, 19, 98-101. http://dx.doi.org/10.1002/mrc.1270190211

39. Al-Ravi, J.M.A., Behnam, G.O., Naceur, A. and Kruemer, R. (1985) Carbon-13 Chemical Shift Assignment of Some Or-ganophosphorus Compounds. IV—2-oxo- and 2-thio-2-phenoxy-1,3,2-Diazaphosphorinanes and Related P(IV) Com- pounds. Magnetic Resonance in Chemistry, 23,728-731. http://dx.doi.org/10.1002/mrc.1260230910

40. Gorenstein, D.G. and Kar, D.J. (1977) Effect of Bond Angle Distortion on Torsional Potentials. Ab Initio and CNDO/2 Calculations on

Dimethoxymethane and Dimethyl Phosphate. Journal of the American Chemical Society, 99, 672-677.

41. Reddy, C.D., Reddy, G.T. and Reddy, M.S. (1993) ^{1}H,^{13}C and ^{31}P NMR Studies of 2-(Substituted Phenoxy) 2, 3-Di- hydro-1H-Naphtho-[1, 8-de]-1, 3,2-Diazaphosphorine-2-Oxides. Asian Journal of Chemistry, 5, 291-295.

42. Clercq, E.D. (2002) New Developments in Anti-HIV Chemotherapy. Biochemical et Biophysica Acta, 1587, 258-275.

43. Xie, L., Takeuchi, Y., Mark, L. and Lee, K.H. (1999) Anti-AIDS Agents. 37. Synthesis and Structure-Activity Relationships of (3'R,4'R)-(+)-cis-Khellactone Derivatives as Novel Potent Anti-HIV Agents. Journal of Medicinal Chemistry, 42, 2662-20877.http://dx.doi.org/10.1021/jm9900624

44. Connoly, K.J. and Hammer, S.M. (1992) Comparative Pharmacokinetics, Distributions in Tissue, and Interactions with Blood Proteins of Conventional and Sterically Stabilized Liposomes Containing 2',3'-Dideoxyinosine. Antimicrobial Agents and Chemotherapy, 36, 245-254.

45. Mitsuya, H., Weinhold, K.J., Fuman, P.A., Clair, M.H., Lehrman, S.N., Gallo, R.C., Bolognesi, D., Barry, D.W. and Broder, S. (1985) 3'-Azido-3'-deoxythymidine (BW A509U): An Antiviral Agent That Inhibits the Infectivity and Cytopathic Effect of Human T-lymphotropic Virus Type III/Lymphadenopathy-Associated Virus in Vitro. Proceedings of the National Academy of Sciences of the United States of America, 82, 709670-707100.

46. Mitsuya, H. and Broder, S. (1986) Inhibition of the in Vitro Infectivity Andcytopathic Effect of Human T-Lympho- trophic Virus Type III/Lymphadenopathy-Associated Virus (HTLV-III/LAV) by 2',3'-dideoxynucleosides. Proceed- ings of the National Academy of Sciences of the United States of America, 83, 1911-1915. http://dx.doi.org/10.1073/pnas.83.6.1911

47. Bozzette, S.A. and Richman, D.D. (1990) Salvage Therapy for Zidovudine-Intolerant HIV-Infected Patients with Al- ternating and Intermittent Regimens of Zidovudine and Dideoxycytidine. The American Journal of Medicine, 88, S24- S26.http://dx.doi.org/10.1016/0002-9343(90)90418-D

48. Dunkel, L., Cross, A., Martin, R., Brown, M. and Murray, H., (1990) Dose-Escalating Study of Safety and Efficacy of Dideoxydidehydrothymidine

(d4T) for HIV Infection. Antiviral Research, 13, 116. http://dx.doi. org/10.1016/0166-3542(90)90217-U

49. Romero, D.L., Busso, M., Tan, C.K., Reusser, F., Palmer, J.R., Poppe, S.M., Aristoff, P.A., Downey, K.M., So, A.G., Resnick, L. and Tarpley, W.G. (1991) Non-Nucleoside Reverse Transcriptase Inhibitors That Potently and Specifically Block Human Immunodeficiency Virus Type 1 Replication. Proceedings of the National Academy of Sciences of the United States of America, 88, 8806-8810. http://dx.doi.org/10.1073/ pnas.88.19.8806

50. Srivastava, A.K., Khan, A.A. and Shakil, M. (2001) Quantitative Structure Activity Relationship (QSAR) Studies on Anti-HIV-1 and Cytotoxic Arylpyrrolylsulfones. Journal of the Indian Chemical Society, 78, 154-157.

51. Lu, C. and Li, A.P. (2010) Enzyme Inhibition in Drug Discovery and Development: The Good and the Bad. Wiley, OU143, E 605.

52. Sergei, V., Gulni, K., Elena, A. and Michael, E. (2010) Enzyme Inhibition in Drug Discovery and Developmental Edited, Lu & Li. Wiley & Sons, Inc., New York, 749.

53. Weislow, O.W., Kiser, R., Fine, D., Bader, J., Shoemaker, R.N. and Boyd, M.R. (1989) New Soluble Formazan Assay for HIV-1 Cytopathiceffects, Application to High Flux Screening of Synthetic and Natural Products for AIDS-Anti- viral Activity. Journal of the National Cancer Institute, 81, 577-586. http://dx.doi.org/10.1093/jnci/81.8.577

54. Chaouni, B.A., Galtier, C., Allouchi, H., Kherbeche, A., Chavignon, O., Teulade, J.C., Witvrouw, M., Pannecouque, C., Snoeck, R., Andrei, G., Balzarini, J., De Clercq, E., Fauvelle, F., Enguehard, C. and Gueiffier, A. (2001) 3-Benzamido, Ureido and Thioureidimidazo[1,2-a]Pyridine Derivatives as Potential Antiviral Agents. Chemical and Pharmaceutical Bulletin, 49, 1631-1635. http://dx.doi.org/10.1248/cpb.49.1631

Chapter 9

PHYSICOCHEMICAL AND FUNCTIONAL CHARACTERIZATION OF MUCUNA PRURIES DEPIGMENTED STARCH FOR POTENTIAL INDUSTRIAL APPLICATIONS

Maira Rubi Segura-Campos, Sonia Marina López-Sánchez, Arturo Castellanos-Ruelas, David Betancur-Ancona, Luis Chel-Guerrero*

Facultad de Ingeniería Química, Universidad Autónoma de Yucatán, Mérida, México

ABSTRACT

Starch is a very important biopolymer in the food industry. The velvet bean (M. pruriens) is an excellent potential starch source containing approximately 520 g starch per kg. The objective of this study was to evaluate the physicochemical and functional properties of velvet bean depigmented starch. The starch granules appear oval and spherical shaped. The colour registered L*, a*, b* values of 44.9, 0.324 and 0.341 respectively. The chemical composition registered values of moisture, ash, fat, protein, fibre and NFE of 110.5, 5.8, 5.7, 0.0, 34 and 954.5 g/kg respectively, as well as amylose levels of 215.3 g/kg. Gelatinization onset (T_o), peak (T_p) and final (T_f) temperatures were of 74.23°C, 80.57°C and 86.39°C. The solubility (3.1% - 16.2%), swelling power (SP) (2.86% - 16.17%) and water absorption capacity (WAC) (2.67 - 15.95 g water/g starch) were directly correlated to temperature (60°C - 90°C). The enthalpy values (4.10 - 13.47 j/g) were directly correlated to the time (1 - 21 days). The retrogradation increased as time increased. The viscosity of M. pruriens depigmented starch decreased slightly during the heating stages and then increased during cooling and the refrigeration and freezing stability registered synersis ranges from 17.65 to 23.18 mL/50mL and from 16.4 to 22.6 mL/50mL respectively, indicating that the depigmented starch was unstable in heating-cooling processes.

INTRODUCTION

Starch is a naturally occurring, biodegradable, cheap, renewable, and abundantly available polysaccharide mole- cule. The different botanical sources of starches are cereal (wheat, corn, rice, barley, oat, sorghum, millet, and rye), legume (lima bean, garbanzo bean, lentil bean, red kidney bean, navy bean, faba bean, mung bean, pinto bean, adzuki bean, field pea, cowpea, beach pea, green pea, grass pea, soybean, and groundnut), some under-utilized legume (sword bean, jack bean, and pigeon pea), root and tuber (cassava, potato, yam, cocoyam, and sweet potato), and unripe fruit (banana, plantain, mango, and pawpaw). Starch granules are mainly found in seeds, roots and tubers, as well as in stems, leaves, fruits and even pollens. The granules occur in all shapes and sizes (spheres, ellipsoids, polygon, platelets, and irregular tubules). The two main components of starch are amylose (AM) and amylopectin (AP) and they differ significantly in their properties and functionality. AM has a high tendency to retrograde and produce tough gels and strong films. In contrast, AP, when dispersed in water, is more stable and produces soft gels and weak films. It is possible for entanglements to occur between AM and AP, along with the presence of minor components (proteins, PLs, lipids), which all also have important impacts on the physicochemical properties of the starches from different botanical origin [1] .

Starch is a very important biopolymer in the food industry, where it performs various functions as thickener, binder, stabilizer, texture modifier, gelling and bulking agent. Starch varies greatly in form and functionality between and within botanical species, and even from the same plant cultivar grown under different conditions. Its different properties are utilized for making diverse food products [2] .

A growing demand for starches from the food industry has created the need for new sources of this polysaccharide. A potential new starch source is the legume M. pruriens (Velvet bean), a native of Southeast Asia. This legume has been shown to have high agricultural potential, with yields up to 1000 kg/ha, even under adverse tropical conditions. This grain has a high starch (515 g/kg) and protein (279 g/kg) content, but as with all legumes, it also contains antinutritional components. In particular, the velvet bean contains the non-protein amino acid L-3,4-dihydroxyphenylalanine (L-DOPA) in quantities of 31 g/kg, which limits its direct consumption in food and feed [3] . M. pruriens belongs to the family Fabaceae and is grown primarily as a soil-improving crop to control weeds. It grows even in poor soils and produces abundant seeds. The seeds are eaten by a few ethnic groups such as Igbos in Nigeria, either as a condiment or as part of a main dish. The mature seeds are also known to be eaten by Indian tribal sects, Mundari and Dravidian groups [4] . The process to obtain a protein concentrate helps to reduce or eliminate non-nutritive factors [5] . Ascorbic acid is used in the extraction due to its positive effect in terms of reduction of the dark colour [6] . Usually a starchy coloured by- product is obtained, which depending on the variety of seed may be slightly or highly coloured, and could be more or less acceptable as a food ingredient.

The objective of this study was to investigate the diverse physicochemical and functional properties of depigmented starch isolated from the legume M. pruriens.

MATERIALS AND METHODS

Materials

M. pruriens seeds were obtained from the 2010 harvest in the state of Yucatán, México. Reagents were analytical grade and purchased from J.T. Baker (Phillipsburg, NJ, USA), Sigma Chemical Co. (St. Louis, MO, USA) and Merck (Darmstadt, Germany).

M. pruriens Flour

Impurities and damaged seeds were removed. Sound seeds were milled in a Mykros impact mill until passing through a 20-mesh screen (0.85 mm), and then in a Cyclotec 1093 (Tecator, Sweden) mill until passing through a 60-mesh screen (0.24 mm).

M. pruriens Starch

A single starch extraction was done with the legume flour. This was processed using the wet fractionation meth- od of Hoover et al. [7] . Briefly, whole flour was suspended in distilled water at a 1:6 (w/v) ratio, pH adjusted to 11 with 1 mol/L NaOH, and the dispersion stirred for 1 h at 400 rpm with a mechanical agitator (Caframo Rz-1, Heidolph Schwabach, Germany). This suspension was wet-milled with a Kitchen-Aid® mill and the fibre solids separated from the starch and protein mix by straining through 80- and 150-mesh sieves followed by five washings of the residue with distilled water. The protein-starch suspension was allowed to sediment for 30 min at room temperature to recover the starch and protein fractions. The starch was oven drying at 60°C. The pH of the separated solubilized protein was adjusted to its isoelectric point (4.6) with 1 mol/L HCl. The suspension was then centrifuged at $1317 \times g$ for 12 min (Mistral 3000i, Curtin Matheson Scientific Inc., Morris Plains, NJ, USA), the supernatants discarded and the precipitates freeze-dried at $-47°C$ and 13×10^{-3} mbar.

Starch Discoloration Process

M. pruriens starch was suspended in distilled water at a 1:8 (w/v) ratio containing sodium hypochlorite at 25 g/L, stirred for 10 min and centrifuged at 2500 rpm for 12 min. This step was repeated three times. The precipitate thus obtained was then suspended in distilled water at a 1:8 (w/v) ratio containing HCl 1M to neutralize, stirred for 10 min and centrifuged at 2500 rpm for 12 min. Finally, the starch was suspended in distilled water at a 1:8 ratio (w/v), stirred and centrifuged at 2500 rpm for 12 min. The discoloured starch thus obtained was oven drying at 60˚C.

Physicochemical Characterization of M. pruriens Starch

Starch Shape

The microscopic appearance of the starch granules was assessed using the method of Mc Master [8] by direct microscopic observation using a Leica optical microscope. Starch suspensions at 10 g/L were prepared for this assay.

Colour

The colour of the M. pruriens depigmented starch was assessed with a colour analyser Chroma-meter CR- 200b (Minolta Camera Co., Ltd. Japan), taking five readings and reporting the average value of the CIELAB parameters (L*, a*, b*).

Chemical Composition

Standard AOAC procedures were used to determine nitrogen (method 954.01), fat (method 920.39), ash (method 925.09), crude fibre (method 962.09) and moisture (method 925.09) contents in the M. pruriens starch [9] . Nitrogen (N_2) content was quantified with a Kjeltec Digestion System (Tecator, Höganäs, Skåne län, Sweden) using cupric sulphate and potassium sulphate as catalysts. Protein content was calculated as nitrogen × 6.25. Fat content was obtained from a 1 h hexane extraction. Ash content was calculated from sample weight after burning at 550˚C for 2 h. Moisture content was measured based on sample weight loss after oven-drying at 110˚C for 2 h. Carbohydrate content was estimated as nitrogen-free extract (NFE) by difference from the sum of the protein, fat, ash and crude fibre content. Apparent amylose content was estimated after iodine complexation using the method of Morrison and Laignelet [10] . Amylopectin content was calculated by the difference of total starch minus amylose content.

Functional Characterization of M. pruriens Starch

Gelatinization. Differential Scanning Calorimetry (DSC)

Starch gelatinization was determined with a DSC-7 (Perkin-Elmer Corp., Norwalk, CT) using the technique described by Ruales and Nair [11] . The DSC was calibrated with indium and the data were analysed using the Pyris software program. Two milligrams (d.b.) of starch was placed in an aluminium pan and moisture level adjusted to 700 g/L by adding de-ionized water. The pan was then hermetically sealed and left to equilibrate for 1 h at room temperature. It was then placed in the calorimeter and heated from 30°C to 120°C at a rate of 10°C/ min by using an empty pan as reference. Gelatinization temperature was determined by automatically computing onset temperature (T_o), peak temperature (T_p), final temperature (T_f) and gelatinization enthalpy (ΔH) from the resulting thermogram.

Solubility, Swelling Power (SP) and Water Absorption Capacity (WAC)

Solubility, SP and WAC patterns at 60°C, 70°C, 80°C and 90°C were determined using a modified version of Sathe et al. [12] method. Briefly, 40 mL of a 10 g/L starch suspension was prepared in a previously tared 50 mL centrifuge tube. A magnetic agitator was placed in the tube, and it was kept at a constant temperature (60°C, 70°C, 80°C or 90°C) in a water bath for 30 min. The suspension was then centrifuged at 2120 x g for 15 min, the supernatant decanted and the swollen granules weighed. From the supernatant, 10 mL was dried in an air convection oven (Imperial V) at 120°C for 4 h in a crucible to constant weight. Percentage solubility and SP were calculated using the following formulas:

% Solubility = dry weight at 120°C × 400 sample weight

% SP = weight of swollen granules × 100/sample weight × (100 − % solubility).

WAC was measured using the same conditions as above, but was expressed as weight of the gel formed per sample, divided by treated sample weight.

Retrogradation

Retrogradation was determined with a DSC-7 (Perkin-Elmer Corp., Norwalk, CT) using the technique described by Gudmundsson and Eliasson [13] . The DSC was calibrated with indium and the data were analysed using the Pyris software program. Two milligrams (d.b.) of starch was placed in an aluminium pan and moisture level adjusted to 700 g/L by adding de-ionized water. The pan was then hermetically sealed, left to heat at 105°C for 15 min and stored for 1, 2, 3, 7, 14 and 21 days at 4°C. After that, it was then placed in the calorimeter and heated from 30°C to 120°C at a rate of 10°C /min, using an empty pan as reference.

Pasting Properties

These properties were evaluated in a Viscoamylograph (Brabender PT-100, Duisburg, Germany) according to Wiesenborn et al. [14] . Briefly, 400 ml of 60 g/L (d.b.) starch suspension was heated to 95°C at a rate of 1.5°C/min, held at this temperature for 15 min, then cooled to 50°C at the same rate and held at this second temperature for another 15 min. Maximum viscosity, consistency, breakdown and setback were calculated in Brabender Units (BU) from the resulting amylograms.

Breakdown: peak viscosity [BU] ? viscosity at 95°C × 15 min [BU]

Consistency: viscosity at 50°C [BU] ? peak viscosity [BU].

Setback: viscosity at 50°C [BU]) ? viscosity at 95°C × 15 min [BU].

Starch Clarity

Starch clarity was measured using the method of Bello-Pérez et al. [15] , determining transmittance of a 10 g/L starch paste at 650 nm using a spectrophotometer (Beckman DU-650, Fullerton, CA). Starch suspensions (10 g/L) in tubes with threaded caps were placed in a water bath at 100°C for 30 min, agitated using a Vortex every 5 min, and left to cool at room temperature. Percentage of transmittance (%T) was determined from these suspensions.

Refrigeration and Freezing Stability

Stability under refrigeration and freezing conditions was evaluated using a modified version of Eliasson and Kim [16] method. Pastes were prepared in a Brabender viscoamylograph. Briefly, 400 mL of 60 g/L (d.b.) starch suspension was heated to 95°C at a rate of 1.5°C/min, held at this temperature for 15 min, then cooled to 50°C at the same rate and held at this second temperature for another 15 min. Portions of 50 mL were placed in centrifuge tubes, cooled to room temperature and stored at 4°C and −10°C, and then centrifuged at 8000 x g for 10 min in a J2-HS centrifuge (Beckman Instruments). The water separated from the starch gels during 1, 2, 3, 4 and 5 days was measured.

RESULTS AND DISCUSSION

Physicochemical Characterization of M. pruriens Starch

In the micrograph, the granules of M. pruriens depigmented starch appear oval and spherical shaped (Figure 1). The granule size was heterogeneous and their shape was similar to that reported in potato, yucca, pinto bean, navy bean, field pea, as well as jack and velvet bean [17] - [20] . The images showed the presence of pores along the equatorial region of the M. pruriens depigmented starch granules. Similar pore arrangement has also been observed on potato starch granules, and on wheat, rye and barley granules [21] . These pores are thought to support the oxidation process that suffers the granules during the discoloration process. A similar behaviour was reported by Betancur-Ancona et al. [3] in granules of M. pruriens starch, in the micrograph the granules appear oval-shaped with a minimum diameter of 12 mm and a maximum of 45 mm as well as the presence of pores along the equatorial region that suggested the initial attack of enzymes during the germination process.

The colour analysis of M. pruriens depigmented starch registered L*, a*, b* values of 44.9, 0.324 and 0.341, respectively. The L*, a*, b* values registered in velvet bean depigmented starch suggested a more opaque colour than in velvet bean pigmented starch with L*, a*, b* values of 73.4, −1.73 and 3.3, respectively.

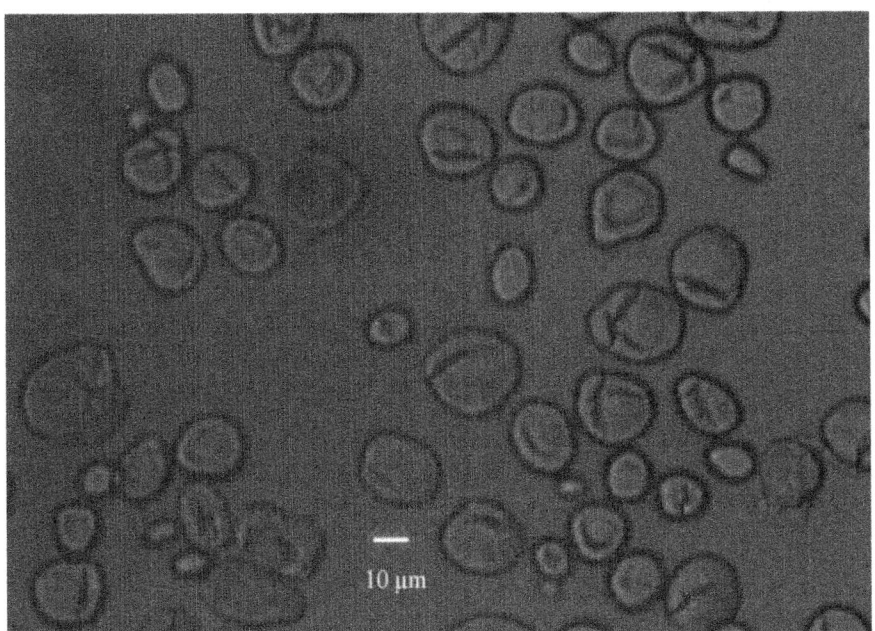

Figure 1: Photomicrograph of M. pruriens depigmented starch.

The proximal composition of M. pruriens depigmented starch registered values of moisture, ash, fat, protein, fibre and NFE of 110.5, 5.8, 5.7, 0.0, 34 and 954.5 g/kg, respectively. Betancur-Ancona et al. [3] suggested that the protein content of M. pruriens starch (7.1 g/kg), when it is found elevated, it should be reduced by means of chemical or enzymatic treatments. The U.S. Food and Drug Administration allows a maximum level of protein in corn starch of 3.5 g/kg prevent formation of undesirable dark syrups resulting from Maillard reactions that may occur during the manufacturing process. Therefore velvet bean depigmented starch could be used in high- glucose syrup production. The apparent amylose level (Table 1), was lower in the velvet bean depigmented starch (215.3 g/kg) than in pigmented starch (392 g/kg) [3] , native cereals and more common tubers starches such as corn (280 g/kg), sorghum (280 g/kg) and wheat (280 g/kg) [20] . This level also is lower than that of other legumes starches such as pinto bean (322 g/kg), navy bean (321 g/kg), field pea (342 g/kg) [19] , mung bean (264 g/kg) [18] and jack bean (375 g/kg) [17] .

Functional Characterization of M. pruriens Starch

Gelatinization onset (T_o), peak (T_p) and final (T_f) temperatures were 74.23°C, 80.57°C and 86.39°C, respectively for the M. pruriens depigmented starch whereas for the pigmented starch. Betancur-Ancona et al. [3] reported 69.4°C, 74.8°C, 81.3°C and 10.70°C, respectively. The gelatinization temperature was higher than that of commercial corn starch (62°C - 73°C) and common food starches such as potato (56°C - 67°C), wheat (58°C - 64°C) and rice (68°C - 78°C) starches [20] [22] . It was similar to the ranges of starches from other legumes such as the 70°C - 80°C range of the lima bean (Phaseolus lunatus), 73°C - 81°C for soybean (Glicine max) [7] and 76°C - 83°C for jack bean (Canavalia ensiformis) [17] . Because of its high gelatinization temperature M. pruriens depigmented starch can be used in thermal food processing at temperatures higher than those in which common starches are used. This characteristic makes it extremely useful in products subjected to high temperatures, such as canned products. Gel enthalpy ΔH for depigmented starch was 10.81 J/g whereas for the pigmented starch it was 10.70, meaning that the latter requires less energy to gelatinize that the former. This coincides with Czuch- ajowska et al. [23] . Who reported that lower gelatinization enthalpy values are linked to higher amylose levels.

Solubility, swelling power and water absorption capacity in depigmented starch were directly correlated to temperature (Figure 2). The solubility increased as temperature increased. Gujska et al. [19] mention that dissolution of pinto bean, navy bean and field pea starches starts at 70°C, probably because the swollen starch granules permit amylose liberation. The native starch (pigmented) [3] , in comparison, exhibited only a moderate increase in solubility as temperature increased, reaching 16.2% at 90°C, less than the 53.6% reached here by depigmented starch at the same temperature. This property would allow greater dispersion of M. pruriens depigmented starch in aqueous solutions and better water reception and retention in the preparations to which it is added.

Table 1: Chemical composition of pigmented and depigmented starch from M. pruriens.

Component (g/kg)	M. pruriens starch	
	Depigmented	Pigmented[a]
Moisture	110.5	98.2
Protein	0	7.1
Fibre	34	5.4
Fat	5.7	4
Ash	5.8	2.8
NFE	954.5	980.6
Amylose	215.3	392
Amylopectin	784.7	608

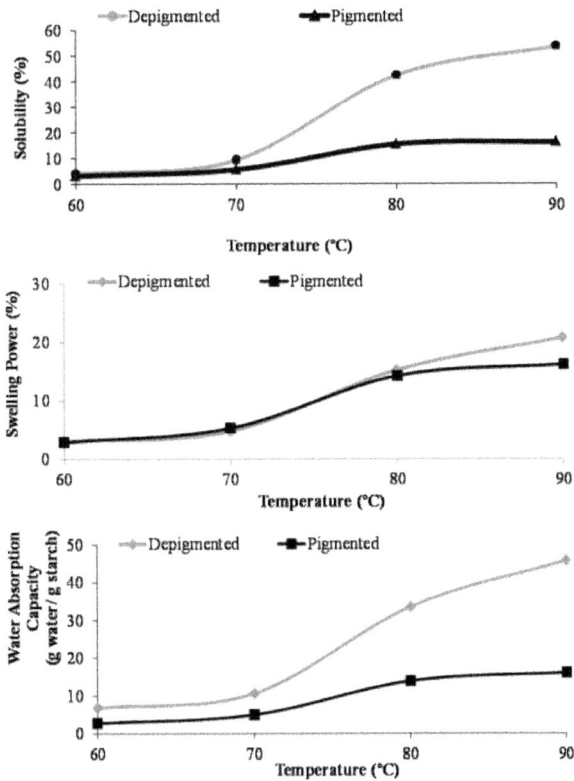

Figure 2: Solubility, swelling power (SP) and water absorption capacity (WAC) of M. pruriens starch. Pigmented starch values from Ref. [3] .

The swelling pattern shows that at temperatures lower than 70°C the velvet bean granules resist swelling, probably due to their high initial gelatinization temperature. From 70°C to 90°C, the granules gradually swell as temperature increases, a result of intermolecular hydrogen bridge rupture in the amorphous areas, which allows irreversible and progressive water absorption. In native starch (pigmented) [3] , the swelling increment was gra- dual, beginning at 60°C but the levels reached were lower than in depigmented starch.

The water absorption pattern showed that in the 70°C to 90°C range the granules swelled gradually as temperature increased due to rupture of the intermolecular hydrogen bridges in the amorphous areas, which allowed progressive water absorption. The native starch (pigmented) [3] , in comparison, exhibited only a moderate increase in WAC as temperature increased, reaching 15.95% at 90°C, less than the 45.75% reached here by depigmented starch at the same temperature.

Table 2 registers the retrogradation parameters of M. pruriens depigmented starch. The enthalpy values were directly correlated to time. The increased of ΔH means that M. pruriens oxided starch retrogrades in higher proportion in days 14 and 21. For the above mentioned, retrogradation increased as time increased.

In paste characterization, corn starch exhibited a classic amylogram curve, whereas the velvet bean starch (pigmented and depigmented) were different since they did not have a defined maximum viscosity peak (Figure 3). To obtain the paste properties of M. pruriens starch, the highest viscosity value developed by the paste was used. Discoloration process reduced viscosity from 256 BU in the native starch [3] to 120 BU in the M. pruriens depigmented starch (Table 3). Velvet bean pigmented starch viscosity was highly stable during a 15 min residence period at 95°C, and remained stable even after gradual cooling. This stability makes this starch suitable for pro- ducts that require cooking at high temperatures during manufacture, since it maintains adequate product consistency during processing. On the other hand, the viscosity of M. pruriens depigmented starch decreased slightly during the heating stages and then increased during cooling, indicating that the depigmented starch was unstable in heating-cooling processes. This must be considered when incorporating this starch into products since upon cooling its paste viscosity will increase, which will be reflected in a higher thickening capacity. The depigmen- ted starch had higher fragility than the pigmented starch, indicating that its viscosity decreased during heating due to swelling of its starch molecules, consequently making them more fragile. The consistency and setback of the depigmented starch was lower than in corn and native M. pruriens starch reported by Betancur-Ancona et al. [3] . This indicated that the depigmented starch was more stable in heating-cooling processes, although it would have thickening capacity upon cooling in foods to which it was added.

Table 2: Retrogradation parameters of M. pruriens depigmented starch.

Day	T_o (°C)	T_p (°C)	T_f (°C)	ΔH (j/g)
1	46.53	56.56	72.14	4.10
2	46.34	58.44	72.70	7.17
3	42.19	56.02	69.11	7.68
7	42.7	56.83	71.53	9.82
14	41.88	57.76	73.34	10.32
21	44.3	58.42	77.55	13.47

Table 3: Pasting properties of pigmented and depigmented starch from M. pruriens.

Parameter	Depigmented starch	Pigmented starch*
Initial gelatinization temperature (°C)	82.5	81
Maximum viscosity (BU)	120	256
Viscosity at 95°C (BU)	120	93
Viscosity at 95°C for 15 min (BU)	104	276
Viscosity at 50°C (BU)	156	350
Viscosity at 50°C for 15 min (BU)	174	350
Breakdown	16	−20
Consistency	36	74
Setback	52	94

The depigmented and pigmented M. pruriens starch transmittance (%T) values were 50.89% and 22.21%, respectively. As it was expected, the depigmented starch was more translucent than the pigmented starch because the discoloration process generated functional properties such as increased paste clarity. Segura-Campos et al. [24] stated that starch clarity was a result of high reflection as the characteristic arrangement of gel molecular chains reduced the intensity of the light transmitted through them. Clarity is a key parameter in starch paste qua- lity because it provides shine and opacity to product colour. The depigmented M. pruriens starch's excellent cla- rity makes it potentially useful in products such as fruit pie fillings and candies.

Compared with the M. pruriens pigmented starch, the depigmented starch had higher stability in refrigeration cycles but lower stability in freezing cycles (Figure 4). This behavior is a disadvantage in the food industry because it means the M. pruriens depigmented starch behaves like a sponge, initially absorbing water and then releasing it when centrifuged.

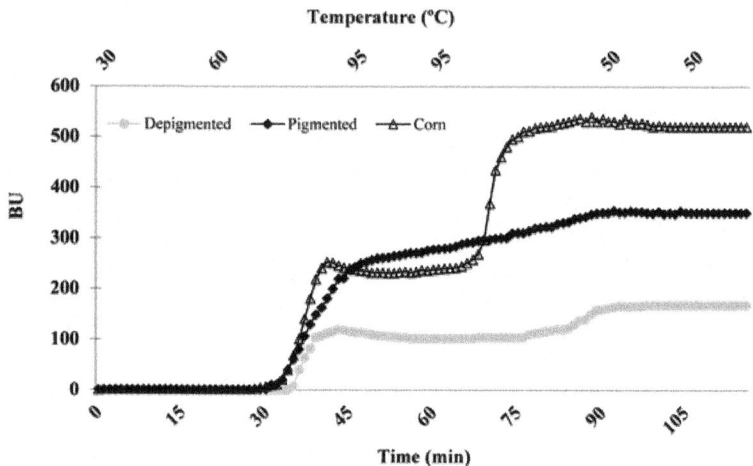

Figure 3: Viscoamylograph of M. pruriens.

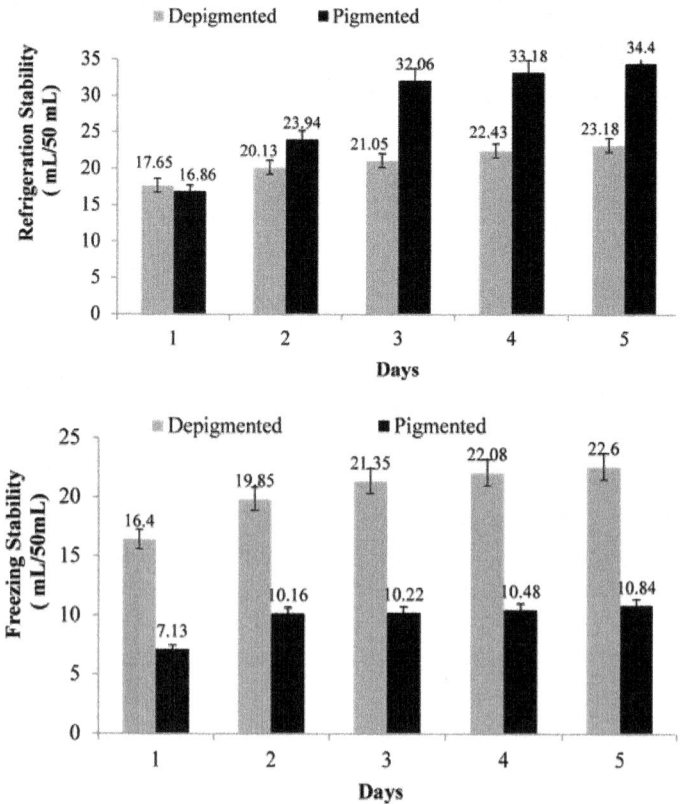

Figure 4: Refrigeration and Freezing Stability of M. pruriens starch.

Its high syneresis can probably be attributed to the organization of the molecules (amylose and amylopectin) that leads to the gel releasing water. The depigmented starch behavior could be caused by the decrease of the starch capacity of the forming hydrogen bond, as well as by the longitude of the amylose chains, as short chains facilitate its lixiviation toward the aqueous means. Successive freezing and thawing of starches can also affect their structure since formation and melting of ice crystals can lead to starch paste redistribution and dilution [25] . This is probably what occurred in the M. pruriens depigmented starch. The retained water would have been released from inter- and intramolecular associations, resulting in two separate phases: one polymer-rich (gel); and another polymer-poor (liquid). Under refrigeration processes the M. pruriens depigmented starch was more stable than its pigmented, but under freezing processes it was unstable. This is vital to consider when using the studied depigmented starch as a food additive since it requires storage at low temperatures.

CONCLUSION

M. pruriens depigmented starch has a wide variety of possible applications as a functional ingredient in food systems and other industrial applications. An example of the above mentioned is the use of starch in the syrups production. On the other hand, the high transmittance makes depigmented starch potentially useful as an additive in jellies and candies to provide brightness. Its high gelatinization temperature, together with its solubility and high WAC, is advantageous for use in sausages, baked products, canned products, sauces, seasonings, jellies, compressed candies, gummy products, etc. However, its low syneresis (i.e., low stability) under refrigeration and freezing conditions makes it inadequate for use as a thickener, stabilizer or gelling agent in refrigerated or frozen foods. The discoloration process of M. pruriens starch produced a derivative with more versatility as a food additive providing some new and improved functional properties. However, the election of depigmented starch as additive in food should take into account its functional properties and characteristics of other ingredients used in the preparation as well as the processing conditions. In sum, the velvet bean is an excellent starch source with many potential applications and is also a promising alternative to corn starch.

REFERENCES

1.   Ashogbon, A.O. and Akintayo, E.T. (2014) Recent Trend in the Physical and Chemical Modification of Starches from Different Botanical

Sources: A Review. Starch-Stärke, 66, 41-57. http://dx.doi.org/10.1002/star.201300106

2. Otegbayo, B., Oguniyan, D. and Akinwumi, O. (2014) Physicochemical and Functional Characterization of Yam Starch for Potential Industrial Applications. Starch-Stärke, 66, 235-250. http://dx.doi.org/10.1002/star.201300056

3. Betancur-Ancona, D.A., Chel-Guerrero, L.A., Bello-Pérez, L.A. and Dávila-Ortiz, G. (2002) Isolation of Velvet Bean (Mucuna pruriens) Starch: Physicochemical and Functional Properties. Starch-Stärke, 54, 303-309. http://dx.doi.org/10.1002/1521-379X(200207)54:7<303::AID-STAR303>3.0.CO;2-2

4. Adebowale, K.O. and Lawal, O.S. (2003) Functional Properties and Retrogradation Behaviour of Native and Chemically Modified Starch of Mucuna Bean (Mucuna pruriens). Journal of the Science of Food and Agriculture, 83, 1541-1546.http://dx.doi.org/10.1002/jsfa.1569

5. Betancur-Ancona, D., Gallegos-Tintoré, S., Delgado-Herrera, A., Pérez-Flores, V., Castellanos-Ruelas, A. and Chel- Guerrero, L. (2008) Some Physicochemical and Antinutritional Properties of Raw Flours and Protein Isolates from Mucuna pruriens (Velvet Bean) and Canavalia ensiformis (Jack Bean). International Journal of Food Science and Technology, 43, 816-823. http://dx.doi.org/10.1111/j.1365-2621.2007.01521.x

6. Adebowale, Y.A., Adeyemi, I.A., Oshodi, A.A. and Niranjan, K. (2007) Isolation, Fractionation and Characterisation of Proteins from Mucuna Bean. Food Chemistry, 104, 287-299. http://dx.doi.org/10.1016/j.foodchem.2006.11.050

7. Hoover, R., Rorke, S.C. and Martin, A.M. (1991) Isolation and Characterization of Lima Bean (Phaseolus lunatus) Starch. Journal of Food Biochemistry, 15, 117-136.http://dx.doi.org/10.1111/j.1745-4514.1991.tb00149.x

8. Mc Master, M.M. (1964) Microscopic Techniques for Determining Starch Granules Properties. In: Whistler, L.R., Smith, J.R. and BeMiller, N.J., Eds., Methods in Carbohydrate Chemistry, Academic Press, London, 233-240.

9. AOAC (1997) Official Methods of Analysis of the Association of Official Analytical Chemists. 20th Edition, Association of Official Analytical Chemists, Washington DC.

10. Morrison, W.R. and Laignelet, B. (1983) An Improved Colorimetric Procedure for Determining Apparent and Total Amylose in Cereal

and Other Starches. Journal of Cereal Science, 1, 9-20. http://dx.doi.org/10.1016/S0733-5210(83)80004-6

11. Ruales, J. and Nair, B.M. (1994) Properties of Starch and Dietary Fibre in Raw and Processed Quinoa (Chenopodium quinoa, Willd) Seeds. Plant Foods for Human Nutrition, 45, 223-246. http://dx.doi.org/10.1007/BF01094092

12. Sathe, S.K., Iyer, V. and Salunkhe, D.K. (1981) Investigations of the Great Northern Bean (Phaseolus vulgaris L.) Starch: Solubility, Swelling, Interaction with Free Fatty Acids, and Alkaline Water Retention Capacity of Blends with Wheat Flours. Journal of Food Science, 46, 1914-1917. http://dx.doi.org/10.1111/j.1365-2621.1981.tb04518.x

13. Gudmundsson, M. and Eliasson, A.C. (1992) Comparison of Thermal and Viscoelastic Properties of Four Waxy Star- ches and the Effect of Added Surfactant. Starch-Stärke, 44, 379-385. http://dx.doi.org/10.1002/star.19920441005

14. Wiesenborn, D.P., Orr, P.H., Casper, H.H. and Tacke, B.K. (1994) Potato Starch Paste Behavior as Related to Some Physical/Chemical Properties. Journal of Food Science, 59, 644-648. http://dx.doi.org/10.1111/j.1365-2621.1994.tb05583.x

15. Bello-Pérez, L.A., Agama-Acevedo, E., Sánchez-Hernández, L. and Paredes-López, O. (1999) Isolation and Partial Characterization of Banana Starches. Journal of Agricultural and Food Chemistry, 47, 854-857. http://dx.doi.org/10.1021/jf980828t

16. Eliasson, A.C. and Kim, A.C. (1992) Changes in Rheological Properties of Hydroxypropyl Potato Starch Pastes during Freeze-Thaw Treatments I. A Rheological Approach for Evaluation of Freeze-Thaw Stability. Journal of Texture Studies, 23, 279-295.http://dx.doi.org/10.1111/j.1745-4603.1992.tb00526.x

17. Betancur, A.D. and Chel, G.L. (1997) Acid Hydrolysis and Characterization of Canavalia ensiformis Starch. Journal of Agricultural and Food Chemistry, 45, 4237-4241.http://dx.doi.org/10.1021/jf970388q

18. Galvez, F.C.F. and Resurreccion, A.V.A. (1993) The Effects of Decortication and Method of Extraction on the Physical and Chemical Properties of Starch from Mung Bean (Vigna radiata (L.) Wilczec). Journal of Food Processing and Preservation, 17, 93-107.http://dx.doi.org/10.1111/j.1745-4549.1993.tb00227.x

19. Gujska, E., Reinhard, W.D. and Khan, K. (1994) Physicochemical Properties of Field Pea, Pinto and Navy Bean Starches. Journal of

Food Science, 59, 634-636.http://dx.doi.org/10.1111/j.1365-2621.1994. tb05580.x

20. Swinkels, J.M. (1985) Sources of Starch, Its Chemistry and Physics. In: Van Beynum, G.M. and Roel, J.A., Eds., Starch Conversion Technology, Marcel Dekker Inc., New York, 15-46.

21. Fannon, J.E., Hauber, R.J. and Bemiller. J.N. (1992) Surface Pores of Starch Granules. Cereal Chemistry, 69, 284-288.

22. Wurzburg, O.B. (1986) Introduction. In: Wurzburg, O.B., Ed., Modified Starches: Properties and Uses, CRC Press, Boca Raton, 24-29.

23. Czuchajowska, Z., Otto, T., Paszczynska, B. and Baik, B.K. (1998) Composition, Thermal Behavior, and Gel Texture of Prime and Tailings Starches from Garbanzo Beans and Peas. Cereal Chemistry, 75, 466-472. http://dx.doi.org/10.1094/CCHEM.1998.75.4.466

24. Segura, M., Chel, L. and Betancur, D. (2010) Effect of Octenylsuccinylation on Functional Properties of Lima Bean (Phaseolus lunatus) Starch. Journal of Food Process Engineering, 33, 712-727.

25. Soni, P.L., Sharma, H., Srivastava, H.C. and Gharia, M.M. (1990) Physicochemical Properties of Canna edulis Starch- Comparison with Maize Starch. Starch-Stärke, 42, 460-464. http://dx.doi.org/10.1002/ star.19900421203

Chapter 10

FECL3 CATALYZED ONE POT SYNTHESIS OF 1-SUBSTITUTED 1H-1,2,3,4-TETRAZOLES UNDER SOLVENT-FREE CONDITIONS

Fatemeh Darvish, Shima Khazraee

Department of Chemistry, K.N.Toosi University of Technology, P. O. Box 15875-4416, Tehran, Iran

ABSTRACT

An efficient procedure for the preparation of 1-substituted-1H-1,2,3,4-tetrazoles via a three-com- ponent condensation of triethyl orthoformate, amine, and trimethylsilyl azide using inexpensive and environment-friendly $FeCl_3$ as catalyst under solvent-free conditions has been reported. The reaction generates the corresponding 1-substituted tetrazole in excellent yields.

INTRODUCTION

Tetrazoles have received considerable attention because of their wide application [1]. They have been used extensively in the synthesis of modified amino acids and peptidomimetic compounds as a metabolically stable equivalent of carboxylic acids [2]. Tetrazole moieties play a major role in material science like propellants and energetic compounds [3]. Furthermore, this nitrogen-rich ring system is an important synthon in organic synthetic and medicinal chemistry [4]. Due to interesting properties of tetrazole, the improvement of known methods for their preparation is still in demand.

The methods reported for the synthesis of 1-substituted tetrazoles involve acid-catalyzed cycloaddition between hydrazoic acid and isocyanides [5] or trimethylsilylazide [6], cyclization between primary amines with an orthocarboxylic acid ester or ethyl orthoformate and sodium azide in the presence of acetic acid [7], acidic ionic liquid [8], ytterbium triflate [9] and natrolite zeolite [10]. Each of these reported methods has at least one or more of the following drawbacks, for instance, the use of expensive, toxic metal catalysts and excess amount of acetic acid or trifluoroacetic acid, utilization of organic solvents, harsh reaction conditions, tedious work-up, low yields, long reaction time and the presence of hydrazoic acid, which is highly toxic and volatile. The few methods that seek to avoid hydrazoic acid liberation during the reaction by avoiding acidic conditions require a very large excess of sodium azide. Thus, the quest for inexpensive, benign catalysts and mild reaction conditions is still a major challenge for the synthesis of 1-substituted tetrazoles.

$FeCl_3$ is an efficient and green catalyst in modern organic synthesis [11]. It has been widely used in several environment-friendly and atom economical organic transformations. Recent reports on $FeCl_3$catalyzed arylation of benzyl alcohols and benzyl carboxylates [12], hydroarylation of styrenes [13], benzylation of 1,3-dicarbonyl compounds [14] and diasteroselective synthesis of cis-oxazolidines [15] have highlighted the applications of $FeCl_3$ in organic synthesis. Herein, we report another remarkable catalytic activity of $FeCl_3$ for the preparation of 1-substituted tetrazoles from a wide variety of primary amines with trimethylsilylazide and trimethylorthoformate under solvent-free conditions (Scheme 1).

RESULTS AND DISCUSSION

Preliminary experiments were carried out in order to determine the best reaction conditions. We examined the reaction of trimethylsilyazide using several different catalysts and solvents as well as neat conditions (Table 1). Fortunately, most of the acid catalysts which were used, afforded the desired product while $FeCl_3$ gave the best result under solvent-free conditions (Table 1, Entry 1).

Further studies showed that the optimum amount of $FeCl_3$ was 0.2 mmol, an excess of $FeCl_3$ did not lead to a substantial improvement in the yield while decreasing the catalyst reduced it (Table 2, Entry 2).

After optimizing the reaction conditions, this process was extended to other substituted anilines. A variety of amines possessing both electron-releasing and electron-withdrawing groups (such as chloro, nitro, bromo, methoxy, ethyl, methyl, and heterocyclic amine like 2-aminopyridine) were employed (Table 3). According to the table, the nature of substituent on the benzene ring did not affect the reaction time and the yields were excellent.

The suggested iron(III) chloride catalyzed transformation mechanism is shown in Scheme 2, in which the conditions: amine (1.0 mmol), triethylorthoformate (1.2 mmol), trimetylsilyazide (1 mmol), $FeCl_3$ (20 mmol%), at 70°C; [b]Isolated yields.

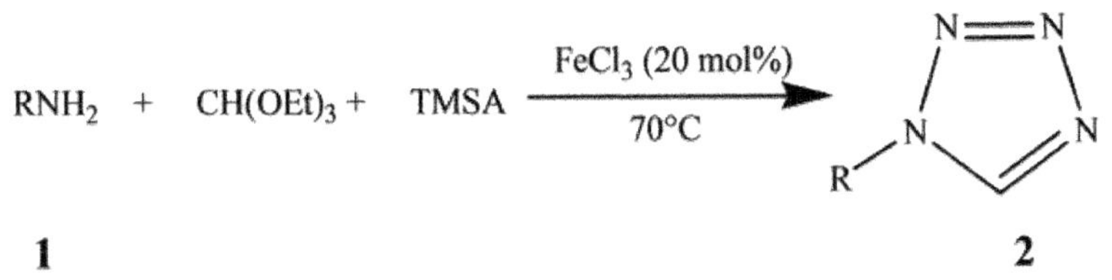

Scheme 1: Synthesis of 1-substituted 1H-1,2,3,4-tetrazoles.

Table 1: Effect of catalyst and solvent on the formation of tetrazole 2a.

Entry	Catalyst	Solvent	Amount of catalyst	Temp/°C	Time/h	Yield[a]/%
1	FeCl$_3$	Neat	0.1 mmol	70	5	74
2	H$_3$PMo$_{12}$O$_{40}$	Neat	0.2 mmol	80	5	40
3	H$_3$PW$_{12}$O$_{40}$	Neat	0.2 mmol	70	5	-
4	SnCl$_2$	Neat	0.2 mmol	80	24	50
5	SnCl$_2$	Methanol	0.2 mmol	60	24	48
6	CuI	Neat	0.2 mmol	70	24	40
7	BiCl$_3$	Neat	0.2 mmol	80	24	55
8	BiCl$_3$	Methanol	0.2 mmol	60	24	50
9	BiCl$_3$	t-BuOH	0.2 mmol	80	48	20
10	BiCl$_3$	DMF	0.2 mmol	80	48	60
11	I$_2$	Methanol	0.2 mmol	60	24	-

Table 2: Effect of the amount of catalyst on the formation of tetrazole 2a.

Entry	FeCl$_3$ [mmol]	Yield[a] [%]	Time/h
1	10	90	7
2	20	95	5
3	30	94	5

Table 3: Synthesis of 1-substituted 1H-1,2,3,4-tetrazoles 2a-i[a].

Entry	R	1	Time [h]	2	Yield[b] [%]	Ref.
1	C$_6$H$_5$	1a	5	2a	95	[16]
2	4-MeO-C$_6$H$_4$	1b	5	2b	88	[16]
3	4-Et-C$_6$H$_4$	1c	5	2c	90	This work
4	3,4-Me$_2$-C$_6$H$_4$	1d	5	2d	93	This work
5	4-Br-C$_6$H$_4$	1e	7	2e	87	[17]
6	4-Cl-C$_6$H$_4$	1f	6	2f	88	[16]
7	3-NO$_2$-C$_6$H$_4$	1g	6	2g	92	[18]
8	4-NO$_2$-C$_6$H$_4$	1h	6	2h	90	[17]
9	2-C$_5$H$_4$N	1i	5	2i	92	[19]

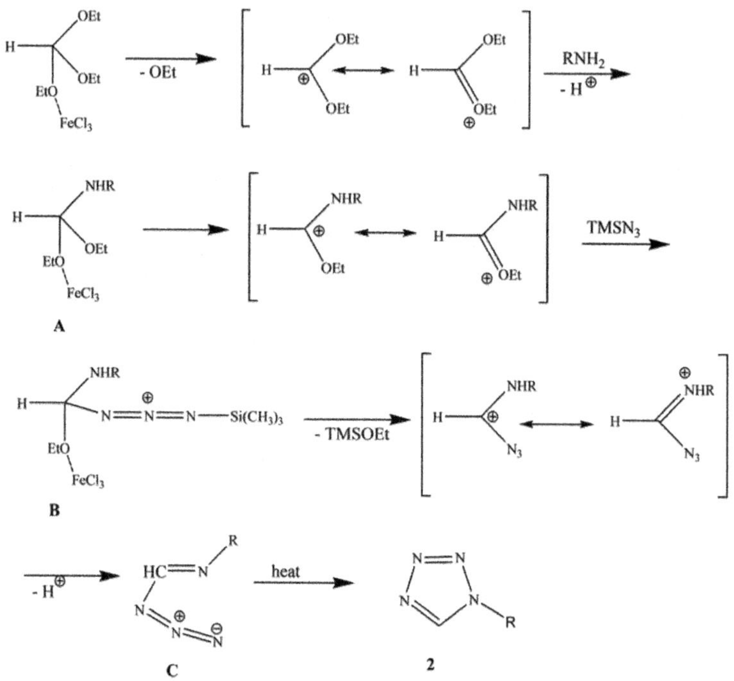

Scheme 2: Plausible mechanism for the formation of 1-substituted 1H-1,2,3,4-tetra-zoles.

Lewis acidity of the catalyst probably has an important role in the promotion of the cyclization process. Apparently the role of $FeCl_3$ is limited to activation of ethoxy groups and to breaking the CO bond, however the possible assistance of trimethylsilyl group in this cleavage should not be neglected. In the first step, carbocations are generated from the cleavage of methoxy group and stabilized by neighboring heteroatom O or N, the following nucleophilic displacements by amine and azide would explain the formation of intermediates A and B. By the assistance of $FeCl_3$ and trimethylsilyl group the elimination of the last methoxy group becomes possible. Finally, 1-substituted tetrazoles will be produced upon the cyclization of intermediate C.

CONCLUSION

We have demonstrated that $FeCl_3$ is an effective catalyst in promoting the reaction between amines, trimethylorthoformate and trimethylsilylazide that affords the corresponding 1-substituted-1H-1,2,3,4-tetrazole products. The process gave rise to excellent isolated yields of 1-substituted-1H-1,2,3,4-tetrazoles under solvent-free conditions and moderate temperature in shorter reaction times than many other reported methods.

EXPERIMENTAL

All the chemicals were purchased from the Merck Company and used without further purification. The melting points were taken in open capillary tubes with Electrothermal 9100 Apparatus. FT-IR (KBr) Spectra were recorded on an ABB FT-IR FTLA 2000 spectrometer. ^{1}H NMR spectra were run on a Bruker DRX-300 (300 MHz) AVANCE instrument using TMS as an internal standard and CDCl$_3$ as solvent. The chemical shifts (d) are reported in ppm relative to the TMS as an internal standard and J values are given in Hz. ^{13}C spectra were recorded at 75 MHz. High-resolution mass spectra were recorded on a Mass-EI-POS (Apex Qe-FT-ICR instrument) spectrometer.

General Procedure

A mixture of amine (1 mmol), triethylorthoformate (1.2 mmol) and trimethylsilylazide (1 mmol) was stirred in the presence of FeCl$_3$ (20 mol%) at 70°C for an appropriate time under inert atmosphere (Table 3). The progress of the reaction was monitored by TLC (EtOAc/n-Hexane, 2:1). After completion, the reaction mixture was extracted with ethyl acetate (10 cm^3 × 3) and washed with brine. The organic layer was dried over magnesium sulfate and the solvent was evaporated. The isolated product was pure (single spot on TLC) for all practical purposes. However, for characterization purposes it was further purified by plate chromatography (silica gel, eluent EtOAc/n-hexane2/1).

Characterization Data

1-Phenyl1-H-1,2,3,4-tetrazole (2a)

Yield: 95%, m.p.: 65°C - 66°C (lit. [16] 65°C - 66°C); pale yellow needle. IR (KBr): v = 3121, 2926, 2844, 1596, 1439, 1463, 1396, 1206, 1093, cm^{-1}. ^{1}H-NMR (CDCl$_3$): d = 7.47 - 7.58 (m, 3H, ArH), 7.70 (d, 2H, J = 7.3 Hz, ArH), 9.07 (s, 1H, CH) ppm. ^{13}C-NMR (CDCl$_3$): d = 121.2, 130.0, 130.2, 133.7, 140.6 ppm.

1-(4-Metoxyphenyl)1-H-1,2,3,4-tetrazole (2b)

Yield: 88%, m.p.: 119°C - 120°C (lit. [16] 116°C - 117°C); Colorless solid. IR (KBr): v = 3132, 3019, 1606, 1597, 1514, 1463, 1389, 1257, 1156, 1093, 1201, 831 cm^{-1}. ^{1}HNMR (CDCl$_3$,): d = 3.86 (s, 3H, OCH$_3$), 7.06 - 7.03 (d, 2H, J = 6.8 Hz, ArH), 7.59 (d, 2H, J = 6.8 Hz, ArH), 8.94 (s, 1H, CH) ppm. ^{13}C NMR (CDCl$_3$): d = 115.2, 122.9, 126.8, 140.6, 160.7 ppm.

1-(4-Ethyllphenyl)1-H-1,2,3,4-tetrazole (2c)

Yield: 90%, m.p. 97°C - 98°C; pale yellow solid. IR (KBr): v = 3137, 2962, 2921, 2880, 1519, 1475, 1367, 1247, 1089, 1024, 838 cm⁻¹; ¹H NMR (CDCl₃): d = 1.25 - 1.30 (3H, t, J = 7.6 Hz, CH₃), 2.78 - 2.70 (2H, q, J = 7.6 Hz, CH₂), 7.41 - 7.38 (2H, d, J = 8.4 Hz, ArH), 7.62 - 7.59 (2H, d, J = 8.4 Hz, ArH), 8.97(1H, s, CH) ppm. ¹³C NMR (CDCl₃): d = 15.3, 28.5, 121.2, 129.5, 131.6, 140.5, 146.7 ppm. HRMS (EI⁺) Calcd for [C₉H₁₀N₄]⁺: 174.0907, found: 174.0901.

1-(4-Dimethyllphenyl)1-H-1,2,3,4-tetrazole (2d)

Yield: 93%. m.p. 57°C - 58°C; colorless solid. IR (KBr): v = 3120, 2957, 2911, 2852, 1619, 1505, 1100 cm⁻¹. ¹H NMR (CDCl₃): d = 2.32 (3H, s, CH₃), 2.34 (3H, s, CH₃), 7.30 - 7.26 (1H, d, J = 8.1 Hz, ArH), 7.41 - 7.38 (1H, d, J = 8.1 Hz, ArH), 7.46 (1H, s, ArH), 8.97 (1H, s, CH) ppm. ¹³C NMR (CDCl₃, d, ppm): 19.5, 19.9, 118.4, 122.2, 130.9, 131.6, 138.9, 139.0, 140.5. HRMS (EI⁺) Calcd for [C₉H₁₀N₄]⁺: 174.0907, found: 174.0896.

1-(4-Bromophenyl)1-H-1,2,3,4-tetrazole (2e)

Yield: 87%. m.p. 184°C - 185°C (lit. [17] 133°C - 134°C); pale yellow solid. IR (KBr): v = 3130, 2917, 2849, 1501, 1462, 1385 cm⁻¹. ¹H NMR (DMSOd₆): d = 7.87 (s, 4H, ArH), 10.11 (s, 1H, CH) ppm.¹³C NMR (DMSOd₆): d = 122.5, 123.1, 132.9, 142.3, 142.4.

1-(4-Chlorophenyl)1-H-1,2,3,4-tetrazole (2f)

Yield: 88%. m.p. 158°C - 159°C (lit. [16] 155°C - 156°C); colorless crystal. IR (KBr): v = 3125, 3105, 2917, 2849, 1505, 1462, 1385, 1201, 1088, 995, 831 cm⁻¹. ¹H NMR (DMSOd₆): d = 7.75 - 7.72 (d.t, 2H, J = 8.8 Hz, ArH), 7.97 - 7.94 (d.t, 2H, J = 8.8 Hz, ArH), 10.11 (s, 1H, CH) ppm;¹³C NMR (DMSOd₆): d = 122.9, 130.1, 132.6, 134.1, 142.3, 142.5.

1-(3-Nitrophenyl)1-H-1,2,3,4-tetrazole (2g)

Yield: 92%. m.p. 110°C - 111°C (lit. [18] 108°C - 109°C); colorless crystal crystal. IR (KBr): v = 3132, 3091, 2911, 2854, 1596, 1519, 1463, 1344, 1211, 1088, 990, 857 cm⁻¹. ¹H NMR (DMSOd₆): 8.26 - 8.22 (m, 3H, ArH), 8.53 - 8.49 (d, 1H, J = 6.9, ArH), 10.20 (s, 1H, CH) ppm;¹³C NMR (DMSOd₆): 122.0, 126.2, 137.5, 141.7, 144.2, 148.3 ppm.

1-(4-Nitrophenyl)1-H-1,2,3,4-tetrazole (2h)

Yield: 90%. m.p. 207°C - 208°C (lit. [16] 202°C - 204°C); pale yellow crystal.

IR (KBr): ν = 3132, 3091, 2911, 2854, 1611, 1596, 1519, 1463, 1344, 1211, 1088, 990, 857 cm⁻¹. ¹H NMR (DMSOd₆): 8.26 - 8.22 (d, 2H, J = 7.0 Hz, ArH), 8.53 - 8.49 (d, 2H, J = 6.9, ArH), 10.30 (s, 1H, CH) ppm; ¹³C NMR (DMSOd₆): 121.9, 125.7, 138.2, 142.7, 142.8, 147.4 ppm.

1-(4-Nitrophenyl)1-H-1,2,3,4-tetrazole (2i)

Yield: 92%. m.p. 128°C - 129°C (lit. [19] 125°C - 126°C); Colorless needle; IR (KBr): ν = 1597, 1576, 1472, 1391, 1212, 1182, 1151, 1090, 1006 cm⁻¹.¹H NMR (DMSOd₆): d = 7.61 - 7.65 (m, 1H), 8.07 - 8.05 (d, 1H, J = 8.1 Hz), 8.21 - 8.15 (td, 1H, J = 1.7 Hz), 8.66 - 8.64 (dd, 1H, J = 0.7 Hz), 10.18 (s, 1H) ppm. ¹³C NMR (DMSOd₆): d = 115.6, 125.3, 140.6, 141.6, 146.5, 149.3 ppm.

REFERENCES

1. Butler, R.N. (1996) Comprehensive Heterocyclic Chemistry: Five-Membered Rings with More Than Two Heteroatoms and Fused Carbocyclic Derivatives. Katrisky, A.R. and Scriven E.F.V., Eds., Academic Press, Elsevier, Waltham, Vol. 4, 621-678.

2. Herr, R. (2002) 5-Substituted-1H-tetrazoles as Carboxylic Acid Isosteres: Medicinal Chemistry and Synthetic Methods. Bioorganic & Medicinal Chemistry, 10, 3379-3393.http://dx.doi.org/10.1016/S0968-0896(02)00239-0

3. Klapötke, T.M., Sabaté, C.M. and Stierstorfer, J. (2009) Neutral 5-Nitrotetrazoles: Easy Initiation with Low Pollution. New Journal of Chemistry, 33, 136-147.http://dx.doi.org/10.1039/B812529E

4. Ichikawa, T., Yamada, M., Yamaguchi, M., Kitazaki, T., Matsushita, Y., Higashikawa, K. and Itoh, K. (2001) Optically Active Antifungal Azoles. XIII. Synthesis of Stereoisomers Metabolites of 1-[(1R,2R)-2-(2,4-Difluorophenyl)-2-hy- droxy-1-methyl-3-(1H-1,2,4-triazol-1-yl)propyl]-3-(4-(1"-1tetrazolyl)phenyl-2-imidazolidinone (TAK-456). Chemical & Pharmaceutical Bulletin, 49, 1110-1119. http://dx.doi.org/10.1248/cpb.49.1110

5. Zimmerman, D.M. and Olofson, R.A. (1969) The Rapid Synthesis of 1-Substituted Tetrazole. Tertrahedron Letters, 58, 5081-5084. http://dx.doi.org/10.1016/S0040-4039(01)88889-4

6. Jin, T., Kamijo, S. and Yamamoto, Y. (2004) Synthesis of 1-Substituted Tetrazoles via the Acid-Catalyzed [3 + 2] Cycloaddition between Isocyanides and Trimethylsilyl Azide. Tetrahedron Letters, 45, 9435-943. http://dx.doi.org/10.1016/j.tetlet.2004.10.103

7. Satoh, Y. and Marcopulos, N. (1995) Application of 5-Lithiotetrazoles in Organic Synthesis. Tetrahedron Letters, 36, 1759-1762. http://dx.doi.org/10.1016/0040-4039(95)00117-U

8. Potewar, T.M., Siddiqui, S.A., Lahoti, R.J. and Srinivasan, K.V. (2007) Efficient and Rapid Synthesis of 1-Substituted- 1H-1,2,3,4-tetrazoles in the Acidic Ionic Liquid 1-n-Butylimidazolium Tetrafluoroborate. Tetrahedron Letters, 48, 1721-1724.http://dx.doi.org/10.1016/j.tetlet.2007.01.050

9. Su, W., Hong, Z., Shan, W. and Zhang, X. (2006) A Facile Synthesis of 1-Substituted-1H-1,2,3,4-tetrazoles Catalyzed by Ytterbium Triflate Hydrate. European Journal of Organic Chemistry, 12, 2723-2726. http://dx.doi.org/10.1002/ejoc.200600007

10. Habibi, D., Nasrollahzadeh, M. and Kamali, T.A. (2011) Green Synthesis of the 1-Substituted 1H-1,2,3,4-Tetrazoles by Application of the Natrolite Zeolite as a New and Reusable Heterogeneous Catalyst. Green Chemistry, 13, 3499- 3504.http://dx.doi.org/10.1039/c1gc15245a

11. Diaz, D.D., Miranda, P.O., Padron, J.I. and Martin, V.S. (2006) Recent Uses of Iron(III) Chloride in Organic Synthesis. Current Organic Chemistry, 10, 457-476.http://dx.doi.org/10.2174/138527206776055330

12. Zhan, Z.-P. and Liu, H.-J. (2006) FeCl$_3$-Catalyzed Coupling of Propargylic Acetates with Alcohols. Synlett, No. 14, 2278-2280. http://dx.doi.org/10.1055/s-2006-949645

13. Kischel, J., Iovel, I., Mertins, K., Zapf, A. and Beller, M.A. (2006) Convenient FeCl$_3$-Catalyzed Hydroarylation of Styrenes. Organic Letters, 8, 19-22.

14. Komeyama, K., Morimoto, T., Nakayama, Y. and Takaki, K. (2007) Cationic Iron-Catalyzed Intramolecular Hydroalkoxylation of Unactivated Olefins. Tetrahedron Letters, 48, 3259-3261. http://dx.doi.org/10.1016/j.tetlet.2007.03.004

15. Cornil, J., Guérinot, A., Reymond, S. and Cossy, J. (2013) FeCl$_3$ · 6H$_2$O, a Catalyst for the Diastereoselective Synthesis of cis-Isoxazolidines from N-Protected δ-Hydroxylamino Allylic Acetates. Journal of Organic Chemistry, 78, 10273- 10287.http://dx.doi.org/10.1021/jo401627p

16. Fallon, F.G. and Herbst, R.M. (1957) Synthesis of 1-Substituted Tetrazoles. Journal of Organic Chemistry, 22, 933- 936. http://dx.doi.org/10.1021/jo01359a020

17. Horwitz, J.P. and Grakauskas, V.A. (1957) The Reactions of Diazonium Salts with Some Substituted Hydrazines. II. 1,6-Bisaryl-3,4-diacetyl-1,5-hexazadienes. Journal of the American Chemical Society, 79, 1249-

1253. http://dx.doi.org/10.1021/ja01562a055

18. Nishiyama, K., Oba, M. and Watanabe, A. (1987) Reactions of Trimethylsilyl Azidewith Aldehydes: Facile and Convenient Syntheses of Diazides Tetrazoles, and Nitriles. Tertrahedron, 43, 693-700. http://dx.doi.org/10.1016/S0040-4020(01)90003-1

19. Grunert, C.M., Weinberger, P., Schweifer, J., Hampel, C., Stassen, A.F., Mereiter, K. and Linert, W. (2005) Synthesis and Characterisation of Tetrazole Compounds: 3 Series of New Ligands Representing Versatile Building Blocks for Iron(II) Spin-Crossover Compounds. Journal of Molecular Structure, 733, 41-52.http://dx.doi.org/10.1016/j.molstruc.2004.07.036

Chapter 11

SYNTHETIC STUDIES OF NAPHTHO [2, 3-B] FURAN MOIETY PRESENT IN DIVERSE BIOACTIVE NATURAL PRODUCTS

Bidyut Kumar Senapati[1*], Dipakranjan Mal[2]

[1]Department of Chemistry, Prabhat Kumar College, Contai, India

[2]Department of Chemistry, Indian Institute of Technology, Kharagpur, India

ABSTRACT

The preparation of several functionalized furan derivatives and attempts to transform them into a derivative containing 6H-furo[3,4-b]furanone skeleton towards the construction of naphtho[2,3-b] furan are described. Attempted Pummerer reaction of a furan sulfoxide derivative produced four interesting furan derivatives. Base promoted annulation between methyl 2-(phenylsulfinylme- thyl)-3-furoate and 2-cyclohexenone proceeded to give dihydro naphtho[2,3-b]furanone derivative in a regiospecific manner.

INTRODUCTION

Functionally embellished naphtho[2,3-b]furan moiety has been widely encountered as a unique sub-structure among a diverse range of bioactive synthetic molecules and natural products. Particularly, the condensed quinone derivatives of naphtho[2,3-b]furans such as furonaphthoquinones have been proved to possess broad anticancer activities [1] . Recently, M. Koketsu and co-workers reported that the synthetic furonaphthoquinones showed moderate cytotoxicity against human leukemia U937 and HL-60 cells [2] . During the past decades, a wide range of furanoid natural products have been isolated from plant sources. Among these, furonapthoquinones (e.g. 1 - 9) are prominent due to their wide biological activities and structural significance (Figure 1). Although some strategies have been used for the construction of furonaphthoquinone skeletons, most of the reported methods employ multistep to secure the target skeletons from readily available precursors [3] . Hence intense research in this area has been carried out in recent years leading to the development of simple and straightforward regiospecific route for the preparation of functionalized furonaphthoquinone compounds [4] .

Our continued interest in the application of anionic [4 + 2] cylcoaddition [5] of isobenzofuranones prompted us to study the preparation of 6H-furo[3,4-b] furanones (10) towards the construction of naphtho[2,3-b]furan skeleton embedded in various biologically important molecules.

RESULTS AND DISCUSSION

Our study began with the preparation of furanosulfoxide derivative 14, following the literature procedure [6] . Bromination of methyl 2-methyl-furan-3-carboxylate (11) with N-bromosuccinimide (NBS) under standard condition gave bromo derivative 12 was prepared in 75% yield. Reaction of compound 12 with sodium methoxide and thiophenol gave compound 13 in 88% yield followed by oxidation with sodium periodate in methanol and water medium provided methyl 2-(phenylsulfinylmethyl)-3-furoate (14a) in 75% yield (Scheme 1). Attempted intramolecular Pummerer reaction [7] of sulfoxide 14a with trimethylsilyl chloride (TMSCl) in dichloromethane for overnight, no reaction took place. But when this was refluxed with acetic anhydride, a polymeric product generated. When compound 14a was refluxed with trifluro acetic anhydride or p-toluenesulfonic acid (PTSA), complex mixtures of products were obtained. Examination of the ^{1}H NMR spectrum of the crude products did not indicate formation of desired 10a. The same result was obtained when the above reactions were performed on acid derivative 14b, prepared by hydrolysis of sulfoxide ester 14a with aqueous NaOH and ethanol.

For Scheme 1. Reagents and conditions: (i) NBS, CCl_4, $(PhCO)_2O$ (cat.), hv, 75%; (ii) PhSH, NaI, MeOH, reflux, 88%; (iii) $NaIO_4$, MeOH/H_2O, rt, 40 h, 75%; (iv) NaOH, ethanol, 85%; (v) TMSCl, CH_2Cl_2, overnight or (vi) Ac_2O, reflux, 10 h or (vii) $(CF_3CO)_2O$, reflux or PTSA, reflux.

Interestingly, treatment of sulfoxide 14a with acetic anhydride and a catalytic amount of sodium acetate under reflux produced four different products instead of 10a. All these products 15, 16, 17 and 18 were separated by column chromatography and characterized by NMR, IR studies. Under the same conditions the acid derivative 14b produced an oily polymeric product, ^{1}H NMR spectrum of which revealed the absence of 10a.

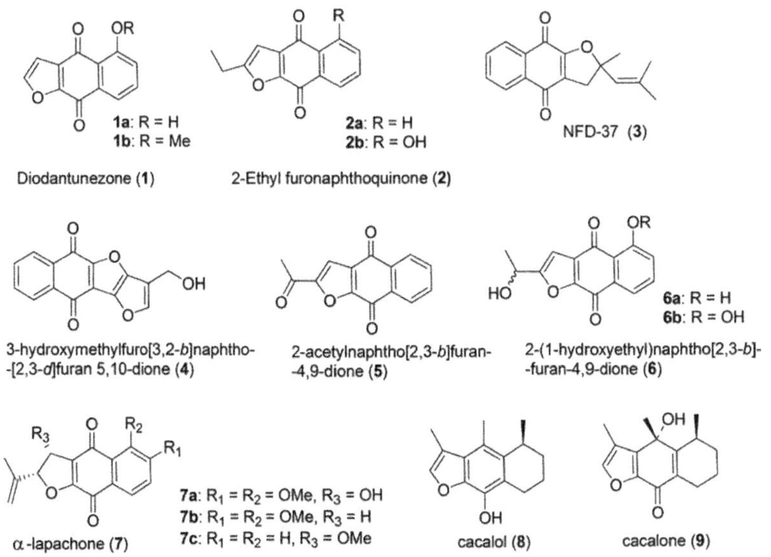

Figure 1: Structure of some biologically active naphtho[2,3-b]furan harbouring natural products.

For Scheme 3. Reagents and conditions: (viii) Ac$_2$O, NaOAc, reflux, 3 h, 28% (for 15), 8% (for 16), 6% (for 17) and 15% (for 18).

Except 18, all these products were expectedly generated through a common Pummerer intermediate 19. Nucleophilic addition of water to the Pummerer intermediate 19 and subsequent expulsion of thiophenol gave aldehydic ester derivative 15 as the major product (28%). Compound 15 could further be added to acetic anhydride to produce furan diacetate derivative 16 in 8% yield (Scheme 2).

Formation of compound 17 (6%) could be explained by addition of one equivalent of thiophenol to the common intermediate 19. On the other hand, the acetate derivative 18 (15%) could be formed by direct nucleophilic displacement of sulfoxide group of 14a by acetate anion.

Having been successful with the above Scheme 3, we focused our attempts to convert ethylsulfoxide 21a to furofuranone 10b via an intramolecular Pummerer reaction. It was presumed that the corresponding intermediate would have favorable geometry for an intramolecular Pummerer reaction [8] . Compound 12 was converted to 20 in 87% yield by the treatment with sodium methoxide and ethanethiol in refluxing methanol. Oxidation of 20 with sodium periodate gave two products which were separated using column chromatography (1:4 mixture of ethyl acetate/petroleum ether). After column chromatography, the desired ethylsulfoxide 21a as isolated in 56% yield and the more oxidized ethylsulfone 21b was isolated in 33% yield as shown in Scheme 4. Both the com pounds 21a and 21b were fully characterized on the basis of spectroscopic (IR, NMR and mass spectral data) analysis. The ^{1}H NMR spectrum exhibited two doublets, one at δ 7.40 (1H) and other at δ 6.73 (1H) for furan ring. It also showed an ABq signal at δ 4.44 (2H) corresponding to two α-hydrogen atoms of ethylsulfoxide group. But, all attempts to effect intramolecular Pummerer reaction of 21a with various reagents such as PTSA in C_6H_6, Ac_2O in toluene, $(CF_3CO)_2O$ in CH_2Cl_2, CF_3CO_2H, pyridinium PTSA in refluxing condition and phenyliodine diacetate (PIDA) in CH_2Cl_2 failed to give the expected furofuranone 10b.

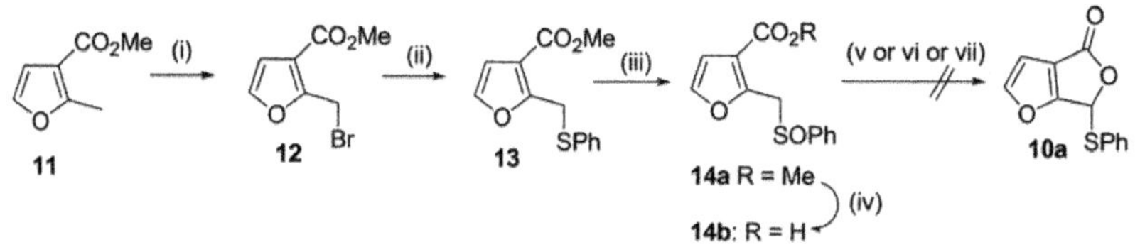

Scheme 1: Synthetic approach to furo[3,4-b]furanone derivative 10a.

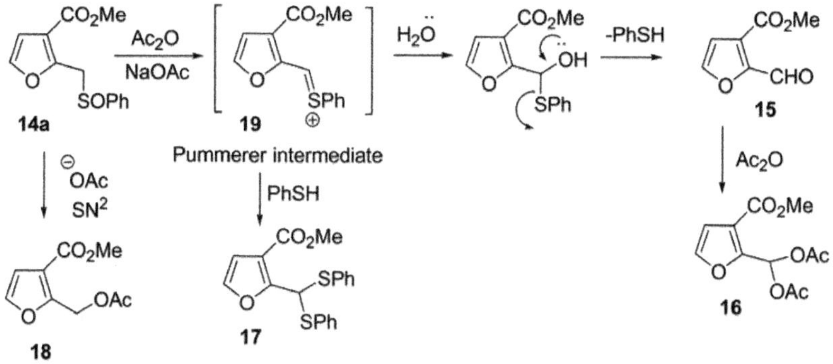

Scheme 2: Possible pathway for the formation of compounds 15, 16, 17 and 18.

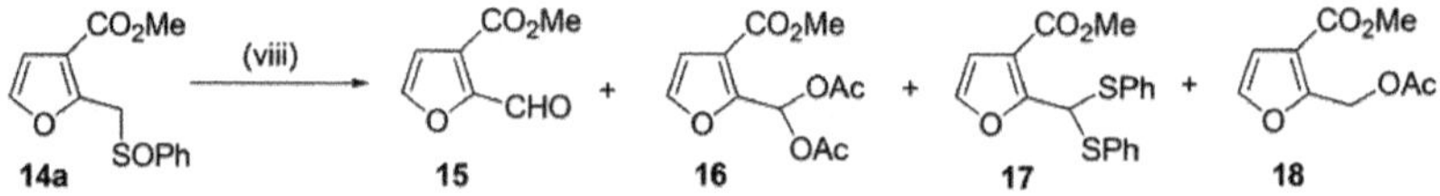

Scheme 3: Intramolecular Pummerer reaction of 14a.

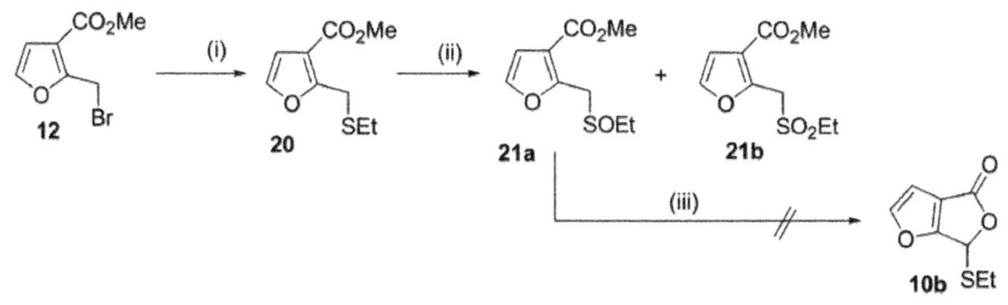

Scheme 4 .Attempted intramolecular Pummerer reaction for the formation of 10b.

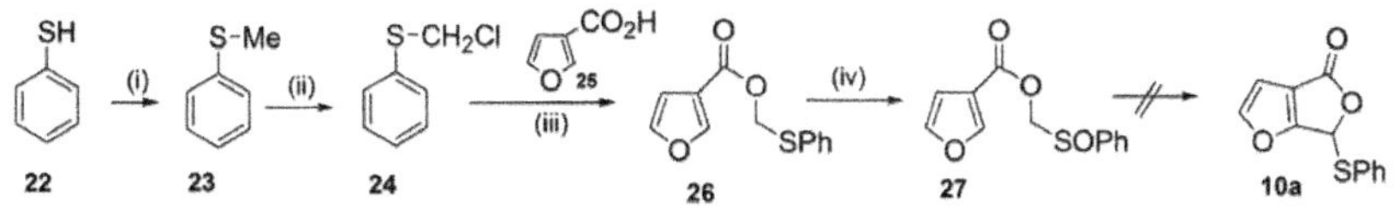

Scheme 5 .Attempted intramolecular Pummerer reaction for the synthesis of 10a.

In all the cases the ^{1}H NMR spectrum of the crude products consisted of broadened signals indicating polymeric materials.

For Scheme 4. Reagents and conditions: (i) EtSH, CHCl$_3$, Et$_3$N, rt, overnight, 87%; (ii) NaIO$_4$, MeOH, 0°C, 2 h, 56%; (iii) PTSA in C$_6$H$_6$, reflux, 10 h or Ac$_2$O in toluene, reflux, 10 h or (CF$_3$CO)$_2$O in CH$_2$Cl$_2$, reflux, 12 h or CF$_3$CO$_2$H, pyridinium PTSA, reflux, 12 h or PIDA in CH$_2$Cl$_2$, reflux, 12 h.

Following the above failures, we turned to preparing furan sulfoxide derivative 26 starting from 3-furoic acid and chloromethylsulfanylbenzene (24) and examining its intramolecular cyclisation via Pummerer reaction to obtain 10a. Methylation of thiophenol with sodium hydroxide and dimehylsulfate in acetone under reflux condition gave 23 in 87% yield. Treatment of 23 with N-chlorosuccinimide in CCl$_4$ produced 24 in 82% yield. Then compound 24 was reacted with 3-furoic acid (25) in the presence of DBU to give 26 (70% yield). This was then transformed to sulfoxide 27 (76% yield) by sodium periodate (NaIO$_4$) oxidation. Both the compounds 26 and 27 gave satisfactory IR, ^{1}H NMR and ^{13}C NMR spectroscopic data. The ^{1}H NMR spectrum showed an ABq signal at δ 5.14 (2H) corresponding to two α-hydrogen atoms of phenylsulfoxide group. Several Pummerer reagents (vide reagents of Scheme 5) were employed for the intramolecular cyclization of 27, but none were effective to give 10a as shown in Scheme 5.

For Scheme 5. Reagents and conditions: (i) aq. NaOH, Me$_2$SO$_4$, reflux, 4 h, 87%; (ii) NCS, CCl$_4$, rt, 11 h, 82%; (iii) DBU, CH$_3$CN, 4 h, 70%; (v) NaIO$_4$, MeOH, 5 h, 76%; (vi) p-TsOH in C$_6$H$_6$, reflux, 10 h or Ac$_2$O in toluene, reflux, 10 h or (CF$_3$CO)$_2$O in CH$_2$Cl$_2$, reflux, 12 h or CF$_3$CO$_2$H, pyridinium PTSA, reflux, 12 h.

As we failed to achieve the preparation of furofuranone 10a by Pummerer procedures, we modified our approach to synthesizing 10a through the desulfanylation of 17, in view of the success of this type of cyclization in benzene system reported by Hauser et al. [8] . Treatment of compound 15 with thiophenol and catalytic amounts of TMSCl in chloroform solvent produced 17 in 85% yield (Scheme 6). Attempted cyclization of 17 in trifluoroacetic acid under reflux condition failed to give expected compound 10a. ^{1}H NMR spectrum of the crude product revealed that starting material decomposed during the course of reaction.

For Scheme 6. Reagents and conditions: (i) PhSH, TMSCl, CHCl$_3$, rt, 85%; (ii) CF$_3$CO$_2$H, H$_2$O, reflux, 12 h.

As we could not carry out the above cyclization of 17, we thought that compound 32 might suit for this de- sulfanylation reaction due to lesser effect of nuclear oxygen atom. It was prepared in good yield starting from commercially available methyl 3-methyl-furan-2-carboxylate (28). The sequence is depicted in Scheme 7. NBS bromination of 28 produced dibromo derivative 29 (60%) along with monobromo derivative 30 in 25% yield. These compounds were separated using column chromatography methods (1:4 mixture of CHCl$_3$/ petroleum eth- er). Hydrolysis of dibromo derivative 29 with silver nitrate in THF/H$_2$O produced furan-3-carboxaldehyde 31 in 42% yield. Finally, treatment of 31 with thiophenol and a catalytic amount of TMSCl provided compound 32. Attempted desulfanylation of 32 with trifluoroacetic acid and water in reflux condition failed to give expected product 10c, starting material was recovered exclusively meaning that no reaction took place.

For Scheme 7. Reagents and conditions: (i) NBS (2 equiv.), CCl$_4$, benzoyl peroxide, hv, 60% (for 29) and 25% (for 30); (ii) AgNO$_3$, THF, H$_2$O, 42%; (iii) PhSH, TMSCl, CHCl$_3$, rt, 80%; (iv) CF$_3$CO$_2$H, H$_2$O, reflux, 12 h.

At this point, we investigated the coupling reaction between diazo derivative of tetronic acid (36) and vinyl acetate for synthesizing 37 from which desired furo[3,4-b]furanone system could be obtained. Compound 36 was prepared from tosyl azide (34) and tetronic acid (35) in the presence of triethylamine according to the literature procedure in 40% yield [9] . ^{1}H NMR of 36 showed only one singlet at δ 4.70 (2H, s) corresponding to -CH$_2$ group. Then we examined its coupling with vinyl acetate under various conditions (Scheme 8) [10] . Butunfortunately, all the attempts failed to give 37. Examination of ^{1}H NMR spectrum of the crude product indicated the exclusive presence of starting material in first two cases (with rhodium diacetate or ceric ammonium nitrate) and unidentifiable product mixture of products with PIDA treatment

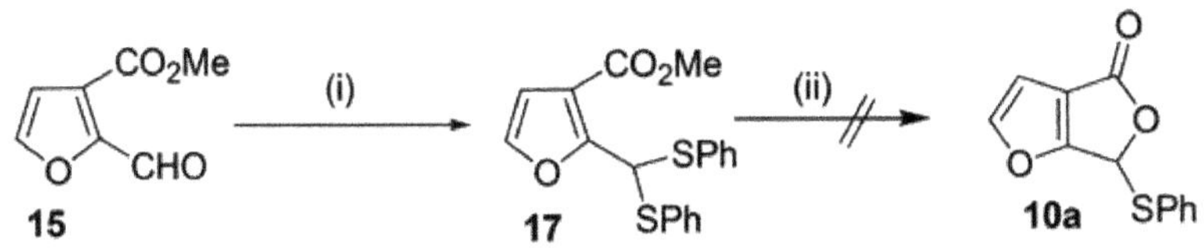

Scheme 6: Attempted intramolecular desulfanylation of 17 to obtain 10a.

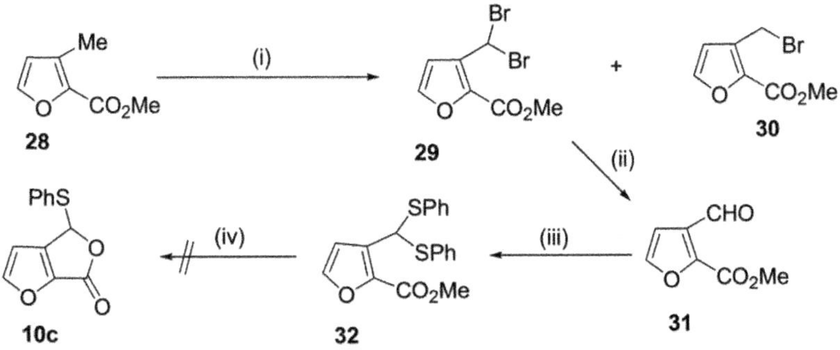

Scheme 7: Investigation of intramolecular desulfanylation of 32.

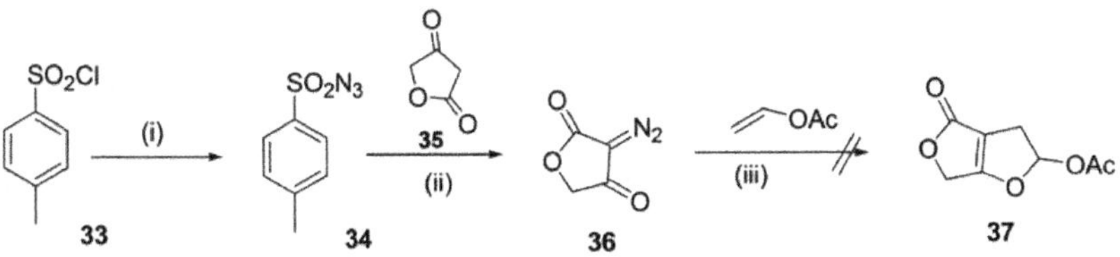

Scheme 8: Attempted routes to synthesis of 37.

For Scheme 8. Reagents and conditions: (i) NaN_3, aq. Acetone, rt, 85%; (ii) Et_3N, CH_3CN, 40%; (iii) $Rh_2(OAc)_2$ or CAN, CH_3CN, 0°C or $PhI(OAc)_2$.

Again we modified our route for the synthesis of furolactone 10d, from which desired compound 10a may be prepared. The hydroxy ester derivative 38 was prepared from 12 by heating it at 80°C in dimetyl sulfoxide and water. The NMR data of 38 matched with literature value [11] . Then compound 38 was transformed to 2-hy- droxymethyl-furan-3-carboxylic acid (39) in 90% yield by the treatment of 40% aqueous solution of KOH solution in methanol (Scheme 9).

For Scheme 9. Reagents and conditions: (i) DMSO, 80°C, 4 h, 92%; (ii) KOH, H_2O, MeOH, 90%; (iii) attempted lactonization with $SOCl_2$ in CH_2Cl_2, DCC in DMF or in CH_2Cl_2, Ac_2O in toluene under refluxing condition and BF_3-ether in C_6H_6.

We then investigated lactonization of compound 39 with various well established literature methods. But, all attempts for lactonization of 39 failed to give the expected furolactone 10d as shown in Scheme 9. In all the cases except with $SOCl_2$, ^1H NMR spectrum of the crude products showed exclusive presence of the starting material 39, meaning no reaction took place. Reaction with $SOCl_2$ produced intractable mixture of products which could not be identified by NMR studies. The above failure of the lactonization may be attributed due to the unfavorable distance between carbonyl "C" and hydroxyl "O" atoms in 39 compared to that in its benzene analog 40 (Figure 2) where lactonization is very facile. Geometries of molecules 39 and 40 were minimized by using density functional theory (DFT) calculations based on the BLYP level of theory with the DND basis set using DMol^3 package program [12] .

Then we thought that sulfoxide 14a itself may serve the purpose of furan annulating agents (i.e. 10a or 10b) for the synthesis of naphtho[2,3-b]furan skeleton. For that purpose, when the sulfoxide 14a was treated with lithium tert-butoxide (t-BuOLi), a light yellow color developed, indicating the generation of carbanion α to -SOPh group, and subsequently the color of the reaction mixture changed to light brown upon addition of 2-cyclohexenone. Work up of the reaction mixture led to formation of tricyclic compound 42 as white solid in 7% yield (Scheme 10). Further transformation of compound 42 to desired naphtho[2,3-b]furan derivative was postponed due to poor yield. We were able to taken only ^1H NMR and IR spectrum of this compound. The ^1H NMR spectrum exhibited three multiplets of six protons at the region of δ 2.18 - 2.97, corresponding to the cyclohexane ring and two doublets, one at δ 7.49 (1H) and other at δ 6.94 (1H) for furan ring. It also showed a ^1H sharp singlet at δ 13.51 corresponding to hydrogen bonded 'OH' group. We repeated the above annulation three times without any improvement in the yield. We also performed this cylcoaddition reaction in presence of lithium diisopropyl amide (LDA). But, ^1H NMR spectrum of the crude indicated the formation of a polymeric material.

CONCLUSION

With the aim of preparing novel naphtho[2,3-b]furan derivatives, an investigation was carried out to synthesize, characterize and study furo[3,4-b] furanones by several approaches. The results showed that the intramolecular Pummerer reaction of furan sulfoxide derivative produced four interesting furan derivatives. The anionic cycloaddition between furan sulfoxide and 2-cyclohexenone produced dihydro naphtho[2,3-b]furanone derivative in poor yield.

39

Total energy, E = - 533.1953949 a.u.

Distance between C-1' and O-2" = 4.326 Å

40

Total energy, E = - 535.3673350 a.u.

Distance between C-1' and O-2" = 3.644 Å

Figure 2: Comparison of distances between carbonyl "C" and hydroxyl "O" atoms in 39 and 40.

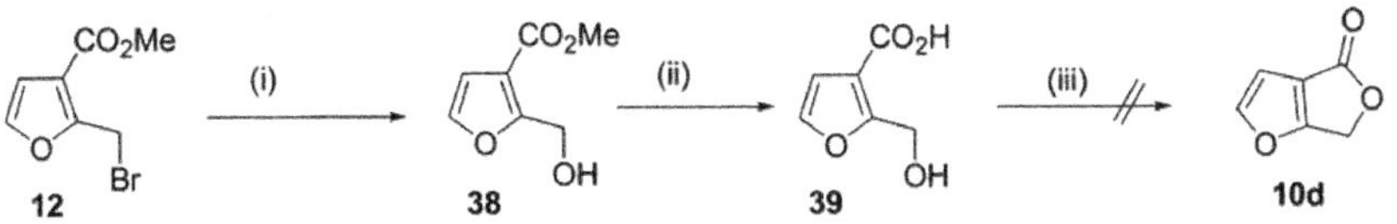

Scheme 9: Attempted lactonization of 2-hydroxymethyl-furan-3-carboxylic acid (39).

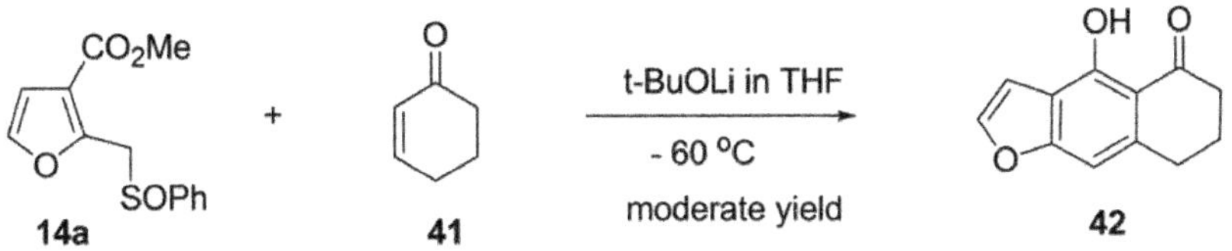

Scheme 10: Synthesis of dihydro naphtho[2,3-b]furanone moiety 42.

This study reveals that synthesis of simple looking furan derivatives (like 10a-d) was elusive and they deserve further study.

EXPERIMENTAL

General

Melting points were determined in open capillary tubes and are uncorrected. Among the spectra, ^{1}H NMR spectra and ^{13}C-NMR spectra were recorded on 200 MHz and 300 MHz spectrometer (Brücker) as solution in ^{2}H- Chloroform with TMS as the internal standard. Chemical shifts are expressed in d unit and ^{1}H-^{1}H coupling constant in Hz. IR spectra were recorded on a Thermo Nicolet Nexus 870 FT-IR spectrophotometers using KBr pellet. EI MS (70 eV) spectra were taken using a VG Autospec M mass spectrometer. Elemental analyses were carried out by using an elemental analyzer VARIO EL instrument. Dry solvents used for reactions were purified, before use, according to the standard protocols. All solvents for chromatography (column and preparative layer chromatography) were distilled prior to use.

Methyl 2-(Phenylsulfinylmethyl)-3-furoate (14a)

To a solution of compound 9 (5 g, 20 mmol) in MeOH (70 mL) containing water (15 mL) was added solid NaIO$_4$ (4.6 g, 21.5 mmol) in portions. The resultant mixture was stirred for 36 h at rt and the solvent was removed under reduced pressure. The resulting thick liquid was purified by column chromatography (3:7 ethyl acetate/petroleum ether, R$_f$ 0.48) over silica gel to furnish the sulfoxide 14a (3.99 g, 75%) as pale yellow solid. mp. 84°C - 85°C (lit. [6] mp. 85°C - 86°C); FT-IR (KBr) cm^{-1} 2935, 1714 (s), 1616, 1440 (m), 1387 (m), 1053, 756; ^{1}H NMR (200 MHz, CDCl$_3$): δ 7.46 - 7.30 (m, 5H), 7.32 (d, 1H, J = 2 Hz), 6.65 (d, 1H, J = 2 Hz), 4.60 (d, 1H, J = 12 Hz), 4.51 (d, 1H, J = 12 Hz), 3.70 (s, 3H); ^{13}C NMR (50 MHz, CDCl$_3$): δ 163.1, 150.4, 142.9, 131.3, 128.9, 123.9, 117.5, 111.0, 55.6, 51.5; MS m/z (EI): 264 (M$^+$), 233, 186, 139 (100%), 125, 109, 97, 77.

2-Phenylsulfinylmethyl-furan-3-carboxylic Acid (14b)

A mixture of methyl 2-(phenylsulfinylmethyl)-3-furoate 14a (1 g, 3.78 mmol), 15 mL of 40% aqueous NaOH solution, 20 mL of MeOH and 15 mL of H$_2$O were stirred for 5 h at ambient temperature. On completion of the reaction, the whole mixture was diluted with water (40 mL) and extracted with ethyl acetate (3 × 30 mL). The combined organic layer was washed with water and 5% of HCl (20 mL), brine (20 mL), dried (Na$_2$SO$_4$) and concentrated. Purification of the crude residue by chromatography on SiO$_2$ (1:1 ethyl acetate/petroleum ether, R$_f$ 0.32) gave compound 14b (0.8 g, 85%) as white solid. mp. 110°C - 112°C; FT-IR (KBr) cm^{-1} 3412, 2362, 1709 (s), 1601 (m), 1444, 1260, 1060, 746; ^{1}H NMR (200 MHz, CDCl$_3$): δ 7.30 (d, 1H, J = 2 Hz), 6.70 (d, 1H, J = 2 Hz), 4.62 (d, 1H, J = 14 Hz), 4.53 (d, 1H, J = 14 Hz); ^{13}C NMR (50 MHz, CDCl$_3$): δ 166.8, 150.6, 143.3, 141.9, 131.7, 129.2, 124.2, 117.9, 111.5, 55.2; HRMS: calcd. for C$_{12}$H$_{10}$O$_4$S [M + Na]$^+$ 251.0380; found 251.0388.

Methyl 2-Formyl-furan-3-carboxylate (15)

To a stirred solution of 14a (1.0 g, 3.78 mmol) Ac$_2$O (10 mL) was added NaOAc (0.31 g, 3.78 mmol) and heated at 110°C for 3 h. After completion of the reaction, the resulting mixture was diluted with water (20 mL) and extracted with ethyl acetate (3 × 20 mL). The organic phases were washed with brine (25 mL), dried (Na$_2$SO$_4$) and concentrated under reduced pressure. Purification of the crude residue by chromatography on silica gel (1:8 ethyl acetate/petroleum ether, R$_f$ 0.52) gave 15 [13] (28%) as white crystalline solid. mp. 76°C - 78°C; FT-IR (KBr) cm^{-1} 3145, 2886, 1719 (s), 1678 (s), 1575, 1404, 1308, 1212 (m), 1073, 1036, 809, 761; ^{1}H NMR (200 MHz, CDCl$_3$): δ 10.21 (s, 1H), 7.63 (d,

1H, J = 0.8 Hz), 6.88 (d, 1H, J = 0.8 Hz), 3.94 (s, 3H); ^{13}C NMR (50 MHz, CDCl$_3$): δ 178.7, 161.9, 152.4, 146.7, 126.2, 112.8, 52.5.

Methyl 2-Diacetoxymethyl-furan-3-carboxylate (16)

This compound was obtained as white solid in 8% yield from 14a on treatment with Ac$_2$O and NaOAc, following the procedure adopted for the preparation of compound 15 from 14a. mp. 96°C; FT-IR (KBr) cm^{-1} 2937, 2388, 1769 (s), 1728 (s), 1623, 1378, 1236, 1200, 1044, 894, 755; ^{1}H NMR (200 MHz, CDCl$_3$): δ 8.18 (s, 1H), 7.42 (d, 1H, J = 2 Hz), 6.74 (d, 1H, J = 2 Hz), 3.85 (s, 3H), 2.12 (s, 6H); ^{13}C NMR (50 MHz, CDCl$_3$): δ 167.8, 162.3, 151.4, 143.0, 117.3, 112.2, 82.5, 51.9, 20.5; Anal. Calcd for C$_{11}$H$_{12}$O$_{17}$: C, 51.57; H, 4.72. Found: C, 51.92; H, 5.04.

Methyl 2-(Bis-phenylsulfanylmethyl)-furan-3-carboxylate (17)

Method 1: This compound was obtained as white solid in 15% yield from 14a on treatment with Ac$_2$O and NaOAc, following the procedure adopted for the preparation of compound 15 from 14a.

Method 2: To a well stirred solution of 15 (200 mg, 1.19 mmol) and thiophenol (132 mg, 1.2 mmol) in dry CHCl$_3$ (10 mL) at rt was added TMSCl (30 mg, 0.28 mmol) and stirring was continued for 5 h. After completion of the reaction, this was washed with 5% NaHCO$_3$ solution (20 mL), diluted with water (50 mL), extracted with ethyl acetate (3 × 30 mL). The combined organic phases were washed with brine (25 mL), dried (Na$_2$SO$_4$) and concentrated under reduced pressure. Purification of the crude residue by chromatography on silica gel (1:2 ethyl acetate/petroleum ether, 0.68) gave 17 (360 mg, 85%) as a thick oil. FT-IR (KBr) cm^{-1} 3140, 1720 (s), 1591, 1475 (m), 1441, 1312 (m), 1162, 1042, 747; ^{1}H NMR (200 MHz, CDCl$_3$): δ 7.35 - 7.43 (m, 5H), 7.24 - 7.30 (m, 5H), 7.33 (d, 1H, J = 2 Hz), 6.35 (s, 1H), 3.66 (s, 3H); ^{13}C NMR (50 MHz, CDCl$_3$): δ 163.0, 156.9, 142.2, 133.3, 128.9, 128.3, 114.2, 110.4, 51.4, 50.8; HRMS: calcd. for C$_{19}$H$_{16}$O$_3$S$_2$ [M + H]$^+$357.0629; found 357.0636.

Methyl 2-Acetoxymethyl-furan-3-carboxylate (18)

This compound was obtained as white solid in 15% yield from 14a on treatment with Ac$_2$O and NaOAc, following the procedure adopted for the preparation of compound 15 from 14a. mp. 47°C; FT-IR (KBr) cm^{-1} 1723 (s), 1633 (s), 1387 (s), 1108 (m), 1041, 754; ^{1}H NMR (200 MHz, CDCl$_3$): δ 7.37 (d, 1H, J = 2 Hz), 6.70 (d, 1H, J = 2 Hz), 5.36 (s, 2H), 3.84 (s, 3H), 2.08 (s, 3H); ^{13}C NMR (50 MHz, CDCl$_3$): δ 170.1, 163.1, 154.3, 142.6, 116.9, 111.0, 56.8, 51.6, 20.6; HRMS: calcd. for C$_9$H$_{10}$O$_5$ [M + H]$^+$ 199.0608; found 199.0615.

Methyl 2-Ethylsulfanylmethylfuran-3-carboxylate (20)

To a stirred solution of ethanethiol (0.16 mL, 2.15 mmol) in dry $CHCl_3$ (5 mL) and triethylamine (217 mg, 2.15 mmol) at rt was added compound 12 (470 mg, 2.15 mmol). After overnight stirring, the resulting mixture was diluted with water (130 mL) and then extracted with chloroform (3 × 40 mL), washed with 5% of HCl (20 mL), brine (30 mL) and dried (Na_2SO_4). The combined organic layer was concentrated under reduced pressure and purified by column chromatography on silica gel (1:10 chloroform/petroleum ether, R_f 0.42) to give 20 (375 mg, 87%) as an oil. FT-IR (KBr) cm^{-1} 3434, 2953, 1722 (s), 1599, 1441, 1308 (m), 1210, 1063, 772; ^1H NMR (200 MHz, CDCl$_3$): δ 7.30 (d, 1H, J = 2), 6.64 (d, 1H, J = 2.4 Hz), 4.07 (s, 2H), 3.82 (s, 3H), 2.55 (q, 2H, J = 8 Hz), 1.32 (t, 3H, J = 8 Hz); ^{13}C NMR (50 MHz, CDCl$_3$): δ 163.8, 158.8, 141.3, 113.9, 110.6, 51.6, 26.5, 25.9, 14.4; HRMS: calcd. for $C_9H_{12}O_3S$ [M + H]$^+$ 201.0587; found 201.0575.

Methyl 2-Ethanesulfinylmethylfuran-3-carboxylate (21a)

To a solution of compound 20 (2 g, 10 mmol) in MeOH (50 mL) containing water (5 mL) was added solid NaIO$_4$ (2.30 g, 10.7 mmol) in portions. The resultant mixture was stirred for 2 h at 0°C and the solvent was removed under reduced pressure. The resulting crude liquid was purified by column chromatography over silica gel (1:5 chloroform/petroleum ether, R_f 0.38) to furnish the sulfoxide 21a (1.20 g, 56%, oily liquid) as the major product along with sulfone derivative 21b (33%). FT-IR (KBr) cm^{-1} 1717 (s), 1654, 1559, 1508, 769; ^1H NMR (200 MHz, CDCl$_3$): δ 7.40 (d, 1H, J = 2 Hz), 6.73 (d, 1H, J = 2 Hz), 4.44 (ABq, 2H, J = 12 Hz), 3.85 (s, 3H), 2.72 (q, 2H, J = 8 Hz), 1.35 (t, 3H, J = 8 Hz); ^{13}C NMR (50 MHz, CDCl$_3$): δ 163.3, 150.6, 143.0, 117.1, 110.9, 51.5, 49.1, 45.2, 6.2; HRMS: calcd. for $C_9H_{12}O_4S$ [M + H]$^+$ 217.0536; found 217.0542.

Methyl 2-Ethanesulfonylmethylfuran-3-carboxylate (21b)

This compound was obtained in the above experiment (for the preparation of 21a) as white solid in 33% yield. mp. 90°C; FT-IR (KBr) cm^{-1} 2940, 1711 (s), 1600, 1508, 1445, 1307, 1042, 827; ^1H NMR (200 MHz, CDCl$_3$): δ 7.46 (d, 1H, J = 2 Hz), 6.75 (d, 1H, J = 2 Hz), 4.74 (s, 2H), 3.86 (s, 3H), 3.02 (q, 2H, J = 8 Hz), 1.38 (t, 3H, J = 8 Hz); ^{13}C NMR (50 MHz, CDCl$_3$): δ 163.2, 148.3, 143.7, 118.1, 111.1, 51.8, 50.8, 47.2, 6.10. Anal. Calcd for $C_9H_{12}NO_5S$: C, 46.54; H, 5.21. Found: C, 46.57; H, 5.04.

Methylsulfanylbenzene (23)

A mixture of thiophenol 22 (5 g, 45.5 mmol) and 20% of aqueous solution NaOH (50 mL) was stirred for 30 min at rt. Then dimethyl sulfate (4.28 mL, 45.5 mmol) was added to the reaction mixture and stirring was continued for 1 h. Afterward, reaction mixture was heated at reflux for 7 h, cooled to rt, extracted with CH_2Cl_2 (3 × 40 mL). The combined extracts were washed with 10% aq. NaOH solution (30 mL), dried (Na_2SO_4) and distilled to give compound 23 [14] (4.9 g, 87%) as colorless oil. ^1H NMR (200 MHz, $CDCl_3$): δ 7.32 - 7.28 (m, 3H), 7.22 - 7.16 (m, 2H), 2.50 (s, 3H).

Chloromethylsulfanylbenzene (24)

To a stirred solution of compound 23 (2 g, 6.10 mmol) in CCl_4 (20 mL) was added N-chlorosuccinimide (2.36 g, 6.71 mmol) at room temperature and stirring was continued for 11 h. The reaction mixture then cooled (0°C) and filtered off. The filtrate was then concentrated under reduced pressure and the residue distilled to give a brownish semisolid of 24 [15] (0.78 g, 82%).^1H NMR (200 MHz, $CDCl_3$): δ 7.58 - 7.48 (m, 2H), 7.40 - 7.14 (m, 3H), 4.97 (s, 2H).

Phenylsulfanylmethyl Furan-3-carboxylate (26)

To a stirred solution of 3-furoic acid (25) (1.0 g, 9.0 mmol) and DBU (1.36 g, 9.0 mmol) in dry acetonitrile (10 mL) under inert atmosphere, was added compound 24 (1.42 g, 9.0 mmol). The resulting mixture was further stirred for 4 h at rt and extracted with ethyl acetate (3 × 30 mL). The combined ethyl acetate extracts were washed with saturated solution of $NaHCO_3$ (20 mL), brine (20 mL) and dried (Na_2SO_4). Concentration of the organic layer gave a light yellow residue. This was purified by column chromatography (1:10 chloroform/petroleum ether, R_f 0.60) to give 26 (1.02 g, 70%) as an oily liquid. FT-IR (KBr) cm^{-1} 2930, 1730 (s), 1431, 1329, 1292, 1150 (s), 1126 (m)1078, 973, 749; ^1H NMR (200 MHz, $CDCl_3$): δ 8.04 (d, 1H, J = 2 Hz), 7.42 - 7.55 (m, 3H), 7.28 - 7.38 (m, 3H), 6.76 (d, 1H, J = 2 Hz), 5.58 (s, 2H); HRMS: calcd. for $C_{12}H_{10}O_3S$ [M + H]$^+$235.0431; found 235.0439.

Benzenesulfinylmethyl Furan-3-carboxylate (27)

To a stirred solution of compound 26 (120 mg, 0.74 mmol) in MeOH (10 mL) containing water (2 mL) was added solid $NaIO_4$ (170 mg, 0.79 mmol) in portions. The resultant mixture was stirred for 5 h at rt and the solvent was removed under reduced pressure. The resulting crude liquid was extracted with ethyl acetate (3 × 20 mL). The combined ethyl acetate extracts was washed with brine (20 mL) and dried (Na$_2$SO$_4$). Concentration of the organic layer gave a solid residue which was purified by column chromatography (1:5 chloroform/petroleum ether, R$_f$ 0.52) to give 27 (140 mg, 76%) as a white solid. mp. 84°C - 85°C; FT-IR (KBr) cm^{-1} 2929, 1747 (s), 1571, 1315, 1169, 1122 (s), 1085 (m), 1049, 757; ^{1}H NMR (200 MHz, CDCl$_3$): δ 8.07 (d, 1H, J = 2 Hz), 7.66 - 7.75 (m, 2H), 7.52 - 7.58 (m, 3H), 7.41 - 7.47 (m, 1H), 5.14 (ABq, 2H, J = 12 Hz); ^{13}C NMR (50 MHz, CDCl$_3$): δ 161.4, 148.8, 144.1, 140.3, 131.8, 129.4, 124.5, 117.5, 109.7, 81.9; HRMS: calcd. for C$_{12}$H$_{10}$O$_4$S [M + H]$^+$ 251.0380; found 251.0388.

Methyl 3-Dibromomethyl-furan-2-carboxylate (29)

A mixture of commercially available methyl 3-methyl-2-furoate (28) (2.0 g, 14.30 mmol), NBS (5.08 g, 28.60 mmol) and a pinch of benzoyl peroxide in CCl$_4$ (150 mL) was heated at reflux for 3.5 h under the exposure of a bulb (100 W). The reaction mixture was then cooled (0°C) and succinimide filtered. The filtrate was concentrated under reduced pressure to give a yellowish residue which was then subjected to column chromatography over silica gel (60 - 120 mesh) using chloroform-petroleum ether mixture (3:7, v/v, R$_f$ 0.58) as eluent to furnish dibromo compound 29 (2.54 g, 60%, white solid) as a main product along with 30 (25%). mp. 80°C - 82°C; FT- IR (KBr) cm^{-1} 3142, 1732 (s), 1608, 1420, 1382, 1252, 1065 (m), 875, 758; ^{1}H NMR (200 MHz, CDCl$_3$): δ 7.51 (d, 1H, J = 2 Hz), 7.36 (s, 1H), 6.92 (d, 1H, J = 2 Hz), 3.95 (s, 3H); HRMS: calcd. for C$_7$H$_6$Br$_2$O$_3$ [M + H]$^+$ 295.8684; found 295.8678.

Methyl 3-Bromomethyl-2-carboxylate (30)

This compound was obtained as white solid in 25% yield and co-product if 30. mp. 51°C (lit. [16] 52°C - 53°C); ^{1}H NMR (200 MHz, CDCl$_3$): δ 7.50 (d, 1H, J = 2 Hz), 6.60 (d, 1H, J = 2 H), 4.65 (s, 2H), 3.92 (s, 3H).

Methyl 3-Formyl-furan-2-carboxylate (31)

To a solution of compound 29 (1.28 g, 4.29 mmol) in THF (20 mL) was added aqueous solution of AgNO$_3$ (1.45 g, 8.58 mmol in 5 mL water) in portions and stirring was continued for overnight at room temperature. The resulting mixture was filtered and after usual work-up of the concentrated filtrate, the residue was purified by column chromatography (1:8 ethyl acetate/petroleum ether, R$_f$0.52) to give 31 (0.28 g, 42%) as white crystalline solid. mp. 72°C - 74°C; FT-IR (KBr) cm^{-1}3264, 2890, 1732 (s), 1682 (s), 1612, 1412, 1320, 1246, 1085, 756; ^{1}H NMR (200 MHz, CDCl$_3$): δ 10.51 (s, 1H), 7.54 (d, 1H, J = 1.8), 6.90 (d, 1H, J = 1.8), 4.0 (s, 1H).

Methyl 3-(1,1-Diphenylsulfanyl)-methylfuran-2-carboxylate (32)

This compound was prepared by reaction of 31 with thiophenol in 80% yield as yellow liquid, according to the procedure described for 17 from 15 (Method 2). FT-IR (KBr) cm^{-1} 3160, 1728 (s), 1592, 1470 (m), 1438, 1310 (m), 1140, 1102, 1046, 746; ^{1}H NMR (200 MHz, CDCl$_3$): δ 7.34 - 7.44 (m, 5H), 7.22 - 7.30 (m, 7H), 6.66 (d, 1H, J = 2 Hz), 6.33 (s, 1H), 3.78 (s, 3H); ^{13}C NMR (50 MHz, CDCl$_3$): δ 158.9, 145.2, 138.9, 133.9, 133.5, 132.8, 128.8, 128.0, 112.5, 51.7, 49.6; MS m/z (EI): [M + H]$^+$ 357.0640.

4-Methyl-benzenesulfonyl Azide (34)

This compound was prepared according to the procedure reported procedure [9] . A mixture of p-toluenesulfonyl chloride (33) (1.35 g, 7 mmol), NaN$_3$ (0.55 g, 8.5 mmol) in aqueous solution of acetone (1:2 mixture of acetone andwater) were stirred for 5 h and then acetone was removed under reduced pressure. After usual work-up, drying (Na$_2$SO$_4$), solvent was evaporated to furnish the desired product 34 [17] as light yellow liquid (1.17 g, 85%), which was sufficiently pure for the next experiment. ^{1}H NMR (200 MHz, CDCl$_3$): δ 7.82 (d, 2H, J = 8), 7.32 (d, 2H, J = 8), 2.45 (s, 3H).

3-Diazotetrahydrofuran-2,4-dione (36)

To a stirred solution of tetrahydrofuran-2,4-dione 35 (2.0 g, 0.02 mol) and p-tosyl azide 34 (3.7 g, 0.02 mol) in acetonitrile (50 mL) was added triethylamine (2 g, 0.02 mol) dropswise over 15 min resulting in a darkening of the solution. After one hour stirring at room temperature the reaction mixture was concentrated and extracted with ether (3 × 50 mL). The combined organic phases were washed with 5% of HCl (20 mL), brine (25 mL), dried (Na$_2$SO$_4$) and concentrated under reduced pressure. Purification of the crude residue by chromatography on silica gel (1:1 ethyl acetate/petroleum ether, R$_f$ 0.61) gave 36 [9] (1.0 g, 40%) as a yellowish solid. mp. 90°C; FT-IR (KBr) cm^{-1} 2166, 1760 (s), 1692 (s); ^{1}H NMR (200 MHz, CDCl$_3$): δ 4.70 (2H, s).

Methyl 2-Hydroxymethylfuran-3-carboxylate (38)

To a solution of DMSO and water (100 mL, 90:10, v/v) at 80°C temperature was added compound 12 (1.0 g, 4.58 mmol) and stirring was continued for 4 h. The resulting reaction mixture was extracted with diethyl ether (3 × 50 mL). The combined extracts were dried (Na$_2$SO$_4$) and the organic phase was evaporated under reduced pressure. The residue was subjected to column chromatography over silica gel (60 - 120 mesh) (1:10 ethyl acetate-petroleum ether, R$_f$ 0.56) to furnish the alcohol 38 [11] (660 mg, 92%) as brownish liquid.

FT-IR (KBr) cm⁻¹ 3448, 2925, 1724 (s), 1438, 1260, 1024, 762; ¹H NMR (200 MHz, CDCl₃): δ 7.27 (d, 1H, J = 1.6 Hz), 6.64 (d, 1H, J = 1.6 Hz), 4.78 (s, 2H), 3.83 (s, 3H); ¹³C NMR (50 MHz, CDCl₃): δ 164.96, 161.25, 141.26, 114.86, 110.75, 57.23, 51.89; MS m/z (EI): [M + 2H]⁺ 158.0267, [M + Na-OMe]⁺ 149.0236, [M + 2H-OH]⁺ 141.0020.

2-Hydroxymethylfuran-3-carboxylic acid (39)

Hydroxy ester compound 38 (0.78 g, 5 mmol) was treated with a mixture of 15% solution of aqueous KOH (15 mL) and methanol (30 mL) for 2 h at rt. On completion of the reaction, 5% of HCl solution (20 mL) was added dropwise till pH 6.5. A white solid separated out from the reaction mixture, which was filtered and washed thoroughly with water to furnish pure acid derivative 39 (640 mg, 90%) as white solid. mp. 76°C - 78°C FT-IR (KBr) cm⁻¹ 3455, 2924, 1686 (s), 1551 (m), 1269, 1166, 1375, 743; ¹H NMR (200 MHz, d₆-DMSO): δ 7.40 (d, 1H, J = 2 Hz), 6.75 (d, 1H, J = 2 Hz), 4.81 (s, 1H), 2.59 (d, 1H, J = 2 Hz); ¹³C NMR (50 MHz, d₆-DMSO): δ 169.54, 165.18, 147.36, 119.82, 116.04, 59.50. HRMS: calcd. for $C_6H_6O_4$ [M + Na]⁺ 165.0156; found 165.0166.

4-Hydroxy-7,8-dihydro-6H-naphtho[2,3-b]furan-5-one (42)

To a stirred solution of lithium tert-butoxide (2.42 mmol) in THF (10 mL) at −60°C (chloroform/liquid N₂ bath) under an inert atmosphere was added a solution of furansulfoxide (200 mg, 0.75 mmol) in THF (1.5 mL). The resulting yellowish solution was stirred at −60°C for 25 min, after which a solution of a 2-cyclohexenone (0.90 mmol) in THF (1.5 mL) was added to it. The cooling bath was removed after about 1 h at −60°C and the reaction mixture was brought to room temperature over a period of 1 h and further stirred for 5 h. The reaction was then quenched with 10% NH₄Cl (10 mL) and the resulting solution was concentrated under reduced pressure. The residue was diluted with ethyl acetate (20 mL) and the layers were separated. The aqueous layer was extracted with ethyl acetate (3 × 15 mL). The combined extracts were washed with brine (3 × 1/3 vol.), dried (Na₂SO₄) and concentrated to provide crude product. The crude solid product was purified by column chromatography on silica gel to give compound 42 (10mg, 7%) as white solid. mp. 118°C - 20°C; FT-IR (KBr) cm⁻¹ 3405, 2940, 2502, 2375, 1982, 1630 (s), 1450, 1450 (m), 1356, 1331, 1284, 11285, 1120 (m), 1014, 814, 747; ¹H NMR (200 MHz, CDCl₃): δ 13.51 (s, 1H), 7.49 (d, 1H, J = 2 Hz), 6.94 (d, 1H, J = 0.8 Hz), 6.84 (s, 1H), 2.97 - 3.10 (m, 2H), 2.66 - 2.74 (m, 2H), 2.05 - 2.18 (m, 2H).

ACKNOWLEDGEMENTS

Financial support from the UGC Minor Research Grant, F. PSW-092/11-12 (ERO) dated 3rd August, 2011, New Delhi is gratefully acknowledged.

REFERENCES

1. Lin, K.I., Su, J.C., Chien, C.M., Tseng, C.H., Chen, Y.L., Chang, L.S. and Lin, S.R. (2010) Naphtho[1,2-b]furan-4,5- dione Induces Apoptosis and S-Phase Arrest of MDA-MB-231 Cells through JNK and ERK Signaling Activation. Toxicology in Vitro, 24, 61-70.http://dx.doi. org/10.1016/j.tiv.2009.09.002 Ito, C., Katsuno, S., Kondo, Y., Tan, H.T.W. and Furukawa, H. (2000) Chemical Constituents of Avicennia Alba. Isolation and Structural Elucidation of New Naphthoquinones and Their Analogues. Chemical & Pharmaceutical Bulletin, 48, 339-343. Hirai, K., Koyama, J., Pan, J., Simamura, E., Shimada, H. and Yamori, T. (1999) Cancer Detection and Prevention, 23, 539-550. Nagata, K., Hirai, K., Koyama, J., Wada, Y. and Tamura, T. (1998) Antimicrobial Agents and Chemotherapy, 42, 700-702. Takegami, T., Simamura, E., Hirai, K. and Koyama, J. (1998) Inhibitory Effect of Furanonaphthoquinone Derivatives on the Replication of Japanese Encephalitis Virus. Antiviral Research, 37, 37-45.

2. Inagaki, R., Ninomiya, M., Tanaka, K., Watanabe, K. and Koketsu, M. (2013) Synthesis and Cytotoxicity on Human Leukemia Cells of Furonaphthoquinones Isolated from Tabebuia Plants. Chemical & Pharmaceutical Bulletin, 61, 670-673.http://dx.doi.org/10.1248/cpb.c13-00011

3. Thomson, R.H. (1997) Naturally Occurring Quinones IV, Recent Advances. Blackie Academics & Professional, London, 112-308. Diaz, F. and Medina, J.D. (1996) Furanonaphthoquinones from Tabebuia ochracea ssp. Neochrysanta. Journal of Natural Products, 59, 423-424. Corral, J.M.D., Castro, M., Oliveira, A., Gualberto, S., Cuevas, C. and San, A.F. (2006) New Cytotoxic Furoquinones Obtained from Terpenyl-1,4-naphthoquinones and 1,4-Anthracenediones. Bioorganic & Medicinal Chemistry, 14, 7231-7240. Kobayashi, K., Uneda, T., Kawakita, M., Morikawa, O. and Konishi, H. (1997) One-Pot Synthesis of Naphtho[2,3-b] furan-4,9-diones by Sequential Coupling/Ring Closure Reactions. Tetrahedron Letters, 38, 837-840.

4. Kobayashi, K., Shimizu, H., Sakai, A. and Suginome, H. (1993) Photoinduced Molecular Transformations. 140. New One-Step General Synthesis of Naphtho[2,3-b]furan-4,9-diones and Their 2,3-Dihydro

Derivatives by the Regioselective [3 + 2] Photoaddition of 2-Hydroxy-1,4-naphthoquinones with Various Alkynes and Alkenes: Application of the Photoaddition to a Two-Step Synthesis of Maturinone. Journal of Organic Chemistry, 58, 4614-4618. http://dx.doi.org/10.1021/jo00069a023 Kobayashi, K., Kanno, Y. and Suginome, H. (1993) Photoinduced Molecular Transformations. Part 141. New One- Step General Synthesis of Benzofuran-4,7-diones by the Regioselective (3 + 2) Photoaddition of 2-Hydroxy-1,4-ben- zoquinones with Various Alkenes. Journal of the Chemical Society, Perkin Transactions, 1, 1449-1452. Lee, Y.R., Suk, J.Y. and Kim, B.S. (2000) One-Pot Construction of Medium- and Large-Sized Ring Substituted Furans. Efficient Conversion to Dibenzofurans, Coumestans, and 4-Pyrones. Organic Letters, 2, 1387-1389.

5.   Hauser, F.M., Dorsch, W.A. and Mal, D. (2002) Total Synthesis of (±)-O-Methyl PD 116740. Organic Letters, 4, 2237-2239. Mal, D., Senapati, B.K. and Pahari, P. (2006) Regioselective Synthesis of 1-Hydroxycarbazoles via Anionic [4 + 2] Cycloaddition of Furoindolones: A Short Synthesis of Murrayafoline-A. Tetrahedron Letters, 47, 1071-1075. http://dx.doi.org/10.1016/j.tetlet.2005.12.048 Mal, D., Senapati, B.K. and Pahari, P. (2007) Anionic [4 + 2] Cycloaddition Strategy in the Regiospecific Synthesis of Carbazoles: Formal Synthesis of Ellipticine and Murrayaquinone A. Tetrahedron, 63, 3768-3781. http://dx.doi.org/10.1016/j.tet.2007.02.060 Mal, D. and Pahari, P. (2007) Recent Advances in the Hauser Annulation. Chemical Reviews, 107, 1892-1918.

6.   Mal, D., Bandhyopadhyay, M., Datta, K. and Murty, K.V.S.N. (1998) Anionic [4 + 2] Cycloaddition Strategy to Linear Furocoumarins: Synthesis of 8-Methoxypsoralen and Its Isoster. Tetrahedron, 54, 7525-7538. http://dx.doi.org/10.1016/S0040-4020(98)00387-1

7.   Feldman, K.S. (2006) Modern Pummerer-Type Reactions. Tetrahedron, 62, 5003-5034. http://dx.doi.org/10.1016/j.tet.2006.03.004 Padwa, A. (2004) Tandem Methodology for Heterocyclic Synthesis. Pure and Applied Chemistry, 76, 1933-1952. Bur, S.K. and Padwa, A. (2004) The Pummerer Reaction: Methodology and Strategy for the Synthesis of Heterocyclic Compounds. Chemical Reviews, 104, 2401-2432.

8.   Padwa, A., Danca, M.D., Hardcastle, K.I. and McClure, M.S. (2003) A Short Diastereoselective Synthesis of the Putative Alkaloid Jamtine, Using a Tandem Pummerer/Mannich Cyclization Sequence. Journal of Organic Chemistry, 68, 929-941. Hauser, F.M., Rhee, R.P. and Prasanna, S. (1980) ortho-Toluate Carbanion Chemistry: Sulfenylation and Selenation. Synthesis, 1, 72-74. http://dx.doi.org/10.1055/s-1980-28963

9. Murphy, P.V., O'Sullivian, T.J., Kenndy, B.D. and Geraghty, N.W. (2000) The Reactions of Diazo Compounds with Lactones. Part 2. The Reaction of Cyclic 2-Diazo-1,3-dicarbonyl Compounds with Diketene: Benzofuranformation. Journal of the Chemical Society, Perkin Transactions, 1, 2121-2126. http://dx.doi.org/10.1039/b001394n

10. Pirrung, M.C. and Lee, Y.R. (1994) Dipolar Cycloaddition of Rhodium Carbenoids with Vinyl Esters. Total Synthesis of Pongamol and Lanceolatin B. Tetrahedron Letters, 35, 6231-6234. http://dx.doi.org/10.1016/S0040-4039(00)73399-5

11. Pevzner, L.M. (2001) Synthesis and Properties of (1,3-Dioxolan-2-yl) furans. Russian Journal of General Chemistry, 71, 1045-1049.http://dx.doi.org/10.1023/A:1013149519992

12. Parr, R.G. and Yang, W. (1989) Density Functional Theory of Atoms and Molecules. Oxford University Press, Oxford.

13. Khatuya, H. (2001) On the Bromination of Methyl 2-Methyl-3-Furoate. Tetrahedron Letters, 42, 2643-2644. http://dx.doi.org/10.1016/S0040-4039(01)00275-1

14. Yamamoto, T. and Sekine, Y. (1984) Condensation of Thiophenols with Aryl Halides Using Metallic Copper as a Reactant. Intermediation of Cuprous Thiophenolates. Canadian Journal of Chemistry, 62, 1544-1547. http://dx.doi.org/10.1139/v84-263

15. Tanikaga, R., Miyashita, K., Ono, N. and Kaji, A. (1982) A Convenient Synthesis of 2-Alkenoic Esters. Synthesis, 1982, 131-132. http://dx.doi.org/10.1055/s-1982-29714

16. Clayden, J., Greeves, N., Warren, S. and Wothers, P. (2001) Organic Chemistry. Oxford University Press Inc., New York, 1133. Vegh, D., Morel, J., Decroix, B. and Zalupsky, P. (1992) A New Convenient Method for Preparation of Condensed Aromatic and Heterocyclic Thiolactones. Synthetic Communications, 22, 2057-2061.http://dx.doi.org/10.1080/00397919208021340

17. Curphey, T.J. (1998) Preparation of p-Toluenesulfonyl Azide. A Cautionary Note. Organic Preparations and Procedures International, 13, 112-115.http://dx.doi.org/10.1080/00304948109356105

Chapter 12

1-(4-(PYRROLIDIN-1-YLSULFONYL)PHENYL)
ETHANONE IN HETEROCYCLIC
SYNTHESIS: SYNTHESIS, MOLECULAR
DOCKING AND ANTI-HUMAN LIVER CANCER
EVALUATION OF NOVEL SULFONAMIDES
INCORPORATING THIAZOLE, IMIDAZO[1,2-A]
PYRIDINE, IMIDAZO[2,1-C] [1,2,4]TRIAZOLE,
IMIDAZO[2,1-B]THIAZOLE, 1,3,4-
THIADIAZINE AND 1,4-THIAZINE MOIETIES

Mahmoud Sayed Bashandy

Chemistry Department, Faculty of Science (Boys), Al-Azhar University, Nasr City, Cairo, Egypt

ABSTRACT

This article describes the synthesis of some novel sulfonamides having the biologically active, thiazole 4-6, 8, 10-12a,b, 20, 22, 34, 35, imidazo[1,2-a] pyridine 14, imidazo[2,1-c][1,2,4]triazole 15, imidazo[2,1-b]thiazole 23, 24, 33, nicotinonitrile 25, 1,3,4-thiadiazine 27, quinoxaline 30 and 1,4- thiazine 31 moieties, starting with 1-(4-(pyrrolidin-1-ylsulfonyl)phenyl)ethanone (1). The structures of the newly synthesized compounds were confirmed by elemental analysis, IR, ^{1}H NMR, ^{13}C NMR and Ms spectral data. All the compounds were tested in-vitro antihuman liver hepatocellular carcinoma cell line (HepG2). Compounds 8, 11, 4, 22, 12a, 33, 35, 27 and 24 with selectivity index (SI) values of 33.21, 30.49, 19.43, 14.82, 10.29, 7.3, 6.87, 6.15 and 4.62, respectively, exhibited better activity than methotrexate (MTX) as a reference drug with SI value of 4.14. Molecular Operating Environment (MOE) performed virtual screening using molecular docking studies of the synthesized compounds. The results indicated that some synthesized compounds are suitable inhibitors against dihydrofolate reductase (DHFR) enzyme (PDB ID: 4DFR) with further modification.

INTRODUCTION

Sulfonamides have been demonstrated to possess antibacterial [1] -[3] , antifungal [4] , insulin-releasing [5] [6] , carbonic anhydrase inhibitory [7] -[9] , hypoglycemic [10] , anesthetic [11] , anti-inflammatory [12] [13] , andanti-carcinogenic [14] [15] activities. Liver cancer (hepatocellular carcinoma) remains one of the most important health problems in the world because it is the third foremost cause of cancer-related deaths worldwide [16] . In view of these reports and as a continuation of previous works [17] -[21] directed towards the synthesis of substituted heterocycles, incorporating benzenesulfonamide with anticipated biological activities, therefore, this article reports new and convenient methods for the synthesis of heterocyclic ring systems that are required to medicinal chemistry utilizing 1-(4-(pyrrolidin-1-ylsulfonyl)phenyl) ethanone (1) as a starting material. Since, the carbonyl and the methyl functions of compound 1 suitably situated to enable reaction with common bi-dentate reagents to form a variety of heterocyclic compounds having sulfonamide function, and investigated their anti- human liver cancer activities.

MATERIAL AND METHODS

Experimental

Melting points (°C, uncorrected) were determined in open capillaries on a Gallen Kemp melting point apparatus (Sanyo Gallen Kemp, Southborough, UK). IR spectra (KBr) were recorded on FT-IR 5300 spectrometer and Perkin Elmer spectrum RXIFT-IR system (v, cm^{-1}). Pre-coated silica gel plates (silica gel 0.25 mm, 60 G F 254; Merck, Germany) were used for thin layer chromatography. The NMR spectra in (DMSO-d$_6$) were recorded at 300 MHz on a Varian Gemini NMR spectrometer (δ, ppm). Mass spectra were obtained on GC Ms-QP 1000 EX mass spectrometer at 70 ev. Elemental analyses were performed on Carlo Erba 1108 Elemental Analyzer (Heraeus, Hanau, Germany). All compounds were within ± 0.4% of the theoretical values. Analyses were carried out by the Micro analytical Research Center, Faculty of Science, Cairo University and Al-Azhar University. 1-(4-(Pyrrolidin-1-ylsulfonyl)phenyl)ethanone (1) was prepared according to the procedures reported in the literature [22] . Yellowish white crystals, Yield, 83%; mp 115°C - 116°C (ethanol). IR (KBr, cm^{-1}): v$_{max}$ = 3070 (CH aromatic), 2973 (CH aliphatic), 1696 (C=O), 1347, 1153 (SO$_2$). ^{1}H NMR (DMSO-d$_6$): δ = 1.87 (t, 4H, CH$_2$-CH$_2$ pyrrolidine), 2.55 (s, 3H, CH$_3$), 3.25 (t, 4H, CH$_2$-N-CH$_2$ pyrrolidine), 7.65, 8.26 (dd, 4H, Ar-H, AB system, J = 9.41 Hz). ^{13}C NMR (DMSO-d$_6$): δ = 24.7 (2C, CH$_2$-CH$_2$ pyrrolidine), 28.3 (CH$_3$), 63.1 (2C, CH$_2$- N-CH$_2$ pyrrolidine), [125.6 (2C), 130.2 (2C), 141.4, 145.2] (6ArC's), 199.3 (C=O). MS m/z (%): 253.11 [M$^+$] (9.07), 183.02 (6.98),

174.09 (4.39), 119.06 (19.80), 104.05 (7.35), 91.09 (11.94), 76.06 (14.71), 70.08 (100.00), 43.08 (36.05). Anal. Calcd. for $C_{12}H_{15}NO_3S$ (253.32): C, 56.90; H, 5.97; N, 5.53; S, 12.66. Found: C, 56.88; H, 5.94; N, 5.61; S, 12.53%.

2-Bromo-1-(4-(pyrrolidin-1-ylsulfonyl)phenyl)ethanone (2)

To a stirred solution of 1-(4-(pyrrolidin-1-ylsulfonyl)phenyl)ethanone (1; 2.53 g, 0.01 mol) in dioxane/diethyl- lether mixture (1:2) (30 mL), the bromine (1.59g, 0.01 mol) was added drop wise with constant stirring.After complete addition, the reaction will left for 1 h, then the reaction mixture poured in cold water (100 mL), the separated solid was filtered off and recrystallized from ethanol to give 2. White crystals, Yield, 90%; mp 96˚C - 98˚C. IR (KBr, cm⁻¹): v_{max} = 3056 (CH aromatic), 2909 (CH aliphatic), 1707 (C=O), 1336, 1161 (SO₂). ¹H NMR (DMSO-d₆): δ = 1.99 (t, 4H, CH₂-CH₂ pyrrolidine), 3.27 (t, 4H, CH₂-N-CH₂ pyrrolidine), 4.56 (s, 2H, CH₂), 7.84, 8.22 (dd, 4H, Ar-H, AB system, J = 8.57 Hz). ¹³C NMR (DMSO-d₆): δ = 22.1 (2C, CH₂-CH₂ pyrrolidine), 32.7 (CH₂), 64.2 (2C, CH₂-N-CH₂ pyrrolidine), [123.0 (2C), 128.8 (2C), 134.0, 143.0] (6ArC's), 193.5 (C=O). MS m/z (%): 333.03 [M⁺+2] (1.59), 332.02 [M⁺+1] (1.58), 331.03 [M⁺] (1.92), 330.02 (1.06), 238.07 (15.70), 196.98 (3.49), 174.10 (22.39), 118.10 (10.24), 116.07 (18.07), 104.06 (14.23), 90.08 (10.25), 89.07 (14.76), 70.09 (100.00), 63.05 (5.49), 42.06 (54.43). Anal. Calcd. for $C_{12}H_{14}BrNO_3S$ (332.21): C, 43.38; H, 4.25; N, 4.22; S, 9.65. Found: C, 43.41; H, 4.10; N, 4.16; S, 9.71%.

2-(1-(4-(Pyrrolidin-1-ylsulfonyl)phenyl)ethylidene)hydrazinecar-bothioamide (3)

A mixture of acetophenone derivative 1 (2.53 g, 0.01 mol) and thiosemicarbazide (0.91 g, 0.01 mol) in ethanol (50 mL) was heated under reflux for 5 h, during the reflux period, a pale yellow crystalline solid was separated. The separated solid was filtered off, washed with ethanol, dried and recrystallized from ethanol/benzene to give 3. White crystals, Yield, 53%; mp 130˚C - 131˚C. IR (KBr, cm⁻¹): v_{max} = 3414, 3310 (NH₂), 3198 (NH), 3077 (CH aromatic), 2956 (CH aliphatic), 1587 (C=N), 1345, 1165 (SO₂), 1280 (C=S). ¹H NMR (DMSO-d₆): δ = 1.83 (t, 4H, CH₂-CH₂pyrrolidine), 2.32 (s, 3H, CH₃), 3.29 (t, 4H, CH₂-N-CH₂ pyrrolidine), 6.40 (br, 2H, NH₂exchangeable with D₂O), 7.70, 8.11 (dd, 4H, Ar-H, AB system, J = 8.72 Hz), 8.90 (br, 1H, NH exchangeable with D₂O). MS m/z (%): 326.34 [M⁺] (0.13), 307.12 (1.25), 226.09 (11.06), 183.05 (12.49), 119.10 (17.22), 101.17 (21.02), 86.13 (100.00), 80.02 (6.74), 72.14 (6.44), 58.10 (73.45). Anal. Calcd. for $C_{13}H_{18}N_4O_2S_2$ (326.44): C, 47.83; H, 5.56; N, 17.16; S, 19.65. Found: C, 47.76; H, 5.42; N, 17.22; S, 19.59%.

4-(4-(Pyrrolidin-1-ylsulfonyl)phenyl)-2-(2-(1-(4-(pyrrolidin-1-ylsulfonyl)phenyl)- ethylidene)hydrazinyl)thiazole (4)

A mixture of thiocarbamoyl derivative 3 (3.26g, 0.01 mol), phenacyl bromide derivative 2 (3.32 g, 0.01 mol) and fused sodium acetate (6.56 g, 0.08 mol) in ethanol (50 mL) was heated under reflux for 4 h, during the reflux period, a yellow crystalline solid was separated. The separated solid was filtered off, washed with ethanol, dried and recrystallized from dioxane to give 4. Yellow crystals, Yield, 42%; mp 170°C - 171°C. IR (KBr, cm^{-1}): v_{max} = 3263 (NH), 3098 (CH aromatic), 2970 (CH aliphatic), 1588 (C=N), 1356, 1172 (SO$_2$). ^{1}H NMR (DMSO-d$_6$): δ = 1.93 (t, 8H, 2CH$_2$-CH$_2$pyrrolidine), 2.51 (s, 3H, CH$_3$), 2.80 (t, 8H, 2CH$_2$-N-CH$_2$ pyrrolidine), 7.28 (s, 1H, CH-thiazole), 7.92-8.11 (m, 8H, Ar-H), 8.87 (s, 1H, NH exchangeable with D$_2$O). MS m/z (%): 559.47 [M$^+$] (0.81), 305.08 (23.78), 291.99 (9.32), 268.11 (19.98), 251.06 (7.89), 199.05 (74.82), 140.05 (100.00), 135.05 (12.84), 91.07 (82.77), 77.05 (79.36). Anal. Calcd. for C$_{25}$H$_{29}$N$_5$O$_4$S$_3$ (559.72): C, 53.65; H, 5.22; N, 12.51; S, 17.19. Found: C, 53.59; H, 5.08; N, 12.34; S, 17.25%.

4-(4-(Pyrrolidin-1-ylsulfonyl)phenyl)thiazol-2-amine (5)

1)   Procedure (A)

Thiourea (1.52 g, 0.02 mole) and I$_2$ (2.53 g, 0.01 mole) were triturated and mixed with acetophenone derivative 1 (2.53 g, 0.01 mol) in dioxane (40 mL). The mixture was refluxed with occasional stirring for 8 h. The obtained solid was washed with aqueous sodium thiosulfate to remove excess iodine and then with water. The crude product was dissolved in hot water, filtered to remove the sulphone, and 2-aminothiazole derivative 5 was precipitated by addition of NH$_3$, H$_2$O. The crude product was dried and recrystallized from dioxane to give 5, (yield 23%).

2)   Procedure (B)

A solution of phenacyl bromide derivative 2 (3.32 g, 0.01 mol) and thiourea (0.76 g, 0.01 mole) in ethanol (40 mL) was refluxed for 2 h. After addition of pyridine (5 mL) and continued reflux for 5 h, the solvent was removed in vacuo. The obtained product collected and recrystallized; mp and mixed mp determined with authentic sample gave no depression. Yellow crystals, Yield, 90%; mp 203°C - 204°C. IR (KBr, cm^{-1}): v_{max} = 3384, 3334 (NH$_2$), 3100 (CH aromatic), 2931 (CH aliphatic), 1573 (C=N), 1364, 1181 (SO$_2$). ^{1}H NMR (DMSO-d$_6$): δ = 1.82 (t, 4H, CH$_2$-CH$_2$ pyrrolidine), 3.33 (t, 4H, CH$_2$-N-CH$_2$ pyrrolidine), 6.99 (s, 1H, CH- thiazole), 7.45 (s, 2H, NH$_2$ exchangeable with D$_2$O), 7.83, 8.12 (dd, 4H, Ar-H, AB system, J = 9.01 Hz). ^{13}C NMR (DMSO-d$_6$): δ = 27.8 (2C, CH$_2$-CH$_2$pyrrolidine), 66.2 (2C, CH$_2$-N-

CH_2 pyrrolidine), 105.2 (thiazole-C_5), [121.5 (2C), 124.1 (2C), 135.0, 140.7] (6ArC's), 153.1 (thiazole-C_4), 170.3 (thiazole-C_2). MS m/z (%): 309.13 [M⁺] (3.44), 307.11 (56.57), 264.08 (75.34), 200.12 (51.86), 157.13 (15.99), 133.13 (100.00), 103.11 (25.94), 77.09 (34.24), 58.10 (54.31), 42.07 (20.69). Anal. Calcd. for $C_{13}H_{15}N_3O_2S_2$(309.41): C, 50.46; H, 4.89; N, 13.58; S, 20.73. Found: C, 50.34; H, 4.92; N, 13.41; S, 20.80%.

2-Cyano-N-(4-(4-(pyrrolidin-1-ylsulfonyl)phenyl)thiazol-2-yl) acetamide (6)

A mixture of 2-aminothiazole derivative 5 (3.09g, 0.01 mol) and ethyl cyanoacetate (1.13 g, 0.01 mol) was heated at 160°C for 30 min. the separated solid was filtered off and recrystallized from ethanol to give 6. Buff solid, Yield, 66%; mp 230°C - 231°C. IR (KBr, cm⁻¹): v_{max} = 3273 (NH), 3052 (CH aromatic), 2928 (CH aliphatic), 2220 (C≡N), 1696 (C=O), 1579 (C=N), 1349, 1159 (SO_2). ¹H NMR (DMSO-d_6): δ = 1.82 (t, 4H, CH_2-CH_2 pyrrolidine), 3.30 (t, 4H, CH_2-N-CH_2pyrrolidine), 4.20 (s, 2H, CH_2), 7.61 (s, 1H, CH-thiazole), 7.85, 8.23 (dd, 4H, Ar-H, AB system, J = 8.67 Hz), 9.15 (s, 1H, NH exchangeable with D_2O). MS m/z (%): 376.96 [M⁺] (1.64), 309.08 (4.00), 284.09 (16.72), 266.04 (8.22), 245.10 (13.46), 238.06 (22.74), 174.09 (100.00), 146.06 (16.59), 134.04 (7.08), 104.06 (23.62), 76.04 (59.65), 67.04 (37.05), 44.02 (29.38). Anal. Calcd. For $C_{16}H_{16}N_4O_3S_2$ (376.45): C, 51.05; H, 4.28; N, 14.88; S, 17.04. Found: C, 50.87; H, 4.13; N, 14.90; S, 16.97%.

N-(4-Fluorobenzylidene)-4-(4-(pyrrolidin-1-ylsulfonyl)phenyl) thiazol-2-amine (8)

A mixture of 2-aminothiazole derivative 5 (3.09 g, 0.01 mol) and 4-fluorobenzaldehyde (1.24 g, 0.01 mol) in (50 mL) ethanol with a few drops of piperidine was heated under reflux for 4 h, during the reflux period, a pale yellow crystalline solid was separated. The separated solid was filtered off, washed with ethanol, dried and recrystallized from ethanol/benzene to give 8. Pale yellow crystals, Yield, 91%; mp 213°C - 214°C. IR (KBr, cm⁻¹): v_{max} = 3066 (CH aromatic), 2940 (CH aliphatic), 1590 (C=N), 1360, 1155 (SO_2), 1170 (C-F). ¹H NMR (DMSO-d_6): δ = 1.93 (t, 4H, CH_2-CH_2pyrrolidine), 3.70 (t, 4H, CH_2-N-CH_2 pyrrolidine), 7.45 (s, 1H, CH-thiazole), 7.36 - 8.07 (m, 8H, Ar-H), 8.57 (s, 1H, CH=N). MS m/z (%): 417.12 [M⁺+2] (1.39), 416.14 [M⁺+1] (1.99), 415.10 [M⁺] (7.28), 281.05 (25.03), 240.00 (5.57), 175.04 (28.47), 148.05 (5.25), 134.04 (21.17), 122.05 (7.01), 105.04 (23.58), 89.05 (66.36), 70.06 (100.00), 42.05 (81.36). Anal. Calcd. For $C_{20}H_{18}FN_3O_2S_2$ (415.50): C, 57.81; H, 4.37; N, 10.11; S, 15.43. Found: C, 57.78; H, 4.26; N, 10.10; S, 15.51%.

2-Hydrazinyl-4-(4-(pyrrolidin-1-ylsulfonyl)phenyl)thiazole (10)

A solution of phenacyl bromide derivative 2 (3.32g, 0.01 mol) in ethanol (30 mL) and thiosemicarbazide (0.91g, 0.01 mol) was refluxed for 1 h. The solid product which obtained after cooling was collected and recrystallized from ethanol/benzene to give 10. White solid, Yield, 42%; mp 163°C - 164°C. IR (KBr, cm^{-1}): v$_{max}$ = 3454, 3341 (NH$_2$), 3269 (NH), 3074 (CH aromatic), 2977 (CH aliphatic), 1596 (C=N), 1344, 1171 (SO$_2$). ^{1}H NMR (DMSO-d$_6$): δ = 1.77 (t, 4H, CH$_2$-CH$_2$ pyrrolidine), 3.41 (t, 4H, CH$_2$-N-CH$_2$ pyrrolidine), 5.06 (br, 2H, NH$_2$ exchangeable with D$_2$O), 7.45 (s, 1H, CH-thiazole), 7.99, 8.47 (dd, 4H, Ar-H, AB system, J = 8.73 Hz), 10.35 (s, 1H, NH exchangeable with D$_2$O). ^{13}C NMR (DMSO-d$_6$): δ = 26.5 (2C, CH$_2$-CH$_2$ pyrrolidine), 66.7 (2C, CH$_2$- N-CH$_2$ pyrrolidine), 109.4 (thiazole-C$_5$), [119.3 (2C), 124.2 (2C), 133.6, 139.4] (6ArC's), 162.0 (thiazole-C$_4$), 182.6 (thiazole-C$_2$). MS m/z (%): 325.09 [M$^+$+1] (1.47), 324.05 [M$^+$] (2.77), 291.98 (4.41), 255.05 (4.63), 227.02 (9.96), 199.04 (15.82), 172.03 (6.34), 134.04 (14.00), 89.04 (34.81), 70.06 (100.00), 42.03 (68.78). Anal. Calcd. For C$_{13}$H$_{16}$N$_4$O$_2$S$_2$ (324.42): C, 48.13; H, 4.97; N, 17.27; S, 19.77. Found: C, 48.22; H, 4.86; N, 17.15; S, 19.80%.

2-(2-(4-Fluorobenzylidene)hydrazinyl)-4-(4-(pyrrolidin-1-ylsulfonyl)phenyl)thiazole (11)

1)   Procedure (A)

A mixture of phenacyl bromide derivative 2 (3.32 g, 0.01 mol) and 2-(4-fluorobenzylidene)hydrazinecarbo- thioamide (1.97 g, 0.01 mol) in ethanol (30 mL) was refluxed for 3h. The solid product which formed on heating collected by filtration and recrystallized from dioxane to give 11, (yield 70%).

2)   Procedure (B)

Amixture of 2-hydrazinylthiazole derivative 10 (3.24 g, 0.01 mol) and 4-fluorobenzaldehyde (1.24 g, 0.01 mol) in ethanol (20 mL) was refluxed for 2 h. The obtained product which formed was collected by filtration and recrystallized to give 11, mp and mixed mp determined with authentic sample gave no depression. Yellowish white solid, Yield, 75%; mp 185°C - 186°C. IR (KBr, cm^{-1}): v$_{max}$ = 3253 (NH), 3062 (CH aromatic), 2908 (CH aliphatic), 1571 (C=N), 1350, 1158 (SO$_2$), 1165 (C-F). ^{1}H NMR (DMSO-d$_6$): δ = 1.88 (t, 4H, CH$_2$-CH$_2$ pyrrolidine), 3.49 (t, 4H, CH$_2$-N-CH$_2$pyrrolidine), 7.52 (s, 1H, CH-thiazole), 7.40-8.10 (m, 9H, Ar-H + CH=N), 11.07 (s, 1H, NH exchangeable with D$_2$O). MS m/z (%): 431.10 [M$^+$+1] (3.02), 430.12 [M$^+$] (5.13), 429.18 (0.70), 383.11 (20.49), 319.14 (16.97), 309.10 (39.38), 265.08 (27.59), 250.09 (87.55), 222.06 (12.96), 175.07 (61.14), 146.05 (63.42), 135.07 (10.22), 125.05

(55.14), 117.10 (54.76), 108.07 (22.07), 104.09 (52.36), 95.07 (52.80), 70.09 (100.00), 42.08 (89.83). Anal. Calcd. For $C_{20}H_{19}FN_4O_2S_2$ (430.52): C, 55.80; H, 4.45; N, 13.01; S, 14.90. Found: C, 55.91; H, 4.33; N, 12.86; S, 14.84%.

General Procedure for the Formation of Compounds (12a,b)

A mixture of phenacyl bromide derivative 2 (3.32 g, 0.01 mol) and thioacetamide (0.75 g, 0.01 mol) and/or phenyl thiourea (1.52 g, 0.01 mol) in ethanol (40 mL) was refluxed for 2 h, the obtained product was collected by filtration and recrystallized to give 12a,b, respectively.

1) 2-Methyl-4-(4-(pyrrolidin-1-ylsulfonyl)phenyl)thiazole (12a)

Brown solid, Yield, 62%; mp 260°C - 261°C (dioxane). IR (KBr, cm^{-1}): v_{max} = 3070 (CH aromatic), 2962 (CH aliphatic), 1589 (C=N), 1339, 1161 (SO$_2$). ^{1}H NMR (DMSO-d$_6$): δ = 1.80 (t, 4H, CH$_2$-CH$_2$ pyrrolidine), 2.82 (s, 3H, CH$_3$), 3.23 (t, 4H, CH$_2$-N-CH$_2$ pyrrolidine), 7.31 (s, 1H, CH-thiazole), 7.91, 8.01 (dd, 4H, Ar-H, AB system, J = 8.43 Hz). ^{13}C NMR (DMSO-d$_6$): δ = 16.8 (CH$_3$), 29.2 (2C, CH$_2$-CH$_2$ pyrrolidine), 56.3 (2C, CH$_2$-N- CH$_2$ pyrrolidine), 99.5 (thiazole-C$_5$), [120.1 (2C), 130.6 (2C), 134.7, 140.1] (6ArC's), 159.4 (thiazole-C$_4$), 176.9 (thiazole-C$_2$). MS m/z (%): 308.08 [M$^+$] (7.24), 238.01 (27.13), 190.06 (36.39), 174.06 (100.00), 134.05 (19.79), 89.03 (41.39), 70.09 (89.05), 63.05 (8.53), 42.06 (30.91). Anal. Calcd. For $C_{14}H_{16}N_2O_2S_2$(308.42): C, 54.52; H, 5.23; N, 9.08; S, 20.79. Found: C, 54.44; H, 5.17; N, 9.21; S, 20.84%.

2) N-Phenyl-4-(4-(pyrrolidin-1-ylsulfonyl)phenyl)thiazol-2-amine (12b)

Yellow solid, Yield, 74%; mp 247°C - 248°C (ethanol/benzene). IR (KBr, cm^{-1}): v_{max} = 3184 (NH), 3091 (CH aromatic), 2995 (CH aliphatic), 1598 (C=N), 1337, 1157 (SO$_2$). ^{1}H NMR (DMSO-d$_6$): δ = 1.94 (t, 4H, CH$_2$-CH$_2$ pyrrolidine), 3.27 (t, 4H, CH$_2$-N-CH$_2$ pyrrolidine), 6.81-7.62 (m, 6H, Ar-H + CH-thiazole), 7.95, 8.33 (dd, 4H, Ar-H, AB system, J = 9.27 Hz), 8.97 (s, 1H, NH exchangeable with D$_2$O). MS m/z (%): 385.14 [M$^+$] (2.79), 308.10 (8.22), 238.05 (6.69), 190.07 (10.22), 174.08 (30.84), 133.06 (14.88), 91.09 (14.31), 89.07 (64.29), 70.11 (100.00), 55.09 (35.61), 42.09 (60.82). Anal. Calcd. For $C_{19}H_{19}N_3O_2S_2$ (385.50): C, 59.20; H, 4.97; N, 10.90; S, 16.64. Found: C, 59.15; H, 4.86; N, 10.77; S, 16.51%.

3-Oxo-3-(4-(pyrrolidin-1-ylsulfonyl)phenyl)propanenitrile (13)

A mixture of phenacyl bromide derivative 2 (3.32 g, 0.01 mol) and potassium cyanide (0.65 g, 0.01 mol) in ethanol (20 mL) was heated under reflux for 4 h, during the reflux period, a yellow crystalline solid was separated. The separated solid filtered off, washed with ethanol/water and recrystallized from ethanol to give the compound 13. Yellow solid, Yield, 46%; mp 105°C - 106°C. IR (KBr, cm^{-1}): v_{max} = 3074 (CH aromatic), 2983 (CH aliphatic), 2218 (C≡N), 1696 (C=O), 1362, 1156 (SO$_2$). ^{1}H NMR (DMSO-d$_6$): δ = 1.92 (t, 4H, CH$_2$-CH$_2$ pyrrolidine), 3.41 (t, 4H, CH$_2$-N-CH$_2$pyrrolidine), 3.67 (s, 2H, CH$_2$), 7.99, 8.20 (dd, 4H, Ar-H, AB system, J = 10.36 Hz). MS m/z (%): 279.06 [M$^+$+1] (1.27), 278.08 [M$^+$] (1.45), 244.13 (17.59), 227.04 (12.72), 197.03 (5.18), 158.03 (5.11), 107.05 (4.40), 91.06 (100.00), 65.05 (17.56). Anal. Calcd. For C$_{13}$H$_{14}$N$_2$O$_3$S (278.33): C, 56.10; H, 5.07; N, 10.06; S, 11.52. Found: C, 56.03; H, 4.88; N, 10.30; S, 11.43%.

3-(4-(Pyrrolidin-1-ylsulfonyl)phenyl)imidazo[1,2-a]pyridine (14)

A mixture of phenacyl bromide derivative 2 (3.32 g, 0.01 mol) and 2-aminopyridine (0.94 g, 0.01 mol) in ethanol (30 mL) was refluxed for 3 h. The solid product collected by filtration and recrystallized from acetic acid to give 14. Brown crystals, Yield, 32%; mp 300°C - 301°C. IR (KBr, cm^{-1}): v_{max} = 3055 (CH aromatic), 2969 (CH aliphatic), 1381, 1148 (SO$_2$). ^{1}H NMR (DMSO-d$_6$): δ = 1.79 (t, 4H, CH$_2$-CH$_2$ pyrrolidine), 3.52 (t, 4H, CH$_2$-N- CH$_2$ pyrrolidine), 6.87 - 7.53 (m, 4H, CH$_{2,6,7,8}$-imidazopyridine), 8.00, 8.41 (dd, 4H, Ar-H, AB system, J = 8.56 Hz), 8.50 (d, 1H, CH$_5$-imidazopyridine). MS m/z (%): 327.16 [M$^+$] (13.96), 263.15 (6.38), 258.06 (10.66), 209.11 (18.04), 193.12 (100.00), 167.11 (9.79), 140.09 (9.74), 97.17 (9.92), 89.08 (48.23), 78.06 (52.22), 70.10 (42.84), 42.07 (68.83). Anal. Calcd. For C$_{17}$H$_{17}$N$_3$O$_2$S (327.40): C, 62.36; H, 5.23; N, 12.83; S, 9.79. Found: C, 62.20; H, 5.18; N, 12.76; S, 9.86%.

5-(4-(Pyrrolidin-1-ylsulfonyl)phenyl)-7H-imidazo[2,1-c][1,2,4] triazole (15)

A mixture of phenacyl bromide derivative 2 (3.32g, 0.01 mol) and 4H-1,2,4-triazol-3-amine (0.84g, 0.01 mol) in ethanol (30 mL) was refluxed for 4 h. The solid product collected by filtration and recrystallized from dioxane to give 15. Brown crystals, Yield, 44%; mp 249°C - 250°C. IR (KBr, cm^{-1}): v_{max} = 3262 (NH), 3010 (CH aromatic), 2914 (CH aliphatic), 1340, 1174 (SO$_2$). ^{1}H NMR (DMSO-d$_6$): δ = 1.92 (t, 4H, CH$_2$-CH$_2$ pyrrolidine), 3.77 (t, 4H, CH$_2$-N-CH$_2$ pyrrolidine), 7.51 (s, 1H, CH$_6$-imidazotriazole), 7.88, 8.16 (dd, 4H, Ar-H, AB system, J = 8.53 Hz), 8.81 (s, 1H, CH$_3$-imidazotriazole), 12.01 (s, 1H, NH exchangeable with D_2O). MS m/z (%): 317.08 [M$^+$] (0.77), 308.06 (10.01), 253.07 (51.19), 239.07 (14.82), 193.09 (100.00), 158.06 (43.55), 139.06 (18.53), 119.08 (21.57), 111.05 (16.00), 92.97 (13.00), 78.08 (22.59), 72.14 (22.44), 44.08 (21.44). Anal. Calcd. For $C_{14}H_{15}N_5O_2S$ (317.37): C, 52.98; H, 4.76; N, 22.07; S, 10.10. Found: C, 52.86; H, 4.65; N, 21.79; S, 10.22%.

2-((4-Chlorophenyl)amino)-1-(4-(pyrrolidin-1-ylsulfonyl)phenyl) ethanone (16)

A Solution of phenacyl bromide derivative 2 (3.32 g, 0.01 mol) and 4-chloroaniline (1.52 g, 0.012 mol) in ethanol (30 mL) was heated under reflux for 3 h, after cooling the solid product which formed, was collected and recrystallized from ethanol to give 16. Yellow solid, Yield, 70%; mp 179°C - 180°C. IR (KBr, cm^{-1}): v$_{max}$ = 3201 (NH), 3030 (CH aromatic), 2924 (CH aliphatic), 1696 (C=O), 1369, 1141 (SO$_2$). ^{1}H NMR (DMSO-d$_6$): δ = 1.90 (t, 4H, CH$_2$-CH$_2$ pyrrolidine), 3.53 (t, 4H, CH$_2$-N-CH$_2$ pyrrolidine), 4.55 (s, 2H, CH$_2$), 6.54, 7.27 (dd, 4H, Ar-H, AB system, J = 8.24 Hz), 7.82, 8.12 (dd, 4H, Ar-H, AB system of benzenesulfonamide, J = 8.54 Hz), 9.51 (s, 1H, NH exchangeable with D$_2$O). ^{13}C NMR (DMSO-d$_6$): δ = 27.4 (2C, CH$_2$-CH$_2$ pyrrolidine), 53.8 (2C, CH$_2$-N- CH$_2$ pyrrolidine), 73.8 (CH$_2$), [110.3 (2C), 119.5, 124.0 (2C), 129.8 (2C), 131.9 (2C), 140.1, 144.9, 152.4] (12 ArC's), 184.1 (C=O). MS m/z (%): 380.67 [M$^+$+2] (6.34), 379.45 [M$^+$+1] (3.10), 378.36 [M$^+$] (50.21), 357.50 (3.82), 309.14 (3.10), 293.20 (11.91), 263.11 (9.74), 244.29 (11.61), 194.12 (100.00), 134.06 (11.83), 106.10 (78.12), 89.07 (73.31), 78.11 (50.34), 72.12 (21.33), 53.08 (39.24). Anal. Calcd. For $C_{18}H_{19}ClN_2O_3S$ (378.87): C, 57.06; H, 5.05; N, 7.39; S, 8.46. Found: C, 57.16; H, 4.94; N, 7.27; S, 8.33%.

2-Oxo-2-(4-(pyrrolidin-1-ylsulfonyl)phenyl)ethyl diethylcarbamo-dithioate (17)

A mixture of phenacyl bromide derivative 2 (3.32 g, 0.01 mol) and ammonium diethylcarbamodithioate (1.66 g, 0.01 mol) in ethanol (20 mL) was heated under reflux for 4 h, during the reflux period, a brown crystalline solid was separated. The separated solid filtered off, washed with ethanol/water and recrystallized from ethanol to give 17. Brown solid, Yield, 63%; mp 119°C - 120°C. IR (KBr, cm^{-1}): v$_{max}$ = 3025 (CH aromatic), 2983 (CH aliphatic), 1696 (C=O), 1345, 1164 (SO$_2$), 1287 (C=S). ^{1}H NMR (DMSO-d$_6$): δ = 1.17 (t, 6H, CH$_3$-CH$_2$), 1.96 (t, 4H, CH$_2$-CH$_2$ pyrrolidine), 3.33 (t, 4H, CH$_2$-N-CH$_2$ pyrrolidine), 3.87 (q, 4H, CH$_3$-CH$_2$), 4.89 (s, 2H, CH$_2$CO), 8.01, 8.52 (dd, 4H, Ar-H, AB system, J = 7.99 Hz). MS m/z (%): 400.15 [M$^+$] (3.09), 369.17 (6.93), 332.05 (4.41), 311.26 (5.49), 238.09 (51.90), 174.13 (20.29), 141.03 (22.52), 139.02 (73.23), 130.10 (17.02), 115.14 (14.00), 105.08 (20.54), 88.06 (82.45), 75.08 (31.74),

70.11 (74.64), 60.03 (100.00), 41.08 (40.75). Anal. Calcd. For $C_{17}H_{24}N_2O_3S_3$ (400.58): C, 50.97; H, 6.04; N, 6.99; S, 24.01. Found: C, 50.82; H, 5.96; N, 7.00; S, 24.13%.

General Procedure for the Formation of Compounds 20, 22

To a stirred solution of a suspension of finely powdered potassium hydroxide (0.56 g, 0.01 mol) in dry dimethylformamide (10 mL), ethyl cyanoacetate (1.13 g, 0.01 mol) and/or malononitrile (0.66 g, 0.01 mol) and then phenyl isothiocyanate (1.35 g, 0.01 mol) was add in portions. The reaction mixture was stirred at room temperature with phenacyl bromide derivative 2 (3.32 g, 0.01 mol)and left at room temperature for 3 h, then it was poured onto ice/water and acidified with 0.1 N HCl. The resulting precipitate filtered off, washed with water, dried and recrystallized to give 20 and 22, respectively.

1) Ethyl   2-cyano-2-(3-phenyl-4-(4-(pyrrolidin-1-ylsulfonyl)phenyl) thiazol-2(3H)-ylidene)acetate (20)

Brown crystals, Yield, 59%; mp 222°C - 223°C (ethanol). IR (KBr, cm^{-1}): v_{max} = 3079 (CH aromatic), 2964 (CH aliphatic), 2222 (C≡N), 1750 (C=O ester), 1371, 1150 (SO$_2$). ^{1}H NMR (DMSO-d$_6$): δ = 1.29 (t, 3H, CH$_3$-CH$_2$), 1.92 (t, 4H, CH$_2$-CH$_2$ pyrrolidine), 3.24 (t, 4H, CH$_2$-N-CH$_2$pyrrolidine), 4.20 (q, 2H, CH$_3$-CH$_2$), 6.73 (s, 1H, CH-thiazole), 6.90 - 8.62 (m, 9H, Ar-H). ^{13}C NMR (DMSO-d$_6$): δ = 15.9 (CH$_3$), 26.3 (2C, CH$_2$-CH$_2$ pyrrolidine), 57.7 (2C, CH$_2$-N-CH$_2$pyrrolidine), 63.2 (CH$_2$), 99.8, 109.4 (thiazole-C$_5$), 115.5 (C≡N), [123.1, 125.3 (2C), 127.9 (2C), 129.0 (2C), 130.9 (2C), 135.8, 138.2, 140.3] (12 ArC's), 149.4 (thiazole-C$_4$), 160.4 (C=O), 175.2 (thiazole-C$_2$). MS m/z (%): 482.11 [M$^+$+1] (2.64), 481.11 [M$^+$] (9.55), 452.08 (32.97), 318.04 (40.10), 274.04 (16.68), 241.08 (34.62), 238.04 (100.00), 214.07 (16.10), 174.09 (22.19), 142.06 (20.81), 105.04 (96.02), 93.05 (28.25), 77.03 (60.31). Anal. Calcd. For $C_{24}H_{23}N_3O_4S_2$(481.59): C, 59.86; H, 4.81; N, 8.73; S, 13.32. Found: C, 59.79; H, 4.68; N, 8.66; S, 13.51%.

2) 2-(3-Phenyl-4-(4-(pyrrolidin-1-ylsulfonyl)phenyl)thiazol-2(3H)- ylidene)malono-nitrile (22)

Yellowish white crystals, Yield, 48%; mp 260°C - 261°C (ethanol/ benzene). IR (KBr, cm^{-1}): v_{max} = 3058 (CH aromatic), 2969 (CH aliphatic), 2225, 2217 (2C≡N), 1371, 1148 (SO$_2$). ^{1}H NMR (DMSO-d$_6$): δ = 1.84 (t, 4H, CH$_2$-CH$_2$ pyrrolidine), 3.26 (t, 4H, CH$_2$-N-CH$_2$ pyrrolidine), 7.21 (s, 1H, CH-thiazole), 6.84-7.65 (m, 9H, Ar-H). MS m/z (%): 434.09 [M$^+$] (12.25), 389.13 (15.04), 382.17 (17.06), 331.19 (18.87), 322.17 (42.95), 320.10 (12.54), 271.29 (35.79), 268.13 (32.23), 253.10 (64.20), 191.09 (58.03), 172.07 (48.74), 151.07 (100.00), 141.07 (62.50), 127.05 (49.90), 97.10 (57.39), 81.07 (70.40), 63.06

(49.12). Anal. Calcd. For $C_{22}H_{18}N_4O_2S_2$(434.53): C, 60.81; H, 4.18; N, 12.89; S, 14.76. Found: C, 60.76; H, 4.09; N, 12.92; S, 14.81%.

5-(4-(Pyrrolidin-1-ylsulfonyl)phenyl)imidazo[2,1-b]thiazole (23)

A mixture of phenacyl bromide derivative 2 (3.32 g, 0.01 mol) and 2-aminothiazole (1.00 g, 0.01 mol) in ethanol (30 mL) was refluxed for 2 h. The product collected and recrystallized from acetic acid to give 23. White solid, Yield, 82%; mp 299°C - 300°C. IR (KBr, cm^{-1}): v_{max} = 3103 (CH aromatic), 2947 (CH aliphatic), 1358, 1151 (SO$_2$). ^{1}H NMR (DMSO-d$_6$): δ = 1.94 (t, 4H, CH$_2$-CH$_2$pyrrolidine), 3.30 (t, 4H, CH$_2$-N-CH$_2$ pyrrolidine), 7.42 - 8.11 (m, 6H, Ar-H + CH=CH of thiazole), 8.33 (s, 1H, CH-imidazole). ^{13}C NMR (DMSO-d$_6$): δ = 18.7 (CH$_3$), 20.8 (CH$_3$), 26.9 (2C, CH$_2$-CH$_2$pyrrolidine), 40.4 (CH$_2$), 71.3 (2C, CH$_2$-N-CH$_2$ pyrrolidine), 107.4, 113.5 (C≡N), 120.4, [124.5 (2C), 129.6 (2C), 134.2, 142.9] (6ArC's), 153.1, 159.7, 165.1, 192.9 (C=O). MS m/z (%): 333.50 [M$^+$] (4.91), 329.18 (1.66), 277.33 (5.39), 251.08 (12.25), 215.10 (52.15), 178.08 (7.58), 127.12 (100.00), 104.05 (7.96), 91.07 (93.91). Anal. Calcd. For $C_{15}H_{15}N_3O_2S_2$ (333.43): C, 54.03; H, 4.53; N, 12.60; S, 19.23. Found: C, 53.96; H, 4.62; N, 12.56; S, 19.18%.

3-(4-(Pyrrolidin-1-ylsulfonyl)phenyl)benzo [d]imidazo [2,1-b] thiazole (24)

A mixture of phenacyl bromide derivative 2 (3.32 g, 0.01 mol) and 2-aminobenzothiazole (1.50 g, 0.01 mol) in ethanol (30 mL) was refluxed for 4 h. The product collected and recrystallized from dioxane to give 24. Yellow crystals, Yield, 81%; mp 318°C - 319°C. IR (KBr, cm^{-1}): v_{max} = 3062 (CH aromatic), 2959 (CH aliphatic), 1371, 1144 (SO$_2$). ^{1}H NMR (DMSO-d$_6$): δ = 1.74 (t, 4H, CH$_2$-CH$_2$ pyrrolidine), 3.41 (t, 4H, CH$_2$-N-CH$_2$ pyrrolidine), 7.59 - 8.22 (m, 9H, Ar-H + CH-imidazole). MS m/z (%): 383.15 [M$^+$] (11.39), 374.12 (5.01), 331.08 (12.14), 313.14 (6.17), 301.12 (12.68), 248.07 (24.67), 197.04 (13.74), 170.05 (20.86), 134.06 (19.80), 111.34 (67.49), 89.07 (80.59), 70.10 (51.61), 41.07 (100.00). Anal. Calcd. For $C_{19}H_{17}N_3O_2S_2$ (383.49): C, 59.51; H, 4.47; N, 10.96; S, 16.72. Found: C, 59.42; H, 4.31; N, 10.85; S, 16.90%.

4,6-Dimethyl-2-((2-oxo-2-(4-(pyrrolidin-1-ylsulfonyl)phenyl) ethyl)thio)nicotinonitrile (25)

A solution of phenacyl bromide derivative 2 (3.32 g, 0.01 mol) in ethanol (50 mL) and 2-mercapto-4,6-dime- thylnicotinonitrile (1.64 g, 0.01 mol) was refluxed for 3 h. The solid product, which formed on heating, collected by filtration and recrystallized from ethanol to give 25. Yellow solid, Yield, 33%;

mp 166°C - 167°C. IR (KBr, cm^{-1}): v_{max} = 3090 (CH aromatic), 2965 (CH aliphatic), 2225 (C≡N), 1696 (C=O), 1566 (C=N), 1373, 1139 (SO$_2$). ^{1}H NMR (DMSO-d$_6$): δ = 1.78 (t, 4H, CH$_2$-CH$_2$ pyrrolidine), 2.43 (s, 3H, CH$_3$), 2.53 (s, 3H, CH$_3$), 3.35 (t, 4H, CH$_2$-N-CH$_2$ pyrrolidine), 4.57 (s, 2H, CH$_2$), 7.51 (s. 1H, CH-pyridine), 7.90, 8.02 (dd, 4H, Ar-H, AB system, J = 8.32 Hz). MS m/z (%): 415.39 [M$^+$] (1.90), 388.15 (3.71), 343.50 (1.15), 251.63 (9.63), 242.53 (9.43), 188.34 (13.51), 182.27 (100.00), 163.14 (2.74), 138.60 (7.92), 81.12 (6.07), 50.31 (77.01). Anal. Calcd. For C$_{20}$H$_{21}$N$_3$O$_3$S$_2$ (415.53): C, 57.81; H, 5.09; N, 10.11; S, 15.43. Found: C, 57.92; H, 5.13; N, 10.22; S, 15.30%.

N-Phenyl-5-(4-(pyrrolidin-1-ylsulfonyl)phenyl)-6H-1,3,4-thiadia-zin-2-amine (27)

A solution of phenacyl bromide derivative 2 (3.32 g, 0.01 mol) in ethanol (50 mL) and N-phenylhydrazinecar- bothioamide (1.67 g, 0.01 mol) was refluxed for 3 h. The solid product that formed on heating collected by filtration and recrystallized from ethanol/benzene to give 27. White solid, Yield, 50%; mp 191°C - 193°C. IR (KBr, cm^{-1}): v_{max} = 3115 (NH), 3062 (CH aromatic), 2910 (CH aliphatic), 1583 (C=N), 1357, 1161 (SO$_2$). ^{1}H NMR (DMSO-d$_6$): δ = 1.77 (t, 4H, CH$_2$-CH$_2$ pyrrolidine), 3.39 (t, 4H, CH$_2$-N-CH$_2$ pyrrolidine), 4.32 (s, 2H, CH$_2$-thiadiazine), 6.43 - 8.11 (m, 9H, Ar-H), 9.68 (s, 1H, NH exchangeable with D$_2$O). ^{13}C NMR (DMSO-d$_6$): δ = 23.9 (2C, CH$_2$-CH$_2$ pyrrolidine), 29.8 (CH$_2$ thiadiazin-C$_6$), 72.3 (2C, CH$_2$-N-CH$_2$ pyrrolidine), [119.0 (2C), 123.8, 125.4 (2C), 127.1 (2C), 130.7 (2C), 135.9, 138.0, 144.4] (12 ArC's), 151.1 (thiadiazin-C$_2$), 168.2 (thiadiazin-C$_5$). MS m/z (%): 400.17 [M$^+$] (2.08), 332.08 (5.74), 313.14 (3.63), 299.14 (23.94), 284.12 (4.11), 253.11 (6.17), 223.05 (4.94), 197.05 (20.47), 170.05 (9.90), 133.05 (23.26), 91.08 (15.20), 77.07 (44.45), 70.09 (69.47), 41.07 (100.00). Anal. Calcd. For C$_{19}$H$_{20}$N$_4$O$_2$S$_2$ (400.52): C, 56.98; H, 5.03; N, 13.99; S, 16.01. Found: C, 56.76; H, 5.21; N, 14.10; S, 15.92%.

General Procedure for the Formation of Compounds 30, 31

A mixture of phenacyl bromide derivative 2 (3.32 g, 0.01 mol) and o-phenylenediamine (1.08 g, 0.01 mol) and/or o-aminothiophenol (1.25 g, 0.01 mol) in ethanol (40 mL) was refluxed for 5 h. The solid product, which formed on heating, collected and recrystallized to give 30 and 31, respectively.

1)   2-(4-(Pyrrolidin-1-ylsulfonyl)phenyl)quinoxaline (30)

Brown solid, Yield, 92%; mp > 360°C (DMF). IR (KBr, cm^{-1}): v_{max} = 3084 (CH aromatic), 2933 (CH aliphatic), 1342, 1160 (SO$_2$). ^{1}H NMR (DMSO-d$_6$): δ = 1.91 (t, 4H, CH$_2$-CH$_2$ pyrrolidine), 3.57 (t, 4H, CH$_2$-N-CH$_2$ pyrrolidine), 7.67 - 8.50 (m, 8H, Ar-H), 8.99 (s, 1H, CH-quinoxaline). ^{13}C NMR (DMSO-d$_6$):

$\delta = 20.9$ (2C, CH_2-CH_2 pyrrolidine), 62.7 (2C, CH_2-N-CH_2 pyrrolidine), [121.0 (2C), 124.4 (2C), 126.8 (2C), 130.4 (2C), 139.3, 140.5, 141.2 (2C), 153.0, 160.2] (14 ArC's + quinoxaline). MS m/z (%): 339.18 [M$^+$] (33.87), 327.15 (18.08), 304.22 (15.92), 269.12 (7.62), 241.10 (15.33), 200.09 (12.35), 193.11 (80.38), 117.10 (11.14), 91.09 (100.00), 78.09 (66.12). Anal. Calcd. For $C_{18}H_{17}N_3O_2S$ (339.41): C, 63.70; H, 5.05; N, 12.38; S, 9.45. Found: C, 63.63; H, 4.82; N, 12.17; S, 9.50%.

2)   3-(4-(Pyrrolidin-1-ylsulfonyl)phenyl)-4H-benzo[b][1,4]thiazine (31)

Yellow solid, Yield, 77%; mp 330°C - 331°C (acetic acid). IR (KBr, cm^{-1}): $v_{max} = 3297$ (NH), 3088 (CH aromatic), 2965 (CH aliphatic), 1346, 1171 (SO$_2$). ^{1}H NMR (DMSO-d$_6$): $\delta = 1.96$ (t, 4H, CH_2-CH_2 pyrrolidine), 3.31 (t, 4H, CH_2-N-CH_2 pyrrolidine), 5.58 (s, 1H, CH-thiazine), 6.00 - 7.82 (m, 8H, Ar-H), 8.66 (br, 1H, NH exchangeable with D$_2$O). MS m/z (%): 358.08 [M$^+$] (2.89), 347.08 (7.75), 254.09 (8.25), 223.10 (14.94), 174.11 (9.38), 134.08 (30.08), 121.09 (13.79), 109.10 (14.10), 105.09 (19.58), 91.09 (36.78), 77.09 (62.50), 70.11 (100.00), 44.06 (94.89), 42.11 (80.17), 41.10 (55.29). Anal. Calcd. For $C_{18}H_{18}N_2O_2S_2$ (358.48): C, 60.31; H, 5.06; N, 7.81; S, 17.89. Found: C, 60.28; H, 4.88; N, 7.72; S, 17.73%.

3-(4-(Pyrrolidin-1-ylsulfonyl)phenyl)benzo[4,5]imidazo[2,1-b] thiazole (33)

A mixture of phenacyl bromide derivative 2 (3.32 g, 0.01 mol) and 1H-benzo[d] imidazole-2-thiol (1.50g, 0.01 mol) in ethanol (40 mL) was heated under reflux for 4 h, during the reflux period, a yellow crystalline solid was separated. The separated solid filtered off, washed with ethanol and recrystallized from acetic acid to give 33. Yellow solid, Yield, 56%; mp 342°C - 343°C. IR (KBr, cm^{-1}): $v_{max} = 3086$ (CH aromatic), 2944 (CH aliphatic), 1381, 1137 (SO$_2$). ^{1}H NMR (DMSO-d$_6$): $\delta = 1.97$ (t, 4H, CH_2-CH_2 pyrrolidine), 3.31 (t, 4H, CH_2-N-CH_2 pyrrolidine), 7.22 - 8.56 (m, 9H, Ar-H + CH-thiazole). MS m/z (%): 383.26 [M$^+$] (0.31), 310.18 (13.60), 309.16 (20.52), 297.15 (15.29), 254.12 (40.57), 241.13 (8.56), 183.17 (31.01), 174.10 (22.15), 137.07 (23.18), 105.08 (21.30), 91.07 (100.00), 67.07 (31.40), 65.07 (31.76), 57.10 (62.06), 40.17 (50.06). Anal. Calcd. For $C_{19}H_{17}N_3O_2S_2$ (383.49): C, 59.51; H, 4.47; N, 10.96; S, 16.72. Found: C, 59.47; H, 4.31; N, 11.01; S, 16.63%.

5-Methyl-2-(4-(4-(pyrrolidin-1-ylsulfonyl)phenyl)thiazol-2-yl)-1H-pyrazol-3(2H)-one (34)

A solution of phenacyl bromide derivative 2 (3.32 g, 0.01 mol) in ethanol (30 mL) and 3-methyl-5-oxo-4,5-di- hydro-1H-pyrazole-1-carbothioamide (1.57

g, 0.01 mol) was refluxed for 2 h. The solid obtained after cooling collected and recrystallized from ethanol/benzene to give 34. Yellow solid, Yield, 88%; mp 219°C - 220°C. IR (KBr, cm^{-1}): v$_{max}$ = 3210 (NH), 3085 (CH aromatic), 2930 (CH aliphatic), 1672 (C=O), 1345, 1164 (SO$_2$). ^{1}H NMR (DMSO-d$_6$): δ = 1.85 (t, 4H, CH$_2$-CH$_2$pyrrolidine), 2.26 (s, 3H, CH$_3$), 3.73 (t, 4H, CH$_2$-N-CH$_2$ pyrrolidine), 5.27 (s, 1H, CH-pyrazole), 7.63 (s, 1H, CH-thiazole), 8.07, 8.40 (dd, 4H, Ar-H, AB system, J = 8.42 Hz), 8.44 (s, 1H, NH exchangeable with D$_2$O). ^{13}C NMR (DMSO-d$_6$): δ = 20.1 (CH$_3$), 29.5 (2C, CH$_2$-CH$_2$ pyrrolidine), 64.7 (2C, CH$_2$-N-CH$_2$ pyrrolidine), 87.4, 93.9, [110.5 (2C), 121.6 (2C), 129.2, 137.4] (6ArC's), 147.8, 153.9, 160.0, 189.2. MS m/z (%): 390.17 [M$^+$] (5.49), 347.11 (26.65), 313.11(14.59), 269.07 (9.33), 173.08 (12.86), 134.06 (52.58), 91.09 (23.50), 89.08 (57.48), 77.07 (68.81), 70.11 (100.00), 44.05 (87.15), 42.10 (73.36). Anal. Calcd. For C$_{17}$H$_{18}$N$_4$O$_3$S$_2$ (390.48): C, 52.29; H, 4.65; N, 14.35; S, 16.42. Found: C, 52.15; H, 4.54; N, 14.22; S, 16.57%.

4-((4-Chlorophenyl)diazenyl)-5-methyl-2-(4-(4-(pyrrolidin-1-yl-sulfonyl)phenyl)- thiazol-2-yl)-1H-pyrazol-3(2H)-one (35)

1)   Procedure (A)

To a cold solution of 34 (3.90 g, 0.01 mol) in pyridine was added 4-chlorobenzenediazonium chloride (0.012 mol) (prepared by diazotization of 4-chloroaniline (1.52 g, 0.012 mol) in concentrated HCl (6 mL) with sodium nitrite (0.69 g in 5 mL H$_2$O) at 0°C) portion wise over 30 min. with constant stirring. After complete addition, the reaction mixture was stirred for a further 3 h at 0°C, the solid product was filtered off, washed with water, dried and recrystallized from ethanol/benzene to give 35, (yield 87%).

2)   Procedure (B)

A mixture of phenacyl bromide derivative 2 (3.32 g, 0.01 mol) and 4-((4-chlorophenyl)diazenyl)-3-methyl-   5-oxo-4,5-dihydro-1H-pyrazole-1-carbothioamide (2.95 g, 0.01 mol) in ethanol (30 mL) was refluxed for 1h. The obtained product collected and recrystallized to give 35, mp and mixed mp determined with authentic sample gave no depression. Brown solid, Yield, 91%; mp 250°C - 251°C. IR (KBr, cm^{-1}): v$_{max}$ = 3165 (NH), 3053 (CH aromatic), 2981 (CH aliphatic), 1666 (C=O), 1346, 1147 (SO$_2$), 715 (C-Cl). ^{1}H NMR (DMSO-d$_6$): δ = 1.80 (t, 4H, CH$_2$-CH$_2$ pyrrolidine), 2.61 (s, 3H, CH$_3$), 3.30 (t, 4H, CH$_2$-N-CH$_2$ pyrrolidine), 7.25, 7.49 (dd, 4H, Ar-H, AB system, J = 7.84 Hz), 7.69 (s, 1H, CH-thiazole), 7.92, 8.25 (dd, 4H, Ar-H, AB system of benzenesulfonamide, J = 8.63 Hz), 8.59 (s, 1H, NH exchangeable with D$_2$O). MS m/z (%): 528.15 [M$^+$] (0.11), 499.14 (15.40), 452.11 (19.31), 318.06 (31.74), 297.12 (66.11), 254.08 (95.92), 218.08 (13.99), 137.02 (41.06), 105.05 (48.74), 77.06 (72.28), 70.08 (100.00), 44.02 (57.38), 43.08 (55.28), 42.06

(54.04), 41.05 (49.51). Anal. Calcd. For $C_{23}H_{21}ClN_6O_3S_2$ (529.03): C, 52.22; H, 4.00; N, 15.89; S, 12.12. Found: C, 52.16; H, 4.13; N, 15.74; S, 12.06%.

DOCKING AND MOLECULAR MODELING CALCULATIONS

Materials

Docking and molecular modeling calculations were carried out in the department of pharmaceutical chemistry, Faculty of pharmacy, Alexandria University. All the molecular studies were carried out on an Intel Pentium 1.6 GHz processor, 512 MB memory with windows XP operating system using Molecular Operating Environment (MOE 2005.06; Chemical Computing Group, Montreal, Canada) as the computational software. All the minimizations were performed with MOE until a RMSD gradient of 0.05 K Cal/mol·Å with MMFF94X force field and the partial charges were automatically calculated.

General Methodology

The coordinates of the X-ray crystal structure of methotrexate (MTX) bound to dihydrofolate reductase (DHFR) enzyme (PDB ID: 4DFR) were obtained from Protein Data Bank (PDB ID: 1BID). Enzyme structures were checked for missing atoms, bonds and contacts. Hydrogen atoms were added to the enzyme structure. Water molecules and bound ligands were manually deleted. The ligand molecules were constructed using the builder molecule and were energy minimized. The active site was generated using the MOE-Alpha site finder. Dummy atoms were created from the obtained alpha spheres. Ligands were docked within the dihydrofolate reductase active sites using the MOE-Dock with simulated annealing used as the search protocol and MMFF94X molecular mechanics force field for 8000 interactions. The lowest energy conformation selected and subjected to an energy minimization using MMFF94X force field.

Docking on the Active Site of Dihydrofolate Reductase (DHFR)

The recent determination of the three dimensional co-crystal structure of dihydrofolate reductase complexed with the potent inhibitor, methotrexate (MTX) (PDB ID: 4DFR) has led to the development of a model for the topography of the binding site of dihydrofolate reductase.

In Vitro Anticancer Screening

Cytotoxicity activity was measured in vitro for the newly synthesized compounds using the Sulfo-Rhodamine-B stain (SRB) assay [23] . Cells were plated in 96-multiwell micro titer plates (10^4cells/well) for 24 h before treatment with the compound(s) to allow attachment of cells to the wall of the plate. Test compounds dissolved in DMSO and diluted with saline to the appropriate volume. Different concentrations of the compound under test (50, 25, 12.5, 6.25 and 3.125 mg/mL) were added to the cell monolayer. Triplicate wells were prepared for each individual dose. Monolayer cells were incubated with the compound(s) for 48 h at 37°C in an atmosphere of 5% CO_2. After 48 h cells were fixed, washed and stained for 30 min with 0.4% (wt/vol) with SRB dissolved in 1% acetic acid. Excess unbound dye was removed by four washes with 1% acetic acid and attached stain was recovered with Tris-EDTA buffer. Color intensity was measured in an ELISA reader. The relation between surviving fraction and drug concentration was plotted to obtain the survival curve for breast tumor cell after the specified time [23]. The molar concentration required for 50% inhibition of cell viability (IC_{50}) was calculated and the results presented in (Table 1). The significant differences in the compounds' cytotoxicity were supported by the results of the selectivity index (SI), which is the ratio of the concentration that causes 50% death in

Table 1: Cytotoxicity of the newly synthesized compounds against human liver hepatocellular carcinoma cell line (HepG2)[a] and mammalian cells of African green monkey kidney cell line (VERO)[a].

Comp. No.	IC_{50}[b] (µg/mL)	IC_{50}[b] (µM)	CC_{50}[c] (µg/mL)	CC_{50}[c] (µM)	SI[d]
1	20.16	79.580	48.53	191.58	02.41
2	40.32	121.37	30.22	90.970	00.75
3	16.77	51.370	50.12	153.54	02.99
4	4.020	7.1800	78.10	139.53	19.43
5	50.13	162.02	27.95	90.330	00.56
6	45.16	119.96	28.14	74.750	00.62
8	3.320	7.9900	110.26	265.37	33.21
10	73.64	226.99	25.33	78.080	00.34
11	3.540	8.2200	107.92	250.67	30.49
12a	7.010	22.730	72.10	233.77	10.29
12b	>100	259.40	20.12	52.190	00.20
13	22.01	79.080	46.12	165.70	02.10
14	38.11	116.40	35.19	107.48	00.92
15	70.26	221.38	27.01	85.110	00.38
16	37.99	100.27	70.32	185.60	01.85
17	23.88	59.610	45.00	112.34	01.88
20	13.11	27.220	50.98	105.86	03.89
22	4.990	11.480	73.94	170.16	14.82
23	37.10	111.27	39.76	119.25	01.07
24	13.89	36.220	64.15	167.28	04.62
25	90.12	216.88	74.46	179.19	00.83
27	10.92	27.260	67.16	167.68	06.15
30	97.52	287.32	51.46	151.62	00.53
31	>100	278.96	33.10	92.330	00.33
33	9.620	25.090	70.19	183.03	07.30
34	30.26	77.490	40.17	102.87	01.33
35	10.11	19.110	69.44	131.26	06.87
MTX	15.26	33.610	63.17	139.14	04.14

[a]Mean of three results obtained from three experiments. [b]IC_{50} value: Concentration causing 50% inhibition of HepG2 cell viability. [c]CC_{50} value: Concentration causing 50% inhibition of VERO cell viability. [d]SI value: selective index = CC_{50}(mg/mL)/IC_{50}(mg/mL).

African green monkey kidney (VERO) (CC_{50}) compared to the concentration that causes 50% death in human liver hepatocellular carcinoma cell line (HepG2) (IC_{50}) [24] -[26] (Table 1).

RESULTS AND DISCUSSION

Chemistry

Treatment of 1-(4-(pyrrolidin-1-ylsulfonyl)phenyl)ethanone (1) with bromine in a mixture of dioxane/diethy- lether afforded the 2-bromo-1-(4-(pyrrolidin-1-ylsulfonyl)phenyl)ethanone (2) in a good yield, (Scheme 1). The IR spectrum of compound 2 showed strong absorption band at $v = 1707$ cm^{-1} assignable to ketonic carbonyl group. Other important bands revealed at $v = 1336$ and 1161 cm^{-1} characterized for sulfonyl group. The ^{1}H NMR spectrum showed two triplet signals at $\delta = 1.99$ and 3.27 ppm corresponding to pyrrolidine protons, and a singlet signal at $\delta = 4.56$ ppm due to active methylene of bromoacetyl moiety. Other important signal appeared at $\delta = 7.84$ and 8.22 ppm due to aromatic protons. Furthermore, the ^{13}C NMR spectrum of compound 2 displayed two important signals at $\delta = 32.7$ and 193.5 ppm corresponding to the active methylene of bromoacetyl and ketonic carbonyl carbons, respectively. The mass spectrum of compound 2 revealed molecular ion peaks at m/z = 331 and 333 reflecting the isotopes of bromine. Condensation of 1 with thiosemicarbazide gave the corresponding thiosemicarbazone derivative 3, which when reacted with phenacyl bromide derivative 2 afforded the corresponding thiazole derivative 4, which exhibited singlet signal in ^{1}H NMR due to CH-thiazole at $\delta = 7.28$ ppm and (D_2O exchangeable) signal at $\delta = 8.87$ ppm due to NH proton (Scheme 1).

2-Aminothiazole derivative 5, was synthesized by two different methods, either starting with an acetophenone derivative 1 (Method 1) or phenacyl bromide derivative 2 (Method 2). Method 1, which involves the reaction of 1 and thiourea in the presence of equivalent amount of iodine. The yield was low, furthermore, iodine had to be recycled because of it is the pollution problems. In order to overcome these drawbacks, the second method was employed. When phenacyl bromide derivative 2, was reacted with thiourea in ethanol (Method 2), the yield could be raised to 90% and the reaction time was decreased (Scheme 2). The IR spectrum of 5 showed, two bi-forked characteristic absorption bands at $v = 3384$ and 3334 cm^{-1} assignable to amino group. Its ^{1}H NMR spectrum revealed a single signal of one proton appeared in aromatic region at $\delta = 6.99$ ppm, corresponding to CH-thiazole, and (D_2O exchangeable) singlet at $\delta = 7.45$ ppm, corresponding to amino protons. The ^{13}C NMR spectrum of compound 5 revealed nine carbon types; for thirteen carbon atoms; the most important signals appeared at $\delta = 105.2$ and 170.3 ppm corresponding to thiazole-C5 and thiazole-C2, respectively. The mass spectrum of 5 showed a molecular ion peak at m/z = 309. The investigation was extended to include the behavior of 2-aminothiazole derivative 5 towards some electrophiles. Thus, treatment of 5 with ethyl cyanoacetate gave acyclic cyanoacetamide derivative 6, rather than the expected cyclic product of thiazolo[3,2-a]pyrimidine derivative 7. The obtained product was established based on elemental analysis and spectral data. Thus, IR spectrum of 6 revealed absorption bands at $v = 3273$, 2220 and 1696 cm^{-1} due to NH, cyano and carbonyl groups, respectively. ^1H NMR spectrum showed singlet signal at $\delta = 4.20$ ppm due to active methylene protons, and (D_2O exchangeable) signal at $\delta = 9.15$ ppm due to NH proton. On other hand, condensation of 5 with 4-flourobenzaldehyde in boiling ethanol gave the corresponding 4-fluorobenzylidene derivative 8 (Scheme 2).

Interaction of phenacyl bromide derivative 2 with thiosemicarbazide afforded 2-hydrazinyl thiazole derivative 10, instead of 2-aminothiadiazine derivative 9, the appearance of NH absorption band at $v = 3269$ cm^{-1}, in IR spectrum, and at $\delta = 10.35$ ppm, in ^1H NMR spectrum, supported the structure 10 and ruled out the other possible structure 9. Cyclocondensation of 2 with 4-fluorobenzylidenethiosemicarbazide gave thiazole derivative 11, an equivocal support for structure 11 was achieved via its synthesis through condensation of 2-hydrazinyl thiazole derivative 10 with 4-fluorobenzaldehyde in refluxing ethanol (Scheme 3).

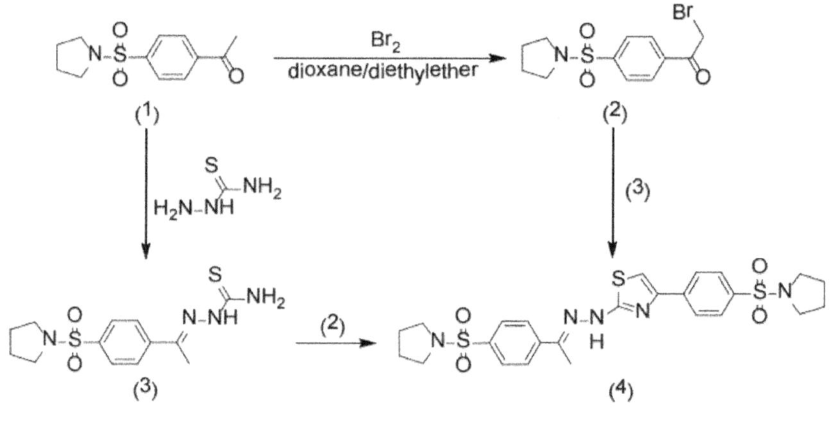

Scheme 1: Synthesis of compounds 1-4.

Cyclocondensation of 2 with thioacetamide derivatives namely thioacetamide and phenylthiourea gave thiazole derivatives 12a,b, respectively (Scheme 4). Thiazole derivatives 12a,b were established on the basis of elemental analysis and spectral data. Thus, IR spectrum of 12a lacked the absorption band of carbonyl function and revealed absorption bands at v = 1339, 1161 cm^{-1} due to sulfonyl group. However, ^{1}H NMR spectrum of 12b showed singlet signal at δ = 8.97 ppm due to NH proton, which discharged with D$_2$O.

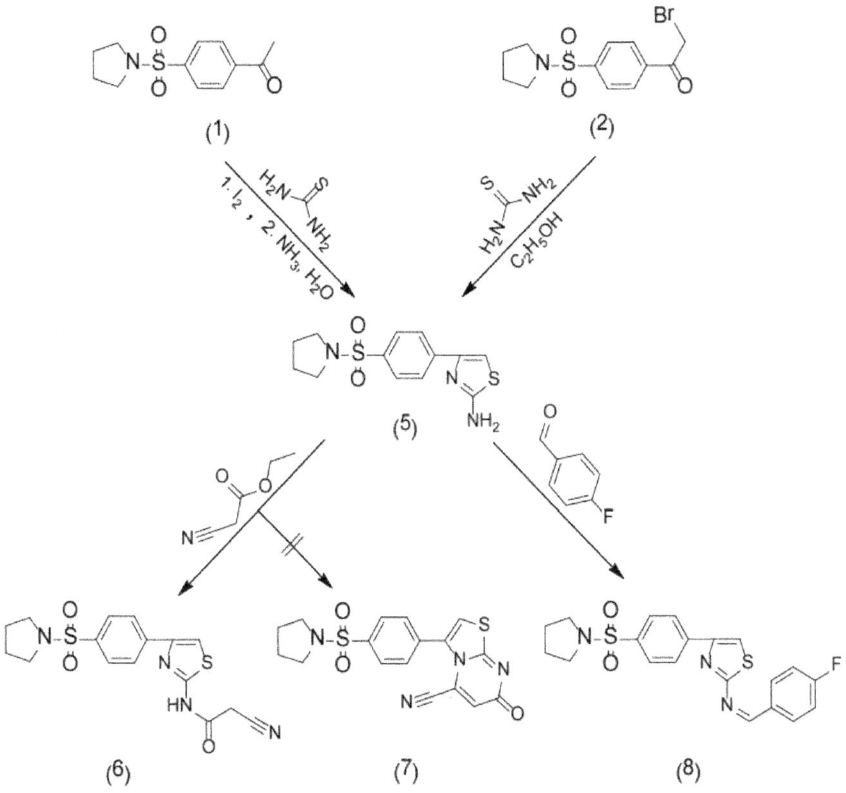

Scheme 2: Synthesis of compounds 5,6,8.

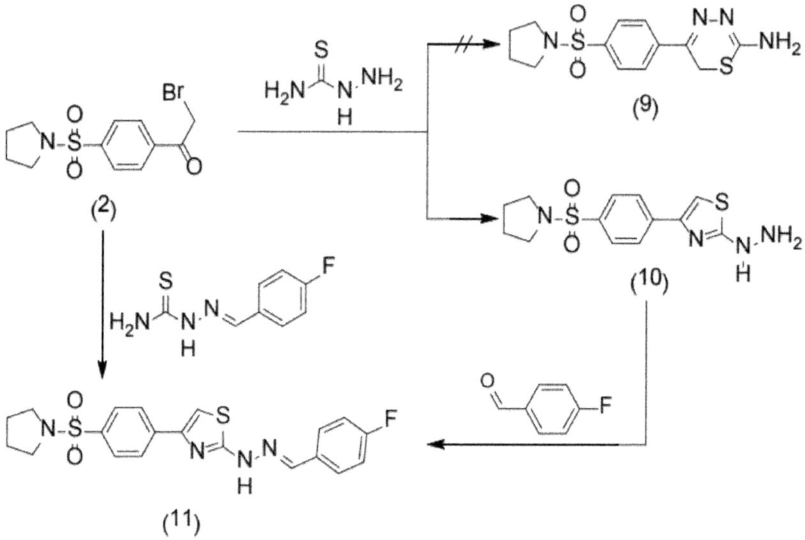

Scheme 3: Synthesis of compounds 10,11.

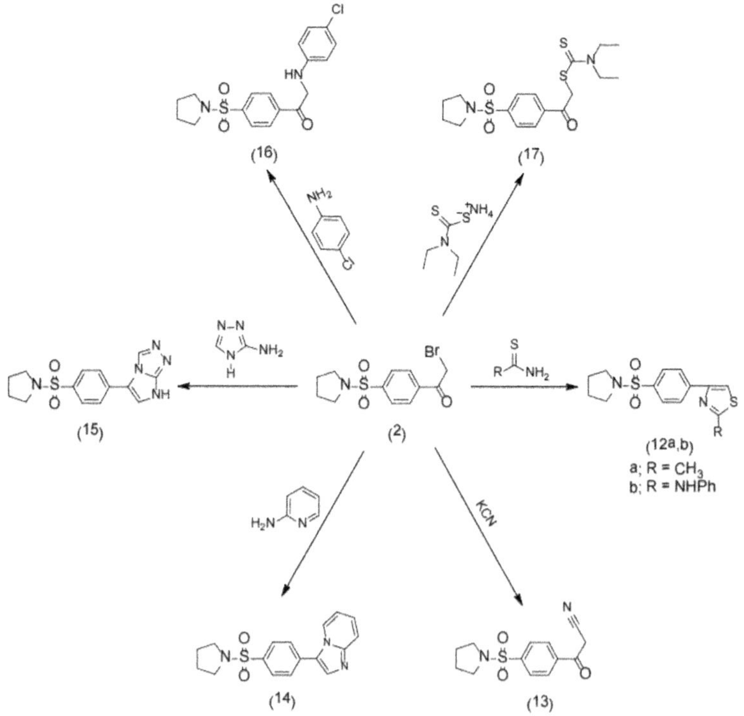

Scheme 4: Synthesis of compounds 12a,b-17.

Also, compound 2 reacted with potassium cyanide in refluxing ethanol to afford acyclic product identified as 3-oxo-3-(4-(pyrrolidin-1-ylsulfonyl)phenyl) propanenitrile (13) which confirmed by elemental analysis and spectral data. Thus, IR spectrum revealed an absorption band at $v = 2218$ cm^{-1}corresponding to cyano group. ^{1}H NMR spectrum showed a singlet signal at $\delta = 3.67$ ppm corresponding to active methylene protons. Additionally, interaction of phenacyl bromide derivative 2 with 2-aminopyridine and 3-amino-1,2,4-triazole afforded fused imidazo derivatives 14, 15, respectively. IR spectrum of 14 lacked the absorption band of carbonyl function of bromoacetyl moiety and its mass spectrum was compatible with molecular formula $C_{17}H_{17}N_3O_2S$ (M$^+$: 327). ^{1}H NMR spectrum of 15 showed two triplets signals at $\delta = 1.92$ and 3.77 ppm corresponding to pyrrolidine protons. Besides, singlet signals at $\delta = 7.51$ and 8.81 ppm due to CH-6 and CH-3, respectively, of imidazotriazole ring, in addition to, D$_2$O exchangeable signal at $\delta = 12.01$ ppm due to NH proton. Aligned with the aim of synthesis of different substituted pyrrolidine benzenesulfonamide, compound 2 was reacted either with 4-chloroaniline or ammonium diethylcarbamodithioate to afford compounds 16 and 17, respectively (Scheme 4).

Treatment of a solution of ethyl acetoacetate in DMF with phenyl isothiocyanate in the presence of potassium hydroxide, at room temperature followed by the addition of an equimolar amount of phenacyl bromide derivative 2 afforded only one isolable product (TLC) for which three proposed structures 18, 19 or 20 seemed possible (Scheme 5). Structures 18 and 19 were ruled out on the basis of ^{1}H NMR spectrum of the isolated product. Thus, ^{1}H NMR spectrum of 20 showed singlet signal at $\delta = 6.73$ ppm due to CH-thiazole. On the other hand, when potassium salt of malononitrile was treated with 2 furnished only one isolable product (TLC) for which two proposed structures 21 or 22 seemed possible (Scheme 5). Structure 21 was ruled out based on IR, ^{1}H NMR and mass spectral data.

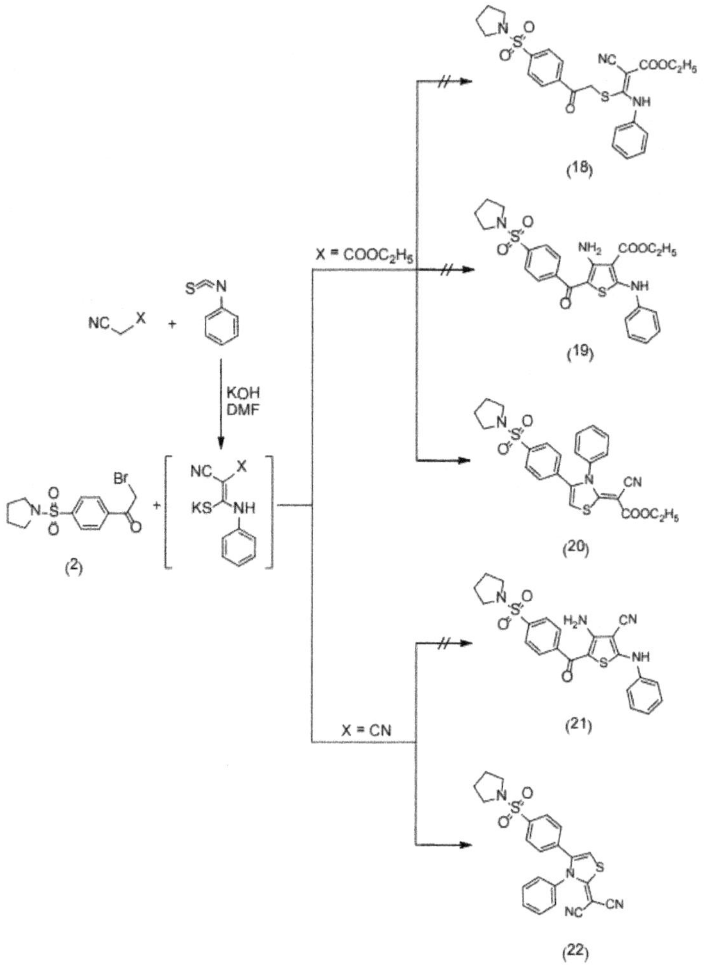

Scheme 5: Synthesis of compounds 20,22.

Thus, IR spectrum of 22 showed no absorption bands for NH, NH_2 or C=O groups, ^1H NMR spectrum showed singlet signal at $\delta = 7.21$ ppm due to CH-thiazole, and the mass spectrum was compatible with the molecular formula $C_{22}H_{18}N_4O_2S_2$ (M^+; 434).

The goal was extended to include the behavior of 2 towards heterocyclic amines for building different fused heterocyclic rings. Thus, treatment of 2 with 2-aminothiazole and 2-aminobenzothiazole in refluxing ethanol yielded imidazo[2,1-b]thiazole 23 and benzo[d]imidazo[2,1-b]thiazole 24, respectively (Scheme 6). Efforts to

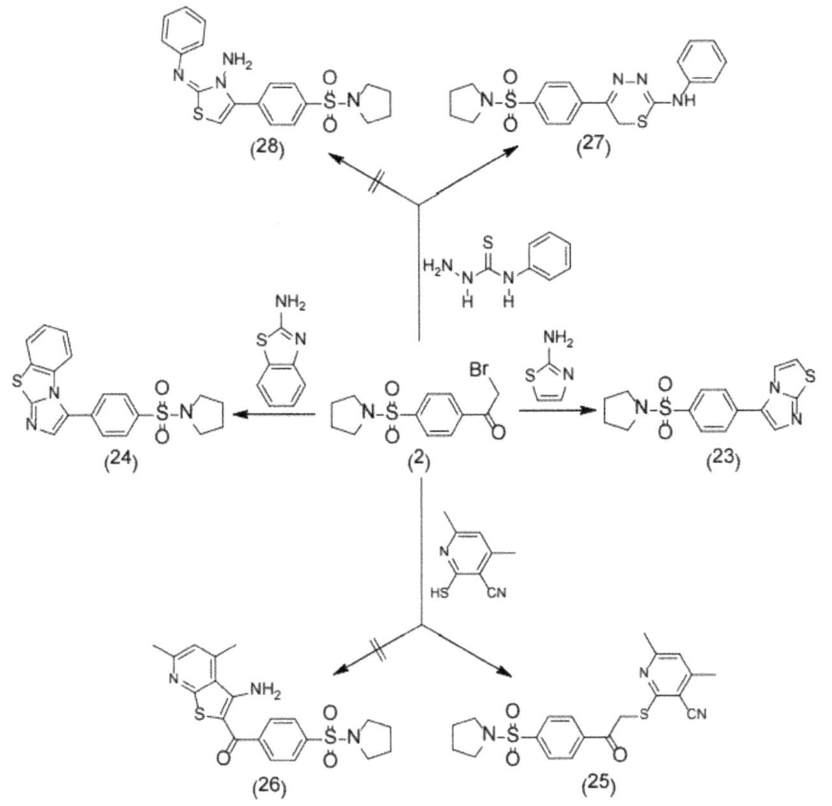

Scheme 6: Synthesis of compounds 23-25,27.

cyclize 2 with 2-mercapto-4,6-dimethylnicotinonitrile [27] to afford thieno[2,3-b]pyridine derivative 26 were not successful, instead the acyclic product 25 was obtained, the latter structure was confirmed based on IR and ^{1}H NMR spectral data. Thus, IR spectrum of 25 revealed absorption band at v = 2225 cm^{-1}, due to cyano group and no absorption band for amino group, ^{1}H NMR spectrum showed singlet signal at δ = 4.57 ppm, for methylene protons. Interaction of 2 with N-phenylhydrazinecarbothioamide afforded thiadiazine derivative 27. The other possible isomeric structure 2-(phenylimino)-4-(4-(pyrrolidin-1-ylsulfonyl)phenyl)thiazol-3(2H)-amine (28) was discarded based on elemental analyses. Among, IR spectrum of 27 lacked absorption bands for amino function and exhibited an absorption band at v = 3115 cm^{-1} corresponding to NH function. Furthermore, ^{1}H NMR spectrum showed singlet signal at δ = 4.32 ppm due to CH$_2$ protons of thiadiazine ring. The presence of ten aromatic carbon types for fourteen aromatic carbon atoms on ^{13}C NMR spectrum between δ = 119.0 and 168.2 ppm, in addition three aliphatic carbon types for five carbon atoms of pyrrolidine moiety at $\delta = 23.9$ and 72.3 ppm, and thiadiazine moiety at $\delta = 29.8$ ppm. Besides, the mass spectrum was compatible with the molecular formula $C_{19}H_{20}N_4O_2S_2$, m/z = 400 confirmed structure 27.

Treatment of phenacyl bromide derivative 2 with o-phenylenediamine in refluxing ethanol afforded a crystalline product identified as 2-(4-(pyrrolidin-1-ylsulfonyl)phenyl)quinoxaline(30) in an excellent yield (Scheme 7). A plausible mechanism may involve the condensation of one of phenylenediamine amino groups with the carbonyl group of bromoacetyl moiety, while the second amino group replaced bromine atom via nucleophilic substitution. The expected product is the dihydroquinoxalinyl derivative 29, however, the spectral data of the isolated product established that the dihydroquinoxalinyl derivative 29 was oxidized under the reaction conditions to give quinoxaline derivative 30.

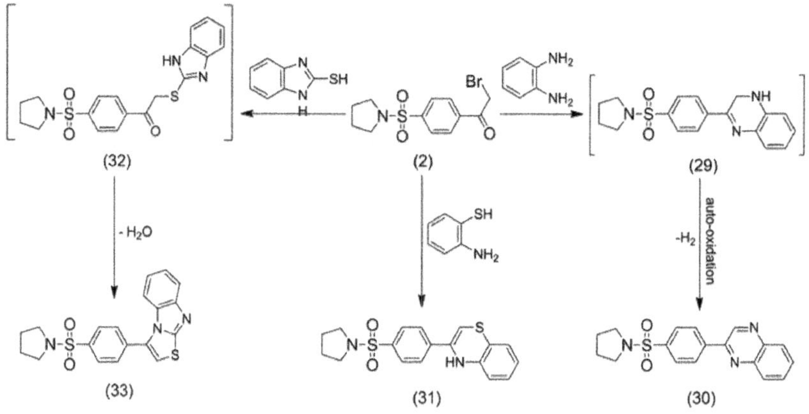

Scheme 7: Synthesis of compounds 30,31,33.

The IR spectrum of 30 showed no absorption band for NH group. The ^{1}H NMR spectrum showed a singlet, of one proton, at $\delta = 8.99$ ppm due to quinoxaline-H$_3$. The presence of nine aromatic carbon types for fourteen aromatic carbon atoms on ^{13}C NMR spectrum of the isolated product between $\delta = 121.0$ and 160.2 ppm, in addition to two aliphatic carbon types for four carbon atoms of pyrrolidine moiety at $\delta = 20.9$ and 62.7 ppm, confirmed structure 30. Cyclocondensation of phenacyl bromide derivative 2 with o-aminothiophenol afforded benzo[b][1,4]thiazine derivative 31. Similarly treatment of 2 with 1H-benzo [d]imidazole-2-thiol afforded benzo[4,5]imidazo[2,1-b]thiazole derivative 33 via cyclization of acyclic intermediate 32 by dehydration under the reaction conditions (Scheme7).

Interaction of phenacyl bromide derivative 2 with 3-methyl-5-oxo-4,5-dihydro-1H-pyrazole-1-carbothioamide [28] in refluxing ethanol gave thiazolylpyrazole derivative 34. Scheme 8 shows three tautomeric structures (a-c) for 34, with tautomeric form-c predominate. IR spectrum of the isolated product revealed absorption bands at $v = 3210$, 1672 cm^{-1} due to NH and C=O groups, respectively. The ^{1}H NMR spectrum showed a singlet, of one proton, at $\delta = 5.27$ ppm due to pyrazole-H$_4$, and (D$_2$O exchangeable) signal at $\delta = 8.44$ ppm due to NH proton. Finally, the methylene group in 34 proved to be highly reactivity, thus compound 34 underwent coupling with equimolar amount of 4-chlorobenzenediazonium chloride in pyridine solution at (0°C - 5°C) to afford a colored product 35, for which the three isomeric structures a, c as azo forms and b as hydrazo form (Scheme 8). IR spectrum of the isolated product revealed absorption bands atv $= 3165$ and 1666 cm^{-1} due to NH and C=O groups, respectively. ^{1}HNMR spectrum showed signals at $\delta = 7.69$ and 8.59 ppm due to thiazol-H$_5$ and NH group, respectively. Structure 35 was further confirmed unequivocally by an independent synthesis from the reaction of compound 2 with 4-((4-chlorophenyl)diazenyl)-3-methyl-5-oxo-4,5-dihydro-1H-pyrazole-1-carbo- thioamide in refluxing dioxane solution (Scheme 8).

Docking and Molecular Modeling

Thymidylate synthase and dihydrofolate reductase are among the main targets involved in anticancer and antimicrobial activity [29] [30] . Molecular modeling study using Molecular Operating Environment (MOE) [31] module was performed in order to rationalize the observed anticancer activity of the newly synthesized compounds. Molecular docking studies further help in understanding the mode of action of the compounds through their various interactions with the active sites of dihydrofolate reductase.

Docking of MTX into DHFR

The active site revealed that hydrogen bond interactions beside hydrophobic interactions were considered responsible for the observed affinity as it acts as a hydrogen bond donor to the backbone Ile 5 and Ile 94 residues and the side chain Asp 27 residue. It also acts as a hydrogen bond accepter to Arg 52 and Arg 57 residues. This beside many hydrophobic interactions with various amino acid residues: ILe 5, Ala 6, Ala7, Asp 27, Leu 28, Phe

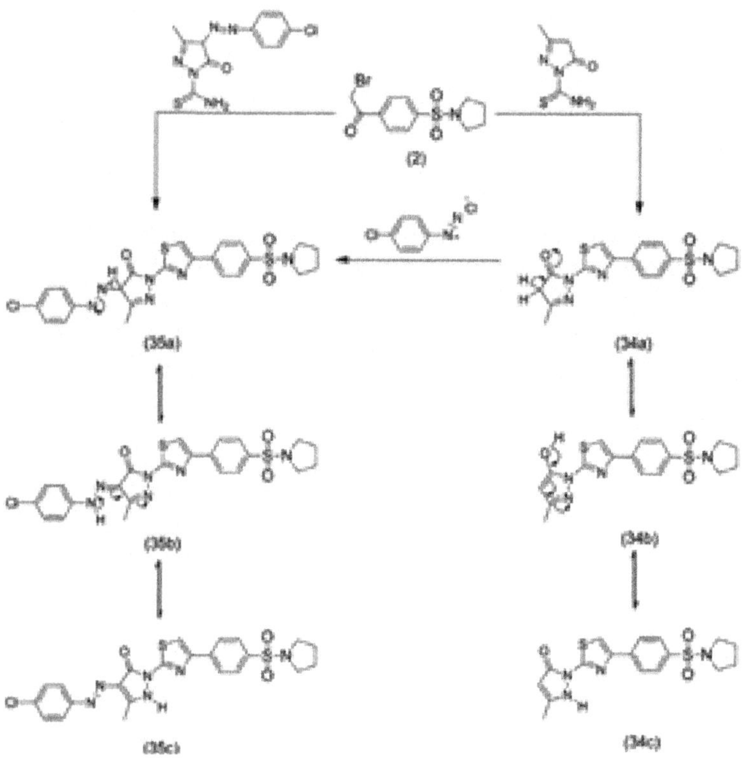

Scheme 8: Synthesis of compounds 34,35.

31, Lys 32, Ser 49, Ile 50, Arg 52, Leu 54, Arg 57, Ile 94, Tyr 100 and Thr 113, as shown in (Figure 1).

Docking Simulation Study of the Synthesized Compounds 1, 2, 4, 5, 8, 11, 14, 20, 24 and 33

MOE docking studies of the inhibitors were performed using dihydrofolate reductase co-crystallized with methotrexate (PDB ID: 4DFR) as a template.

1) Docking of compound 1 into DHFR

The active site revealed the presence of one hydrogen bond interaction as one oxygen atom of SO_2 moiety acted as a hydrogen bond acceptor with the amino acid residue Ser 59 (2.80 Å) with a strength of 72.5%. In addition to, hydrophobic interactions involving carbon atom of carbonyl function, $C_{2,3,5,6}$ of phenyl ring, $C_{2,3,4}$ of pyrrolidine ring and oxygen atom of SO_2 moiety with the following amino acid residues: Val 8, Ala 9, Asp 21, Glu 30, Phe 31, Phe 34, Thr 56 and Ser 59, as shown in (Figure 2).

2) Docking of compound 2 into DHFR

The active site revealed that several molecular interactions were considered responsible for the observed affinity, as the one oxygen atom of SO_2 moiety acted as a hydrogen bond acceptor with the side chain residues; Thr 56 and Ser 59 (3.63 Å and 3.16 Å, respectively) with a strength of 2.4% and 10.9%, respectively. Besides to, hydrophobic interactions involving the bromine atom, oxygen atom of carbonyl function and other carbons as well the second oxygen atom of SO_2 moiety and the following amino acid residues: Ile 16, Leu 22, Phe 31, Ile 60, Pro 61, Val 115 and Tyr 121, as shown in (Figure 3).

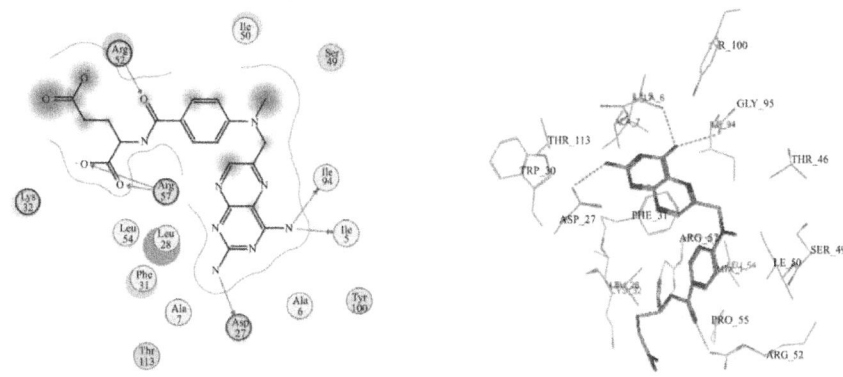

Figure 1: Docking of MTX into DHFR.

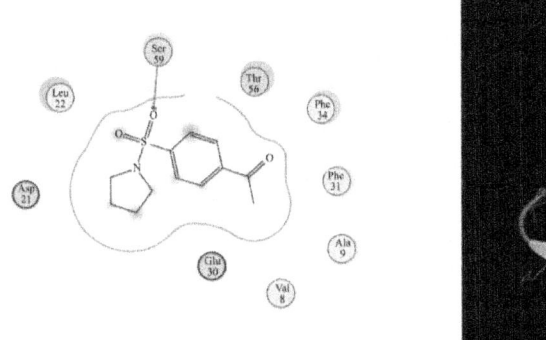
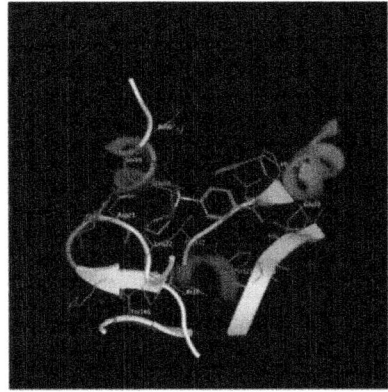

Figure 2: Docking of compound 1 into DHFR.

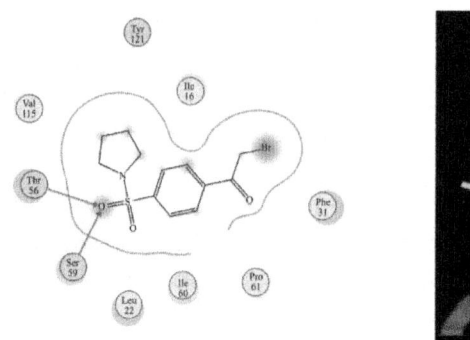
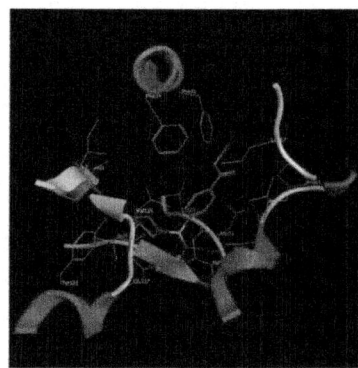

Figure 3: Docking of compound 2 into DHFR.

3) Docking of compound 4 into DHFR

The active site revealed the presence of hydrogen bond interaction between hydrogen atom of NH function as it acted as a hydrogen bond donor with the side chain residue Asp 21 (2.16 Å) with a strength of 23%. Moreover, one oxygen atom of SO_2 moiety acted as a hydrogen bond acceptor with the amino acid residue Thr 56 (2.83 Å) with a strength of 29.7%. Besides to, arene-arene cation interaction between the phenyl ring of benzothiazole moiety and the amino acid residue Phe 31. In addition to, hydrophobic interactions among other atoms of the compound with the following amino acid residues: Ala 9, Ile 16, Gly 17, Asp 21, Leu 22, Gln 35, Thr 56, Ser 59, Ile 60, Pro 61, Asn 64, Lys 68, Arg 70, Val 115 and Tyr 121, as shown in (Figure 4).

4) Docking of compound 5 into DHFR

The active site illustrated the presence of several interactions of the one oxygen atom of SO_2 moiety with different amino acid residues as it acted as a hydrogen bond acceptor with the side chain residues; Thr 56 and Ser 59 (3.57 Å and 3.02 Å, respectively) with a strength of 2.4% and 21.4%, respectively. This beside hydrophobic interaction among the amino function, sulfur atom and C_4 of thiazole moiety, $C_{2,6}$ of benzene ring, oxygen atoms of SO_2 moiety and pyrrolidine C_2, C_3, C_5 and the following amino acid residues: Ile 16, Leu 22, Phe 31, Phe 34, Ile 60, Pro 61, Leu 67, Val 115 and Tyr 121, as shown in (Figure 5).

5) Docking of compound 8 into DHFR

The active site revealed the presence of hydrogen bond interaction between the one oxygen atom of SO_2 moiety as it acted as a hydrogen bond acceptor with the side chain residues; Thr 56 and Ser 59 (3.35 Å and 3.23 Å, respectively) with a strength of 4.1% and 7.4%, respectively. In addition to, hydrophobic interactions involving the other atoms of the compound with the following amino acid residues: Ile 16, Asp 21, Leu 22, Phe 31, Gln 35, Ile 60, Pro 61, Asn 64, Leu 67, Lys 68, Val 115 and Tyr 121, as shown in (Figure 6).

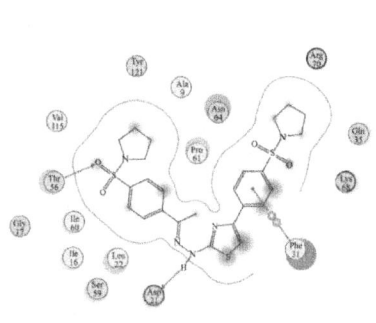
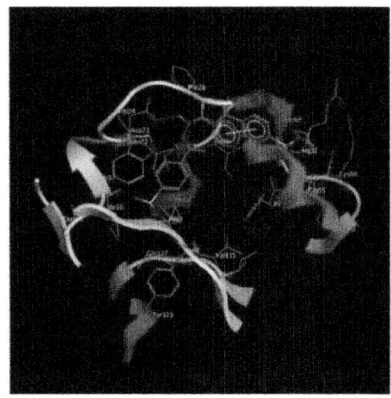

Figure 4: Docking of compound 4 into DHFR.

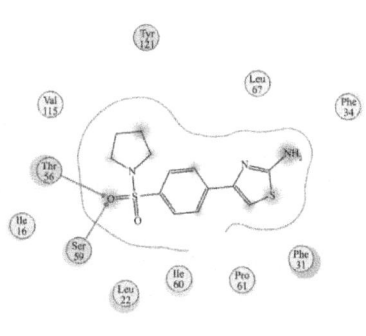
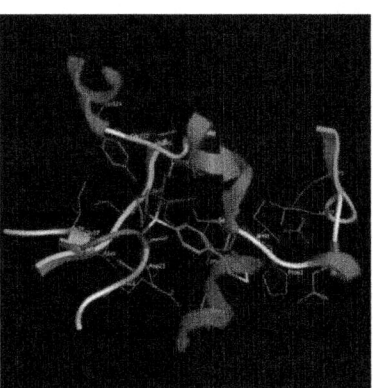

Figure 5: Docking of compound 5 into DHFR.

6) Docking of compound 11 into DHFR

The active site revealed the presence of several molecular interactions in which the one oxygen atom of SO_2 moiety acted as a hydrogen bond acceptor for with the amino acid residue Thr 56 (3.57 Å) with a strength of 2.6%. In addition to, hydrophobic interactions involving other atoms of the compound with the following amino acid residues: Ala 9, Ile 16, Asp 21, Leu 22, Phe 31,

Phe 34, Gln 35, Ser 59, Pro 61, Asn 64, Arg 70, Val 115 and Tyr 121, as shown in (Figure 7).

7) Docking of compound 13 into DHFR

The active site revealed the presence of hydrogen bond interactions between one oxygen atom of SO_2 moiety and the cyano group as they acted as a hydrogen bond acceptor with the side chain residues; Thr 56 and Thr 136 (3.41 Å and 3.25 Å, respectively) with a strength of 1.6% and 13%, respectively. There is also hydrophobic interactions involving the pyrrolidine C_2, C_3, C_4 as well as the oxygen atoms of SO_2 moiety with the following amino acid residues: Ile 7, Val 8, Ala 9, Ile 16, Leu 22, Glu 30, Phe 34, Ser 59, Ile 60, Val 115 and Tyr 121, as shown in (Figure 8).

8) Docking of compound 20 into DHFR

The active site revealed the presence of several molecular interactions, including two hydrogen bonds. In which both oxygen atoms of SO_2 moiety acted as a hydrogen acceptor via two hydrogen bonds with the amino acid residue Asn 64 (2.80 Å and 3.18 Å, respectively) with a strength of 39.1% and 7.2%, respectively). Besides to hydrophobic interactions involving the cyano function as well as the other atoms of the compound with the following amino acid residues: Val 8, Ile 16, Asp 21, Leu 22, Phe 31, Phe 34, Thr 56, Ser 59, Ile 60, Pro 61, Asn 64, Val 115 and Tyr 121, as shown in Figure 9.

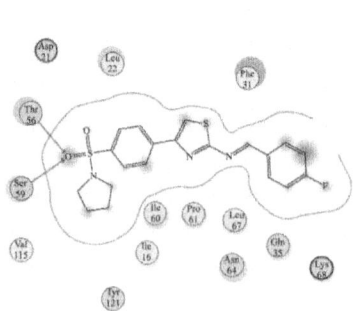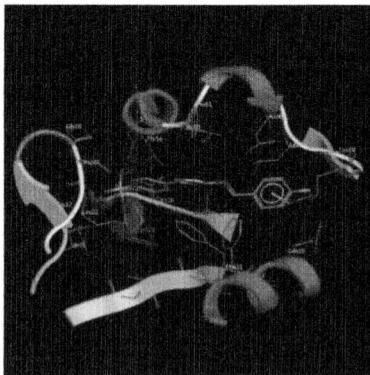

Figure 6: Docking of compound 8 into DHFR.

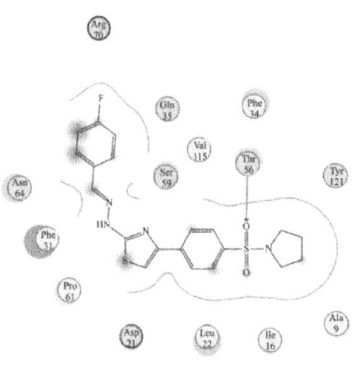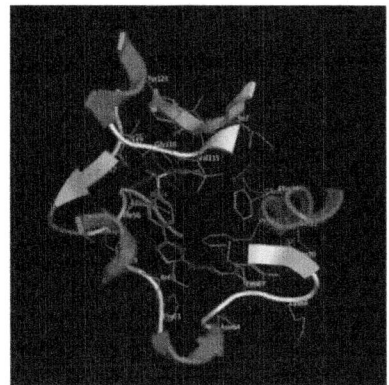

Figure 7: Docking of compound 11 into DHFR.

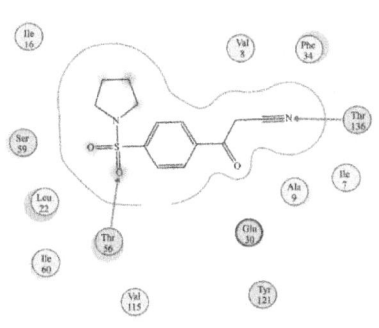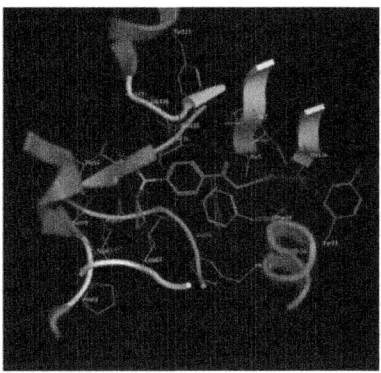

Figure 8: Docking of compound 13 into DHFR.

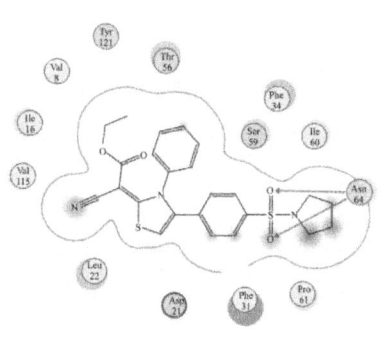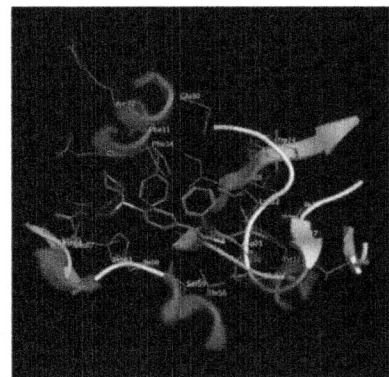

Figure 9: Docking of compound 20 into DHFR.

9) Docking of compound 24 into DHFR

The active site revealed the presence of arene-arene cation interaction between the phenyl ring of benzo[d]imidazothiazole with the amino acid residue Phe 31. In addition to, hydrophobic interactions involving other atoms of the compound with many amino acid residues: Val 8, Ala 9, Ile 16, Leu 22, Phe 31, Thr 56, Ser 59, Ile 60, Pro 61 and Val 115, as shown in (Figure 10).

10) Docking of compound 33 into DHFR

The active site revealed only hydrophobic interactions concerning $C_{2,3,4}$ of pyrrolidine ring and $C_{2,7,8}$ of benzo [d]imidazothiazole with the following amino acid residues: Ile 16, Leu 22, Phe 31, Phe 34, Ser 59, Ile 60, Pro 61, Asn 64, Val 115 and Tyr 121, as shown in (Figure 11).

Docking and Molecular Modeling

Docking was performed for the compounds 1, 2, 4, 5, 8, 11, 13, 20, 24 and 33 on the dihydrofolate reductase in a trial to predict their mode of action as anticancer drugs. The compounds show several interactions with dihydrofolate reductase enzyme. Particularly noteworthy are the compounds 4, 8, 11, 20, 24 and 33, which suggest that they might exert their action through inhibition of the DHFR enzyme (Table 2). It is clear from the present data that the comparison of the docking score energy for tested compounds that the compounds follows the order 4 > 20 > 8 > 11 > 33 > 24 > 5 > 13 > 1 > 2, as shown in Chart 1.

In Vitro Anticancer Activity

The newly synthesized compounds were evaluated for their in-vitro cytotoxicity against human liver hepatocel Score: for all scoring functions, lower scores indicate more poses that are favorable.

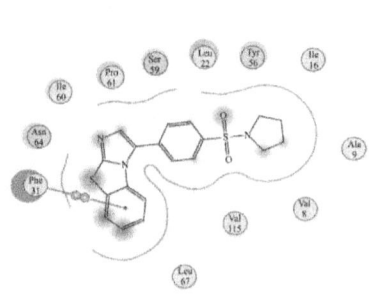
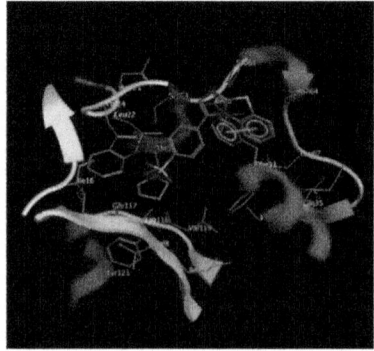

Figure 10: Docking of compound 24 into DHFR.

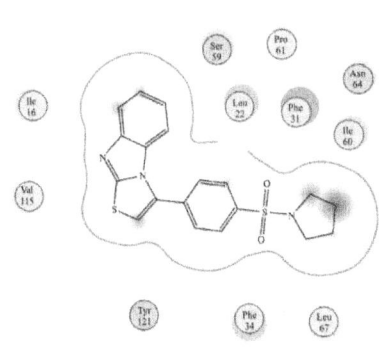 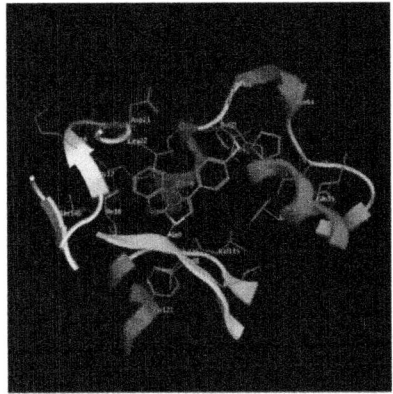

Figure 11: Docking of compound 33 into DHFR.

Table 2: Docking score energy of the selective newly synthesized compounds.

Comp. No.	Score	E-conf	E-place	E-score 1	E-score 2	E-refine
1	−16.87	−05.76	−61.37	−08.38	−16.87	−13.19
2	−16.30	−00.84	−57.58	−08.81	−16.30	−15.73
4	−28.18	08.55	−105.27	−09.29	−28.18	−07.14
5	−16.91	−61.16	−65.08	−09.14	−16.91	−13.89
8	−23.18	−02.87	−93.77	−11.43	−23.18	−15.38
11	−22.42	17.10	−78.04	−10.31	−22.42	−18.02
13	−16.91	−24.24	−61.74	−08.16	−16.91	−08.57
20	−27.21	24.44	−100.42	−09.65	−27.21	−03.40
24	−21.15	26.14	−99.85	−09.28	−21.15	−12.28
33	−21.24	18.53	−65.44	−09.05	−21.24	−07.83

The Unit for all scoring functions is kcal/mol. E-conf: the energy of the conformer. If there is a refinement stage, this is the energy calculated at the end of the refinement. E-place: Score from the placement stage (Placement. A collection of poses is generated from the pool of ligand conformations using one of the placement methods). E-score 1: Score from the first rescoring stage. E-score 2: Score from the second rescoring stage. E-refine: Score from the refinement stage (Refinement: Energy minimization of the system is carried out using the conventional molecular mechanics setup).

lular carcinoma cell line (HepG2). Some of the tested compounds were more potent compared with methotrexate as the reference drug. From the obtained results in Table 1 and Chart 2, observe that compound 8 having 2-(4-fluorobenzylidene)amino thiazole moiety with SI value 33.21, 2-(4-fluorobenzylidene)hydrazinyl thiazole 11 with SI value 30.49,

2-(ethylidene)hydrazinyl thiazole 4 with SI value 19.43, 2-dicyanomethylene thiazole 22 with SI value 14.82, 2-methylthiazole 12a with SI value 10.29, benzo[4,5]imidazo[2,1-b]thiazole 33 with SI value 7.30, 2-(3-oxo-1H-pyrazol-2-yl)thiazole 35 with SI value 6.87, compound 27 having 1,3,4-thiadiazine moiety with SI value 6.15, showed increased activity when compared to methotrexate with SI value 4.14, while compounds 24, 20, 3, 1 and 13 with SI values 4.62, 3.89, 2.99, 2.41 and 2.10, respectively, were found to be nearly as active as methotrexate. While the remaining compounds 17, 16, 34, 23, 14, 25, 2, 6, 5, 30, 15, 10, 31 and 12b with SI values 1.88, 1.85, 1.33, 1.07, 0.92, 0.83, 0.75, 0.62, 0.56, 0.53, 0.38, 0.34, 0.33 and 0.20, respectively showed decreased activity when compared to methotrexate. It is clear from the present data that the comparison of the selective index (SI) for the synthesized compounds against human liver hepatocellular carcinoma cell line (HepG2). Chart 2 has showed that, the cell killing potency follows the order 8 > 11 > 4 >

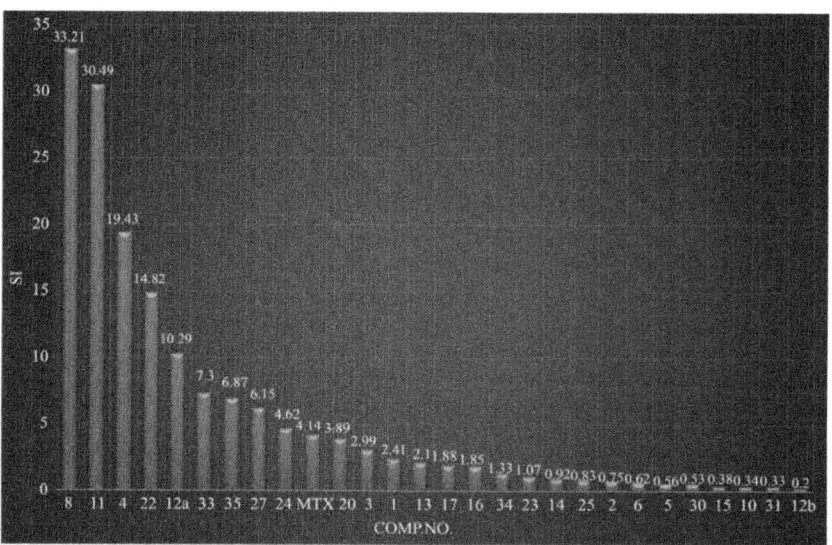

Chart 1: Comparison of the Selective index (SI) for the synthesized compounds against human liver hepatocellular carcinoma cell line (HepG2).

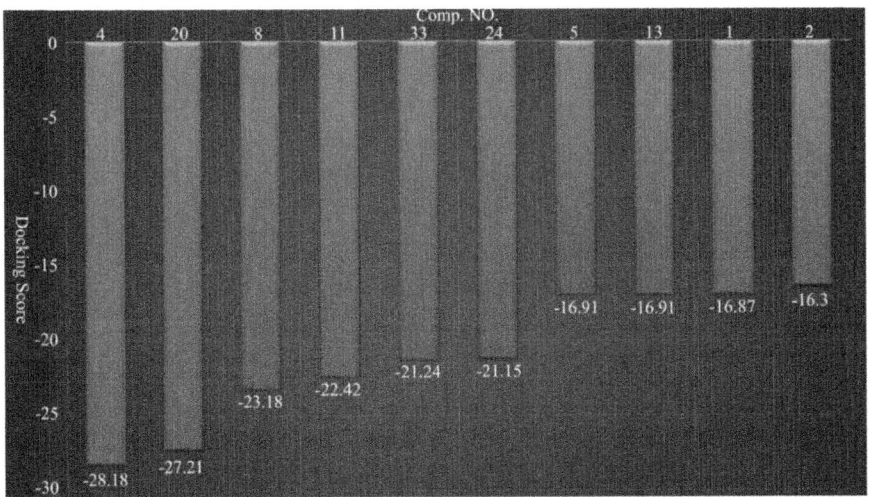

Chart 2: Comparison of the docking Score energy of the selective newly synthesized compounds.

22 > 12a > 33 > 35 > 27 > 24 > MTX > 20 > 3 > 1 > 13 > 17 > 16 > 34 > 23 > 14 > 25 > 2 > 6 > 5 > 30 > 15 > 10 > 31 > 12b. These preliminary results of biological screening of the tested compounds could offer an encouraging framework in this field that may lead to the discovery of potent anticancer agent.

CONCLUSION

This article proved that compounds having pyrrolidine benzenesulfonamide moiety attached to different heterocyclic moieties such as thiazole 4, 8, 11, 12a, 22 and 35, imidazo[2,1-b]thiazole 33 and 1,3,4-thiadiazine 27, showed a significant cytotoxic activity against human liver hepatocellular carcinoma cell line (HepG2) compared to the reference drug Methotrexate.

CITE THIS PAPER

Mahmoud SayedBashandy, (2015) 1-(4-(Pyrrolidin-1-ylsulfonyl)phenyl) ethanone in Heterocyclic Synthesis: Synthesis, Molecular Docking and Anti-Human Liver Cancer Evaluation of Novel Sulfonamides Incorporating Thiazole, Imidazo[1,2-a]pyridine, Imidazo[2,1-c] [1,2,4]triazole, Imidazo[2,1-b] thiazole, 1,3,4-Thiadiazine and 1,4-Thiazine Moieties. *International Journal of Organic Chemistry*,**05**,166-190. doi: 10.4236/ijoc.2015.53018

REFERENCES

1. Konda, S., Raparthi, S., Bhaskar, K., Munaganti, R.K., Guguloth, V., Nagarapu, L. and Akkewar, D.M. (2015) Synthesis and Antimicrobial Activity of Novel Benzoxazine Sulfonamide Derivatives. Bioorganic & Medicinal Chemistry Letters, 25, 1643-1646. http://dx.doi.org/10.1016/j.bmcl.2015.01.026

2. Gamal El-Din, M.M., El-Gamal, M.I., Abdel-Maksoud, M.S., Yoo, K.H. and Oh, C.H. (2015) Synthesis and in Vitro Antiproliferative Activity of New 1,3,4-Oxadiazole Derivatives Possessing Sulfonamide Moiety. European Journal of Medicinal Chemistry, 90, 45-52. http://dx.doi.org/10.1016/j.ejmech.2014.11.011

3. Sadarangani, S.P., Estes, L.L. and Steckelberg, J.M. (2015) Non-Anti-Infective Effects of Antimicrobials and Their Clinical Applications: A Review. Mayo Clinic Proceedings, 90, 109-127.http://dx.doi.org/10.1016/j.mayocp.2014.09.006

4. Farahi, M., Karami, B. and Tanuraghaj, H.M. (2015) Efficient Synthesis of a New Class of Sulfonamide-Substituted Coumarins. Tetrahedron Letters, 56, 1833-1836. http://dx.doi.org/10.1016/j.tetlet.2015.02.087

5. Awadallah, F.M., El-Waei, T.A., Hanna, M.M., Abbas, S.E., Ceruso, M., Oz, B.E., Guler, O.O. and Supuran, C.T. (2015) Synthesis, Carbonic Anhydrase Inhibition and Cytotoxic Activity of Novel Chromone-Based Sulfonamide Derivatives. European Journal of Medicinal Chemistry, 96, 425-435. http://dx.doi.org/10.1016/j.ejmech.2015.04.033

6. Arshad, M.N., Asiri, A.M., Alamry, K.A., Mahmood, T., Gilani, M.A., Ayub, K. and Birinji, A.S. (2015) Synthesis, Crystal Structure, Spectroscopic and Density Functional Theory (DFT) Study of N-[3-Anthracen-9-yl-1-(4-bromophenyl)allylidene]-N-benzenesulfonohydrazine. Spectrochimica Acta Part A: Molecular and Biomolecular Spectroscopy, 142, 364-374. http://dx.doi.org/10.1016/j.saa.2015.01.101

7. Bozdag, M., Carta, F., Vullo, D., Akdemir, A., Isik, S., Lanzi, C., Scozzafava, A., Masini, E. and Supuran, C.T. (2015) Synthesis of a New Series of Dithiocarbamates with Effective Human Carbonic Anhydrase Inhibitory Activity and Antiglaucoma Action. Bioorganic & Medicinal Chemistry, 23, 2368-2376.http://dx.doi.org/10.1016/j.bmc.2015.03.068

8. Grandane, A., Tanc, M., Zalubovskis, R. and Supuran, C.T. (2015) Synthesis of 6-Aryl-Substituted Sulfocoumarins and Investigation of their Carbonic Anhydrase Inhibitory Action. Bioorganic & Medicinal Chemistry, 23, 1430-1436. http://dx.doi.org/10.1016/j.bmc.2015.02.023

9. Yu, Z.Y., Yin, D.Q. and Deng, H.P. (2015) The Combinational Effects between Sulfonamides and Metals on Nematode Caenorhabditis elegans. Ecotoxicology and Environmental Safety, 111, 66-71. http://dx.doi.org/10.1016/j.ecoenv.2014.09.026

10. Karakaya, M., Sert, Y., Sreenivasa, S., Suchetan, P.A. and Cirak, C. (2015) Monomer Spectroscopic Analysis and Dimer Interaction Energies on N-(4-Methoxybenzoyl)-2-methylbenzenesulfonamide by Experimental and Theoretical Approaches. Spectrochimica Acta Part A: Molecular and Biomolecular Spectroscopy, 142, 169-177.http://dx.doi.org/10.1016/j.saa.2015.01.107

11. Booker, S.A., Pires, N., Cobb, S., Silva, P.S. and Vida, I. (2015) Carbamazepine and Oxcarbazepine, but Not Eslicarbazepine, Enhance Excitatory Synaptic Transmission onto Hippocampal CA1 Pyramidal Cells through an Antagonist Action at Adenosine A1 Receptors. Neuropharmacology, 93, 103-115.http://dx.doi.org/10.1016/j.neuropharm.2015.01.019

12. Reddy, N.D., Shoja, M.H., Jayashree, B.S., Nayak, P.G., Kumar, N., Prasad, V.G., Pai, K.S.R. and Rao, C.M. (2015) In Vitro and in Vivo Evaluation of Novel Cinnamyl Sulfonamide Hydroxamate Derivative against Colon Adenocarcinoma. Chemico-Biological Interactions, 233, 81-94. http://dx.doi.org/10.1016/j.cbi.2015.03.015

13. Angel, M., Nieto, A., Apan, T. and Delgado, G. (2015) Anti-Inflammatory Effect of Natural and Semi-Synthetic Phthalides. European Journal of Pharmacology, 752, 40-48. http://dx.doi.org/10.1016/j.ejphar.2015.01.026

14. Chen, Z., Wang, Z.C., Yan, X.Q., Wang, P.F., Lu, X.Y., Chen, L.W., Zhu, H.L. and Zhang, H.W. (2015) Design, Synthesis, Biological Evaluation and Molecular Modeling of Dihydropyrazole Sulfonamide Derivatives as Potential COX-1/COX-2 Inhibitors. Bioorganic & Medicinal Chemistry Letters, 25, 1947-1951.http://dx.doi.org/10.1016/j.bmcl.2015.03.022

15. Tavares, M.T., Pasqualoto, K.F.M., Streek, J., Ferreira, A.K., Azevedo, R.A., Damiao, M.C., Rodrigues, C.P., Junior, P.L., Barbuto, J.A.M., Filho, R. and Ferreira, F.F. (2015) Synthesis, Characterization, in Silico Approach and in Vitro Antiproliferative Activity of RPF151, a Benzodioxole Sulfonamide Analogue Designed from Capsaicin Scaffold. Journal of Molecular Structure, 1088, 138-146. http://dx.doi.org/10.1016/j.molstruc.2015.02.019

16. Medici, S., Peana, M., Nurchi, V.M., Lachowicz, J.I., Crisponi, G. and Zoroddu, M.A. (2015) Noble Metals in Medicine: Latest Advances.

Coordination Chemistry Reviews, 284, 329-350. http://dx.doi. org/10.1016/j.ccr.2014.08.002

17. Bashandy, M.S., Alsaid, M.S., Arafa, R.K. and Ghorab, M.M. (2014) Design, Synthesis and Molecular Docking of Novel N,N-Dimethylbenzenesulfonamide Derivatives as Potential Antiproliferative Agents. Journal of Enzyme Inhibition and Medicinal Chemistry, 29, 619-627. http://dx.doi.org/10.3109/14756366.2013.833197

18. Bashandy, M.S., Al-Said, M.S., Al-Qasoum, S.I. and Ghorab, M.M. (2011) Design and Synthesis of Some Novel Hydrazide, 1,2-Dihydropyridine, Chromene Derivatives Carrying Biologically Active Sulfone Moieties with Potential Anticancer Activity. Arzneimittelforschung / Drug Research, 61, 521-526.

19. Al-Said, M.S., Bashandy, M.S. and Ghorab, M.M. (2011) Novel Quinolines Bearing a Biologically Active Trimethoxyphenyl Moiety as a New Class of Antitumor Agents. Arzneimittelforschung / Drug Research, 61, 527-531.

20. Bashandy, M.S., Hassan, S.M., Fathalla, O.A., Eweas, A.F. and Khalel, A.H. (2012) Synthesis and Biological Evaluation Studies of Some Novel Quinazoline and Quinazolinone Derivatives as Anti-Liver Cancer Agents. Egyptian Journal of Chemistry, 55, 659-676.

21. Al-Said, M.S., Bashandy, M.S., Al-Qasoumi, S.I. and Ghorab, M.M. (2011) Anti-Breast Cancer Activity of Some Novel 1,2-Dihydropyridine, Thiophene and Thiazole Derivatives. European Journal of Medicinal Chemistry, 46, 137- 141. http://dx.doi.org/10.1016/j.ejmech.2010.10.024

22. El-Kashef, H.S., Bayoumy, B.E. and Aly, T.I. (1986) Selenium Heterocycles. I. Synthesis and Antibacterial Activity of Some New 1,2,3-Thia- and 1,2,3-Selenadiazoles Containing Sulfonamides. Egyptian Journal of Pharmaceutical Sciences, 27, 27-30.

23. Skehan, P., Storegng, R., Scudiero, D., Monks, A., McMahon, J., Vistica, D., Warren, J.T., Bokesch, H., Kenney, S. and Boyd, M.R. (1990) New Colorimetric Cytotoxicity Assay for Anticancer-Drug Screening. Journal of the National Cancer Institute, 82, 1107-1112. http://dx.doi.org/10.1093/ jnci/82.13.1107

24. Al-Qubaisi, M., Rozita, R., Yeap, S., Omar, A., Ali, A. and Alitheen, N.B. (2011) Selective Cytotoxicity of Goniothalamin against Hepatoblastoma HepG2 Cells. Molecules, 16, 2944-2959.http://dx.doi.org/10.3390/ molecules16042944

25. Chellaian, J.D. and Johnson, J. (2014) Spectral Characterization, Electrochemical and Anticancer Studies on Some Metal (Li) Complexes

Containing Tridentate Quinoxaline Schiff Base. Spectrochimica Acta Part A: Molecular and Biomolecular Spectroscopy, 127, 396-404. http://dx.doi.org/10.1016/j.saa.2014.02.075

26. Suga, A., Narita, T., Zhou, L., Sakagami, H., Satoh, K. and Wakabayashi, H. (2009) Inhibition of NO Production in LPS-Stimulated Mouse Macrophage-Like Cells By Benzo[b]cyclohept [e] [1,4]oxazine and 2-Aminotropone Derivatives. In Vivo, 23, 691-697.

27. Ho, Y.W. and Wang, I.J. (1995) Studies on the Synthesis of Some Styryl-3-cyano-2(1H)-pyridinethiones and Polyfunctionally Substituted 3-Aminothieno [2,3-b]pyridine Derivatives. Journal of Heterocyclic Chemistry, 32, 819-825.http://dx.doi.org/10.1002/jhet.5570320323

28. Harode, R. and Sharma, T.C. (1989) Synthesis of 2-(3'-Methylpyrazol-5-one-1'-yl)-4-arylthiazoles. Journal of Indian Chemical Society, 66, 282-284.

29. Du, Q.R., Li, D.D., Pi, Y.Z., Li, J.R., Sun, J., Fang, F., Zhong, W.Q., Gong, H.B. and Zhu, H.L. (2013) Novel 1,3,4- Oxadiazole Thioether Derivatives Targeting Thymidylate Synthase as Dual Anticancer/Antimicrobial Agents. Bioorganic & Medicinal Chemistry, 21, 2286-2297. http://dx.doi.org/10.1016/j.bmc.2013.02.008

30. Rao, K.N. and Venkatachalam, S.R. (1999) Dihydrofolate Reductase and Cell Growth Activity Inhibition by the β- Carboline-Benzoquinolizidine Plant Alkaloid Deoxytubulosine from Alangium lamarckii: Its Potential as an Antimicrobial and Anticancer Agent. Bioorganic & Medicinal Chemistry, 7, 1105-1110. http://dx.doi.org/10.1016/S0968-0896(98)00262-4

31. Vilar, S., Cozza, G. and Moro, S. (2008) Medicinal Chemistry and the Molecular Operating Environment (MOE): Application of QSAR and Molecular Docking to Drug Discovery. Current Topics in Medicinal Chemistry, 8, 1555-1572. http://dx.doi.org/10.2174/156802608786786624

Chapter 13

SYNTHESIS OF NEW FLUORINATED 1,2,4-TRIAZINO [3,4-B][1,3,4] THIADIAZOLONES AS ANTIVIRAL PROBES-PART II-REACTIVITIES OF FLUORINATED 3-AMINOPHENYL-1,2,4-TRIAZINOTHIADIAZOLONE

Mohammed Saleh Tawfek Makki, Reda Mohammady Abdel Rahman, Ola Ahmad Abu Ali

Chemistry Department, Faculty of Science, King Abdul Aziz University, Jeddah, Kingdom of Saudi Arabia

ABSTRACT

Some new fluorinated 3-N-acyl/3-N-alkylaminophenyl-1,2,4-triazino[3,4-b][1,3,4]thiadiazolones (2-12) have been obtained from treatment of 2-(4'-fluorophenyl)-6-(2'-amino-5'-fluorophenyl)- 1,2,4-triazino[3,4-b][1,3,4]thiadiazol-4-one(1) with active functional oxygen, sulfur and halogen compounds in different conditions. Former structures of the products have been characterized from elemental and spectral data (UV, IR, NMR and Mass). The new products were evaluated as potential anthelmintic drugs.

INTRODUCTION

The treatment of infectious diseases still remains an important and challenging problem because of a combination of many factors including emerging infectious diseases and the increasing number of multi-drug resistant microbial pathogens [1] -[4] . There is real perceived need for the discovery of new compounds endowed with biocidal activity. Through the various molecules designed and synthesized for this aim, it was demonstrated that fluorinated 1,2,4-triazine fused with 1,3,4-thiadiazole systems. The introduction of fluorine atom to the heterocyclic systems improves or enhances the medicinal properties [5] -[9] . On the other hand, most of heterocyclic nitrogen systems bearing an amino-groups exhibit a wide spectrum of biological activities [10] . And their use is as starting material.

Recently, synthesis and chemistry of 1,3,4-thiadiazoles as biocidal agents have been reviewed [11] [12] . Also, 1,2,4-triazine derivatives have been synthesized and evaluated as biological and pharmacological probes [13] -[15] .

Abdel-Rahman et al. [16] , reported that 1,2,4-triazino[3,4-b][1,3,4] thiadiazolones (Figure 1) used as anti HIV and anticancer drugs. In contamination of our work in these researches for new biocidal agents [17] , the present investigation reports the preparation of fluorinated 3-substitutedamino-1,2,4-triazino[3,4-b][1,3,4]thiadiazolones starting from the corresponding 3-amino analogus, as potential anthelmintic drugs.

RESULTS AND DISCUSSION

3-Substituted-1,2,4-triazines and their azole-fused analogs reacts with bifunctional compounds give a more stable polycyclic systems depends on the triazine substrate nature. Search for new bioactive compounds, the main aim of the present work is preparation of fluorinated 3-substituted amino-1,2,4-triazino[3,4-b][1,3,4]thiadiazo- lones in view of their pharmacological properties.

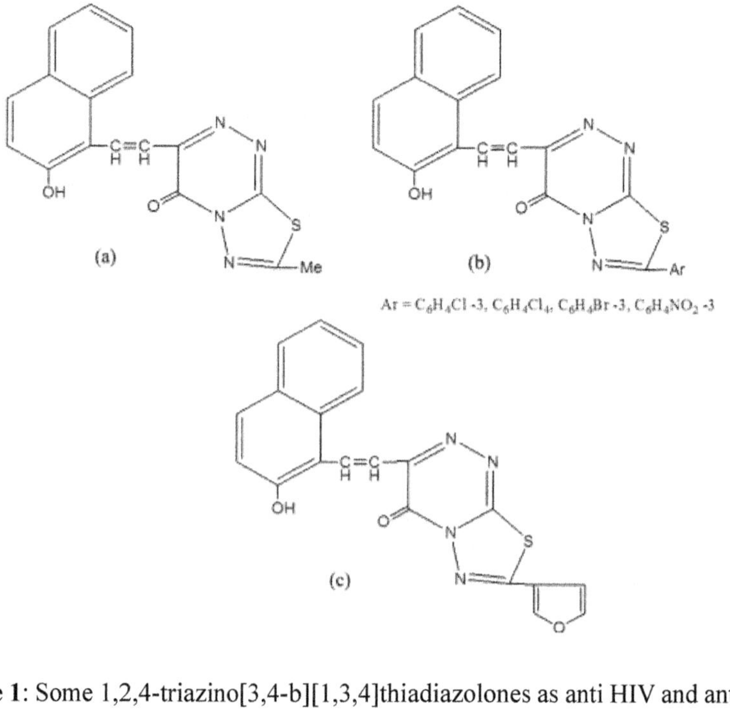

Ar = C_6H_4Cl -3, $C_6H_4Cl_4$, C_6H_4Br -3, $C_6H_4NO_2$ -3

Figure 1: Some 1,2,4-triazino[3,4-b][1,3,4]thiadiazolones as anti HIV and anti-cancer drugs.

Thus, addition of cyclohexyl isocyanate and/or 4-fluorophenyl-isothiocyanate to 3-(2'-amino-5'-fluoro-phenyl) -7-(4'-fluorophenyl)-1,2,4-triazino[3,4-b][1,3,4]thiadiazol-4-one (1) [17] , (Scheme 1) in warm DMF afforded N- (cyclohexyl)-N'-(4'-oxo-1,2,4-triazino[3,4-b][1,3,4] thiadiazol-7'-(4'-fluorophenyl)-3-(4'-fluorophenyl) urea (2) and/or N-(4'-fluorophenyl)-N'-(4'-oxo-1',2',4'-triazino[3,4-b][1,3,4]thiadiazole-7'-(4''-fluoro-phenyl)-3'-(4''-flu- orophenyl)thiourea (3) (Scheme 2). Formation of compound 3 may be tack's place via a nucleophilic attack of NH_2 group to a more electrophilic carbon of isothiocyanate.

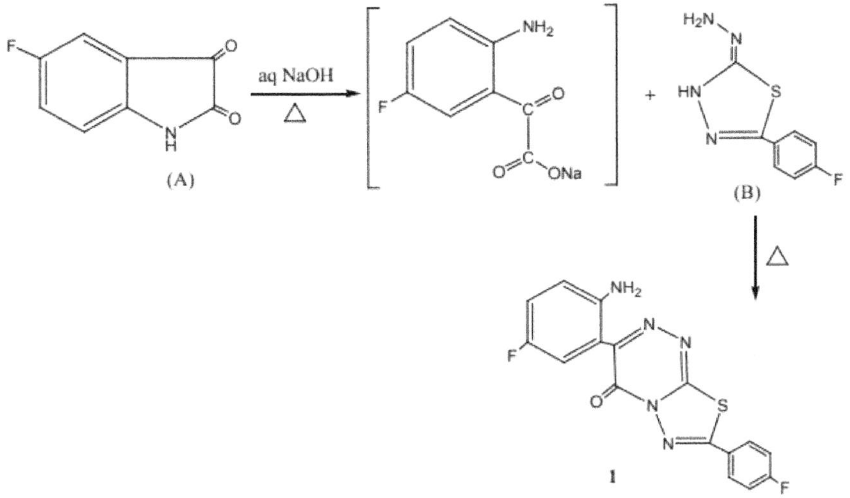

Scheme 1: Synthesis of compound 1.

Acylation of compound 1 using acetyl chloride (DMF) [18] and / orethyltrifluoroacetate in reflux (THF) [19] yielded 3-[(2'-acetylamino)-5'-fluorophenyl]-7-(4'-fluorophenyl)-1,2,4-triazino[3,4-b][1,3,4]thiadiazol-4-one (4) and/or 3-(2'-trifluoroacetylamino-5'-fluoro-phenyl-7-(4'-fluorophenyl)-1,2,4-triazino[3,4-b][1,3,4]thiadiazol-4-one (5) (Scheme 3). Also, self cyclo-condensation of compound 1 via boiling with DMF furnished thiadiazolo-1,2,4- triazinoindole derivative 6 (Scheme 3).

Bonded phosphorus atoms with S, O, N and C-atoms of heterocyclic system enhance their biocidal properties as herbicides, pesticides, insecticides and molluscicidal agents [20] -[22] . With this observations, the present work aims to synthesize of new fused heterobicyclicbearing fluorine and phosphorus atoms through phosphorylation of compound 1 with diphenyl phosphoryl chloride in warm DMF to give 3-(2'-diphenylphosphatoamino-5'-flu-orophenyl)-7-(4'-fluorophenyl)-1,2,4-triazino[3,4-b][1,3,4]thiadiazol-4-

one (7) (Scheme 4).

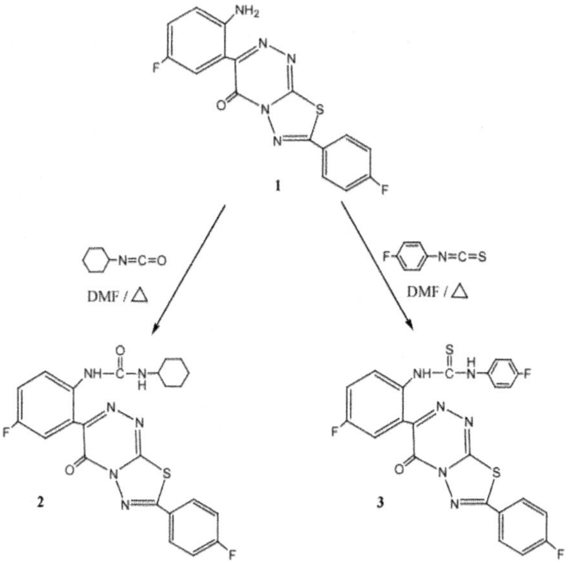

Scheme 2: Synthesis of compounds 2 and 3.

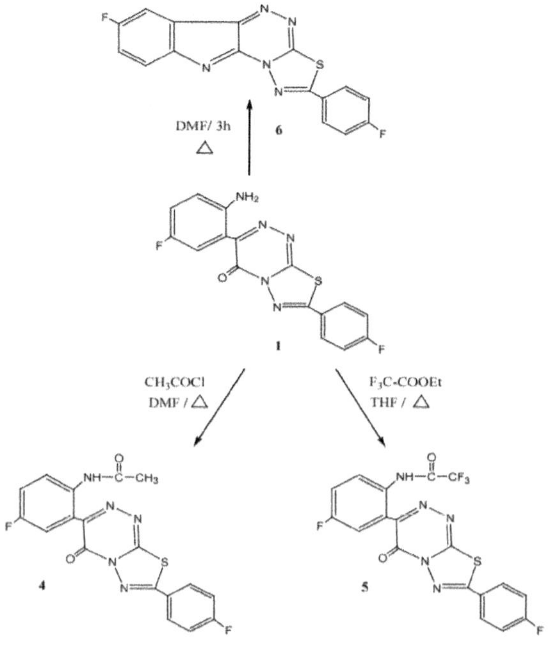

Scheme 3: Synthesis of compounds 4-6.

Due to a highly with drown of P of phosphate moiety, the chlorine atom is very labile. Thus, simple Nu⁻ attack of NH₂ to P atom afforded the aminophosphate derivative.

Full fluorinated 3-[5ʹ-fluoro-2ʹ-(4»-fluorobenzoylamino)phenyl]-7-(4ʹ-fluorophenyl)-1,2,4-triazino[3,4-b][1,3,4] thiadiazole-4-one (8) was obtained from treatment of compound 1 with 4-fluorobenzoylchloride in warm DMF (Scheme 4).

Due to a higher nucleophilicity of amino-group and the better displacement of labile chlorine atom of halo acids the interested point in this investigation is a simple nucleophilic attack of amino-group of compound 1 to a higher electrophilic carbons of α-haloacids as monochloroacetic acid and/or 1,1-dichloroacetic acid in warm DMF, yielded [23] 3-(5»-fluoro-2ʹ-carboxymethylanilino)-7-(4ʹ-fluorophenyl)-1,2,4,-triazino[3,4-b][1,3,4] thia-diazol-4- one (9) and/or 3-(5ʹ-fluoro-2ʹ-carboxymithinicanilino)-7-(4ʹ-fluorophenyl)-1,2,4-triazino[3,4-b][1,3,4] thiadiazol- 4-one (10) (Scheme 5).

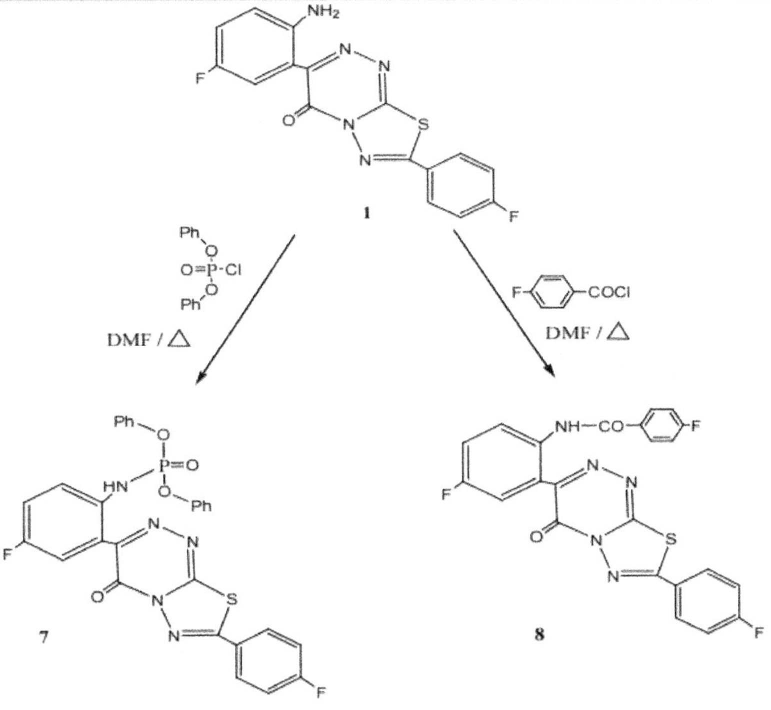

Scheme 4: Synthesis of compounds 7 and 8.

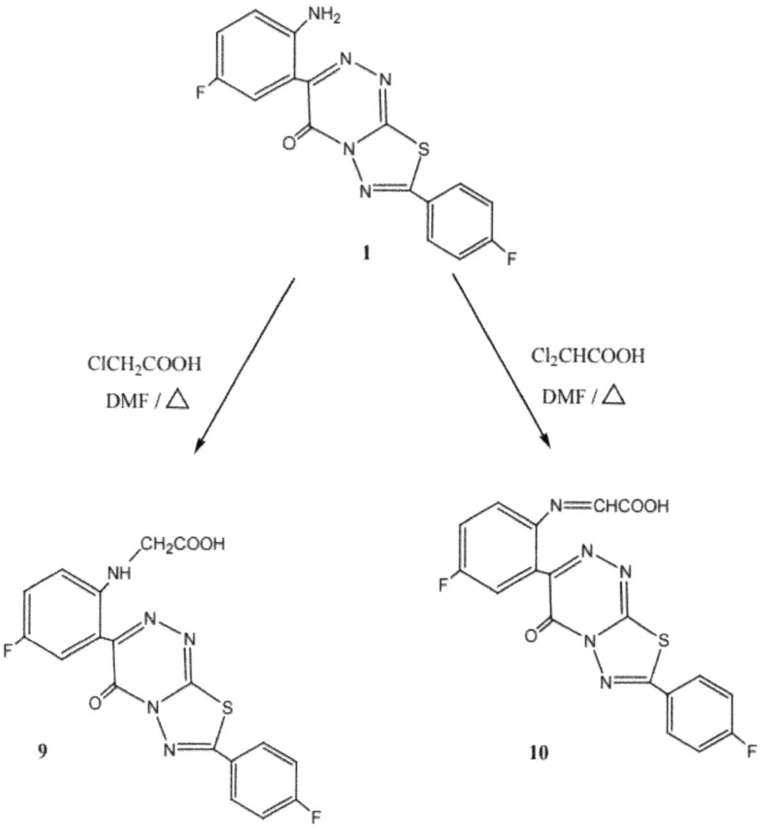

Scheme 5: Synthesis of compounds 9 and 10.

Alkylation, reaction of 3-amino-1,2,4-triazinothiadiazolone 1with chloroacetonitrile in warm DMF produced [24] the 3-(5'-fluoro-2'-(cyanomethylanilido)-7-(4'-fluorophenyl)-1,2,4-triazino[3,4-b][1,3,4] thiadiazol-4-one (11) (Scheme 6). Acidic hydrolysis of 11 by reflux with dil. HCl afforded compound 9. Decarboxylation of 9 via warm with sodium bicarbonate solution, 3-(5'-fluoro-2'-(methylanilido)-7-(4'-fluorophenyl)-1,2,4-triazino[3,4-b] [1,3,4] thiadiazol-4-one (12) was isolated. The compound 12 was also, obtained from stirring of compound 1 with MeI in 1% KOH solution (Scheme 6).

The adducts formed by reactions of nitrogen containing aromatic heterocycles with various electrophilic carbon, may be stable systems, or they can undergo further transformation, such as aromatization.

Abdel-Rahmanetal reported [1] [25] , fluorine substituted thiobarbituric acid derivatives use as anti HIV-1 and cyclin dependent kinase 2 (CDK2) for cell tumor division, thus ring closure reaction of compound 3 with malonic acid in boiling with glacial acetic acid afforded the fluorine substituted N,N'-disubstitutedthiobarbituric acid 13 (Scheme 7). Formation of compound 13 was deduced from ring closure reaction of substituted thiourea 3 with malonic acid (Figure 2).

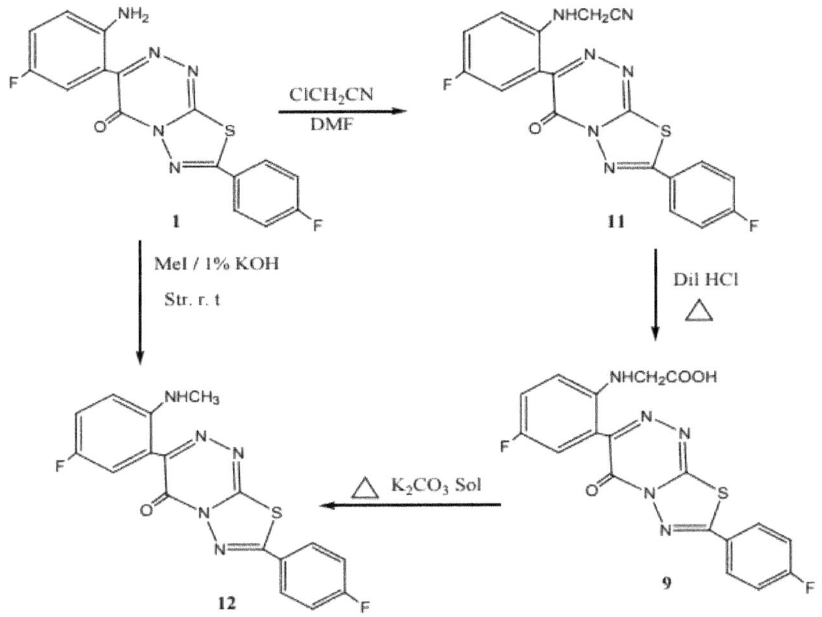

Scheme 6: Synthesis of compounds 11 and 12.

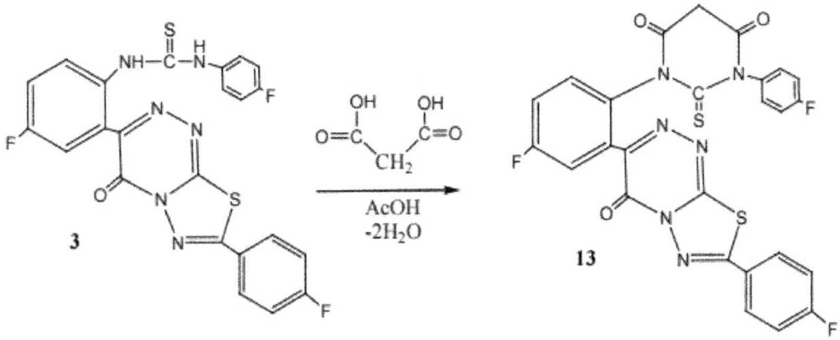

Scheme 7: Synthesis of compounds 13.

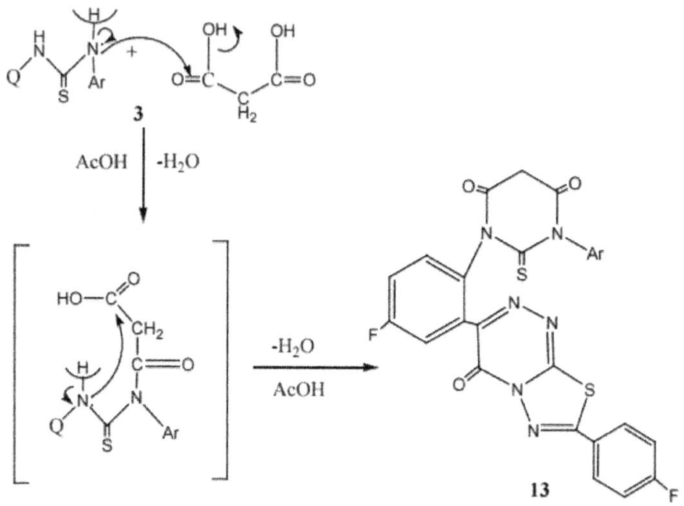

Figure 2: Formation of compound 13 from 3.

Former structures of the fluorinated 1,2,4-triazino[3,4-b][1,3,4] thiadiazolonederivatives have been established by help of their correct elemental analysis and spectral measurements:

1) UV absorption spectra of most N-acyl/phosphoryl derivatives for example 4, 6 and 7 recorded λ_{max} 304 nm as parent amino-derivatives 1, which is may be that electronic inhibition over NH_2 by high electronic acceptor acyl, and/or phosphoryl. On the other hand, UV absorption spectra of compound 5 and 8 showed λ_{max} at 311, and 375 nm respectively, which is may be the introduction of $COCF_3$ and/or COC_6H_4F-p to an amino group of 1. Thus NH proton of these compounds is highly acidic character. In addition, UV absorption spectrum of 13 showed an additive λ_{max} at 410 nm, which attribute to formation of fluorinated thiobarbituric acid bearing of 1,2,4-triazino-1,3,4-thiadiazinone moiety.

2) IR-spectra of all the obtained compounds (expected 6, 10, and13) recorded the absorption bands at 3200 - 3100 cm for NH functional group, while, that of compounds showed an two C=O of NH acyl and 1,2,4-triazinone at 1690 - 1650 cm^{-1}. All the synthesized showed a charactic bands of stretching and bending of C-F at γ 1250 and 720 cm^{-1}. Only the compounds 7, exhibited the presence of P=O and C-O-Ar at 1097 and 1016 cm^{-1}, while the compounds 4, 9, 11, and 12 showed the absorption bands of aliphatic groups at 2880 and 1440 cm^{-1}. Some compounds as 5, 7, 8, and 11recorded a lack's of NH functional group, which is may be formation a type of H-bonding (Figure 3).

3)   NMR spectra of the new synthesized compounds was confired that structures.

a) ^{1}H NMR spectra of all the obtained systems exhibited the presence of a resonated signals at δ 8.73 ppm for NH proton (s), in addition, δ at 7.97 - 7.94 and 7.38 - 7.35 ppm for 7 aromatic protons (m). On the other hand, compounds 2, 4, 9, 11, 12, 13 showed signals at 2.51 ppm for an aliphatic protons (COCH$_3$, CH$_2$CN, CH$_2$- COOH, CH$_2$CO) (J = 8.5). Only the compound 10 recorded δ at 8.8 ppm for N=CH proton (s).

4) ^{13}C NMR spectra of all the prepared compounds showed a resonated signals of fluorinated 1,2,4-triazi- no[3,4-b][1,3,4]thiadiazinone carbons at δ 164-163 (C=O), 135 (C-F), 130 - 120 (aromatic C), 116 (C=N) ppm. Only the compound 3 showed a resonated signals at 180 ppm attribute to C=S. On the other hand most of the new compounds showed an additional δ 160 ppm for new NHCO carbons.

5)   Mass fragmentation study of some new systems 5 showed the molecular ion peak, with a base peak at m/e 95 and 190 (100%).

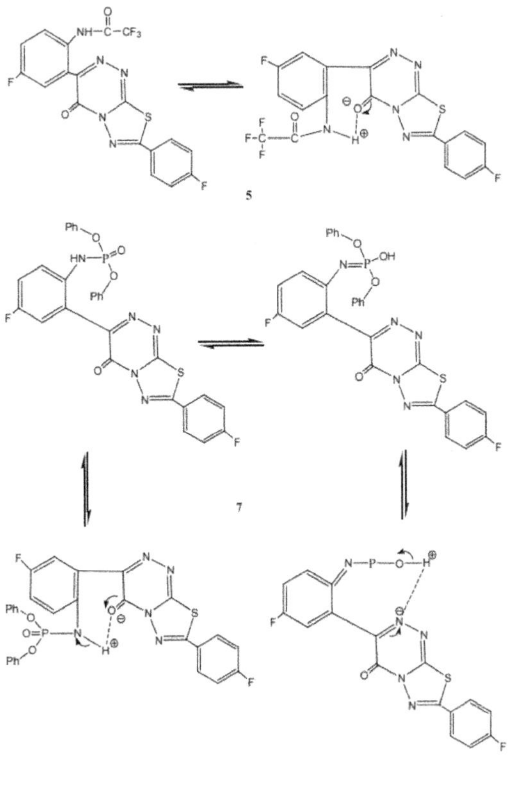

Figure 3: A possible structural formula of compounds 5, 7.

The first base peak is 4-fluorophenylradical while the other is 4,4'-difluorobiphenyl radical.. Also present M^{+2} is attributed to S and F isotopic (Figure 4).

On the other hand, mass fragmentation study of compound 7 recorded a molecular ion peak at m/e 590 with a base peak at m/e 248 as $C_{12}H_{11}NPO_3$iminophosphato ion radical with 4-fluorophenyl cation at 95%. Stability of a base peak is may be due to a higher stability of N = P group, which supported by donation and back-donation between N and P atoms (Figure 5).

CONCLUSION

The present work describes a facile and simple nucleophilic attack of amino group bearing fluorine substituted 1,3,4-thiadiazolo[2,3-c][1,2,4]triazine(1) to active electrophilic reagents via addition, fluorinated acylation/ aroylation, phosphorylation and/or alkylation reactions. Substituted amino derivatives were obtained coupling with electronic modifications overall the molecule which led to potential anthelmintic activity. Among the new compounds 5 > 7 > 1 exhibit a higher activity (55% - 60%), while other 13 > 9 > 6 > 4 (45% - 48%) showed a moderate activity towards N. brasiliensis virus. A higher activity of compounds 5 and 7 may be attributed to bourdation of $COCF_3$ and/ or phosphoryl groups with amino groups of start 1.

EXPERIMENTAL

Melting points of the products were determined on Stuart SMP_3 (UK) and uncorrected. UV absorption spectra (λ_{max} nm) were recorded in DMF on Shimadzu UV and visable 310 IPC-spectro-photometer. A Perkins Elmer Model RXI-FT IR system 55529 used for recording IR spectra of the prepared compounds γ cm^{-1}. A Brucker advanced D P X 400 MHz model using TMS as internal standard used for recording the ^{1}H and ^{13}C NMR spectra of the compounds on deuterated ($CDCl_3$, d$_6$, δ ppm). Mass spectrum was measured on GCMS Q 1000 Ex at 70 eV. Elemental microanalysis were performed by the microanalytical at Cairo-University, Egypt.

3-(2'-Amino-5'-fluorophenyl)-7-(4'-fluorophenyl)-1,2,4-triazino[3,4-b][1,3,4] thiadi-azin-4-one (1)

Equimolar amounts of 5-fluoroisatin (in 5% aqueous NaOH), and 2-hydrazino-2-(4'-fluorophenyl)-4H-1,3,4-thiadiazole and reflux for 3 h, cooled then poured on to ice-HCl.

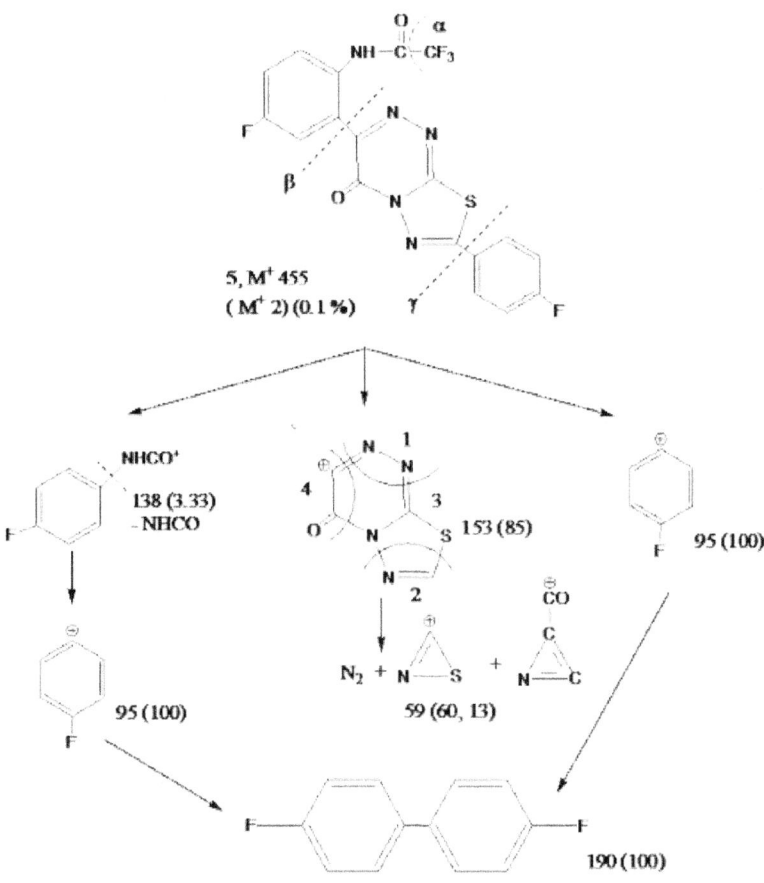

Figure 4: Mass fragmentation pattern of compound 5.

The solid thus obtained filtered off and crystallized from dioxan to give 1 faint yellow crystals, yield 77%, m.p. 189°C - 195°C. UV (EtOH) λ_{max} 303nm. IR (γ cm^{-1}): 3264 and 1631 (NH$_2$), 1681 (C=O), 1601, 1507 (C=N), 1320 (cyclic NCSN), 1225 (C-F), 1156 (C-S), 826 (p-substituted phenyl), 618 (C-F). ^{1}H NMR (DCCl$_3$-d$_6$) δ (ppm): 8.61 - 7.84, 7.84 - 7.83, 7.82 - 7.81, 7.818 - 7.812, 7.24, 7.15, 7.14, 7.13, 7.115, 7.110 (aromatic CH), 3.5 (s, 2H, NH$_2$). ^{13}C NMR (DCCl$_3$-d$_6$) δ (ppm): 165, 163, 160, 130.58, 130.49, 130.37, 116.15, 115.93, 77.34, 77.03, 76.71. M/Z: (Int.%): 357 (57% M$^+$H$_2$O) 244 (1.0), 206 (15.0), 178 (100), 134 (45), 95 (25.0). CHNSF analysis for C$_{16}$H$_9$N$_5$SF$_2$O (357), Calcd: C, 53.78; H, 2.52; N, 19.60; S, 8.63; F, 10.64%. Found: C, 53.58; H, 2.32; N, 19.55; S, 8.33; F, 10.43%.

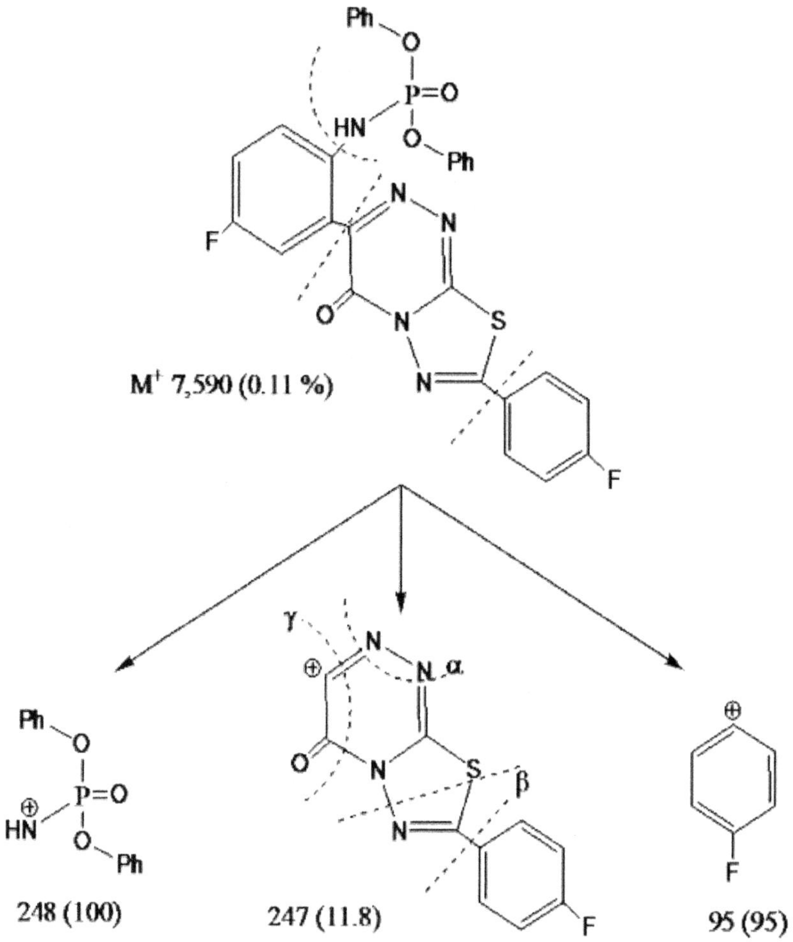

Figure 5: Mass fragmentation pattern of compound 7.

N-(Cyclohexyl)-N'-[4'-oxo-1,2,4-triazino[3,4-b][1,3,4]thiadia-zol-7'- (4"-fluorophenyl)-3'-(4"-fluorophenyl)]-urea (2).

A mixture of compound 1 (0.01 mol) and cyclohexyl isocyanate (0.01 mol) in DMF (50 ml) warmed for 2h, cooled then poured onto ice. The solid produced filtered off and crystallized from dioxan to give 2 as faint yellow. Yield 60%, m.p. 178°C - 180°C. IR (γ) cm⁻¹: 3325 (NH), 2928, 2851 (aliphatic CH, CH$_2$) 1700 (C=O), 1629 (CONH), 1602 (C=N), 1413 (deformation CH$_2$), 1320 (cyclic NCSN), 1225 (C-F), 1155 (C-S), 826, 796 (p-substituted phenyl), 635 (C-F). ¹H NMR (DMSO-d$_6$) δ (ppm): 8.73 (s, 2H, NH, NH), 7.97, 7.96, 7.95, 7.94 (4H, aryl protons), 7.38, 7.37, 7.35 (3H, aryl protons), 2.511, 1.8, 1.7, 1.6,

1.2, (each s, 11H, aliphatic protons). ^{13}C NMR (DMSO-d$_6$) δ (ppm): 168.4, 163.4 (2 C=O), 160 (C-S), 130.71, 130.65, 130.38, 130.36 (aromatic carbons), 116, 116.1 (C-F), 33.32, 25.28, 24.45 (aliphatic carbons). CHNSF analysis for C$_{23}$H$_{20}$N$_6$SF$_2$O$_2$ (482). Calcd: C, 57.26; H, 4.14; N, 17.42; S, 6.6; F, 7.88%. Found: C, 56.96; H, 4.05; N, 17.11; S, 6.33; F, 7.75%.

N-(4'-Fluorophenyl)-N'-[(4'-oxo-1',2',4'-triazino[3,4-b][1,3,4] thiadiazol-7'- (4''-fluoro-phenyl)-3'-(4''-fluorophenyl)]thiourea (3)

A mixture of 1 (0.01 mol) and 4-fluorophenyl isothiocyanate (0.01 mol) in DMF (20 ml) was refluxed for 1h, cooled then poured onto ice. The yielded solid filtered off and crystallized from EtOH to give 3 as yellow crystals. Yield 72%, m.p. 164°C - 165°C. IR (γ) cm^{-1}: 3272 (NH), 3016 (aromatic CH), 1695 (C=O), 1612 (C=N), 1600 (C=C), 1328 (cyclic NCSN), 1225 (C-F), 1208 (C=S), 1154 (C-S), 829,732, 719 (p-substituted phenyl), 639 (C-F).^{1}H NMR (DMSO-d$_6$) δ (ppm): 9.76, 8.73 (each s, 2H, NH), 7.97, 7.95, 7.47, 7.46, 7.45, 7.44, 7.38, 7.35, 7.18, 7.14, 7.11 (m, m, 11H, aromatic protons). ^{13}C NMR (DMSO-d$_6$) δ (ppm): 180.29 (C=S), 164.7 (C=O), 135.98, 135.62, 135.61 (C-N), 130.71, 126.24 (aromatic carbons), 119.97, 119.92 (C-N), 116.14, 115.32, 114.98 (C-F). CHNSF analysis for C$_{23}$H$_{13}$N$_6$S$_2$F$_3$O (510). Calcd: C, 54.11; H, 2.54; N, 16.47; S, 12.54; F, 11.17%. Found: C, 53.98; H, 2.48; N, 16.24; S, 12.38; F, 10.27%.

3-[(2'-Acetylamino)-5'-fluorophenyl]-7-(4'-fluorophenyl)-1,2,4-triazino[3,4-b] [1,3,4]thiadiazol-4-one (4)

A mixture of 1 (0.200 gm) and glacial acetic acid (5 ml) was warmed for 5 min, cooled then poured onto ice. The solid produced filtered off and crystallized from AcOH to give 4 as pall yellow crystals. Yield 55%, m.p. 182°C - 183°C. IR (γ) cm^{-1}: 3123 (NH), 1632 (C=O), 1603 (C=N), 1414 (deformation CH$_3$), 1321 (cyclic NCSN), 1225 (C-F), 1155 (C-S), 827, 795 (p-substituted phenyl), 635 (C-F). ^{1}H NMR (DMSO-d$_6$) δ (ppm): 8.73 (s,1H, NH), 7.97, 7.96, 7.95, 7.94 (m, 4H, aromatic protons), 7.38, 7.37, 7.3 (m, 3H, aromatic protons) 2.51 (s, 3H, CH$_3$). ^{13}C NMR (DMSO-d$_6$) δ (ppm): 164.71 (C=O), 160.50 (C=O), 130.71, 130.65, 130.36, (aromatic carbons), 116.15, 116.0 (C-F), 39.03 (CH$_3$). CHNSF analysis for C$_{18}$H$_{11}$N$_5$SF$_2$O (399). Calcd: C, 54.13; H, 2.75; N, 17.54; S, 8.0; F, 9.52%. Found: C, 53.88; H, 2.69; N, 17.32; S, 7.75; F, 9.40%.

3-(2'-Trifluoroacetylamino-5'-fluorophenyl)-7-(4'-fluorophenyl)-1,2,4-triazino [3,4-b][1,3,4]thiadiazol-4-one (5)

Equimolar mixture of 1 and trifluoroethyl acetate in THF (50 ml) refluxed for 2 h, cooled. The solid thus obtained filtered off and crystallized from dioxan to give 5 as white crystals. Yield 82%, m.p. 179°C - 180°C. IR (γ) cm^{-1}: 1632 (C=O), 1603 (C=N), 1321 (cyclic NCSN), 1225 (C-F), 1155 (C-S), 827, 795 (p-substituted phenyl), 635 (C-F). ^{1}H NMR (DMSO-d$_6$) δ (ppm): 8.73 (s,1H, NH), 7.97, 7.96, 7.95, 7.94 (4H, aromatic protons), 7.38, 7.37, 7.35 (3H, aromatic protons). ^{13}C NMR (DMSO-d$_6$) δ (ppm): 164.71 (C=O), 160.50 (C=O), 130.71, 130.65, 130.38, 130.36 (aromatic carbons), 116.14, 116.00 (C-F). M/S (Int.%): 455 (M$^+$2, 0.11%), 190 (100), 153 (85.0), 138 (3.33), 95 (100), 66 (1.181, 60, 13.0). CHNSF analysis for C$_{18}$H$_8$N$_5$SF$_2$O$_2$ (455, M+2). Calcd: C, 47.68; H, 1.76; N, 15.45; S, 7.06; F, 20.97%. Found: C, 47.38; H, 1.73; N, 15.19; S, 6.88; F, 20.67%.

10-(4'-Fluorophenyl)-4-fluoro-1,3,4-thiadiazolo[2,3-c][1,2,4] triazino[5,6-b]indole (6)

Compounds 1 (0.20 gm) in DMF (20 ml) refluxed for 3h, cooled then poured onto ice. The yielded solid filtered off and crystallized from EtOH to give 6 as yellowish crystals. Yield 60%, m.p. 184°C - 185°C. IR (γ) cm^{-1}: 1603 (C=N), 1321 (cyclic NCSN), 1226 (C-F), 1155 (C-S), 827, 796 (p-substituted phenyl), 635 (C-F). ^{1}H NMR (DMSO-d$_6$) δ (ppm): 7.97, 7.96, 7.95, 7.94 (4H, aromatic protons), 7.38, 7.37, 7.35 (3H, aromatic protons).^{13}C NMR (DMSO-d$_6$) δ (ppm): 130.71, 130.65, 130.38, 130.36 (aromatic carbons), 116.14, 116.0 (C-F). CHNSF analysis for C$_{16}$H$_7$N$_5$SF$_2$ (339). Calcd: C, 56.63; H, 20.6; N, 20.6; S, 9.43; F, 11.20%. Found: C, 56.55; H, 2.03; N, 20.18; S, 9.14; F, 10.98%.

3-(2'-Diphenylphosphatoamino-5'-fluorophenyl)-7-(4'-fluorophenyl)-1,2,4- triazino[3,4-b][1,3,4]thiadiazol-4-one (7)

An equimolar mixture of 1 (0.01 mol) and diphenyl phosphoryl chloride (0.01mol) in DMF (20 ml) refluxed for 30 min, then, cooled and poured onto ice. The produced solid filtered off and crystallized from THF to give 7 as yellowish crystals, yield 60%, m.p. 173°C - 175°C. IR (γ) cm^{-1}: 3061(aromatic CH), 1632 (C=O), 1602 (C=N), 1320 (cyclic NCSN), 1225 (C-F), 1155 (C-S), 1097 (P=O), 1016 (Ph-O-P), 962, 936, 870, 827, 795 (substituted phenyl), 635 (C-F). ^{1}H NMR (DMSO-d$_6$) δ (ppm): 8.73 (s,1H, NH), 7.97, 7.96, 7.95, 7.94 (4H, aromatic protons), 7.38, 7.37, 7.35 (3H, aromatic protons), 7.2-6.8 (m,10H, phenyl protons). ^{13}C NMR (DMSO-d$_6$) δ (ppm): 160.50 (C=O), 130.71, 130.65, 130.38, 130.36 (aromatic carbons), 116.14, 116.0 (C-F). M/S

(Int.%):590 (M⁺, 0.11%), 248 (100), 247 (11.8), 95 (95.0). CHNSF analysis for $C_{28}H_{18}N_5SF_2PO_4$(589). Calcd: C, 57.04; H, 3.05; N, 11.88; S, 5.43; F, 6.45%. Found: C, 57.01; H, 2.98; N, 11.60; S, 5.37; F, 6.14%.

3-[5′-Fluoro-2′-(4″-fluorobenzoylamino)-phenyl]-7-(4′-fluorophenyl)-1,2,4- triazino[3,4-b][1,3,4]thiadiazol-4-one (8)

A mixture of 1 (0.01 mol) and 4-fluorobenzoyl chloride (0.01 mol) in DMF (20 ml) refluxed 1 h, cooled. The reaction mixture poured onto ice. The solid produced filtered off and crystallized from dioxan to give 8 as yellowish crystals, yield 65%, m.p. 149°C - 150°C. IR (γ) cm⁻¹: 3200-3100 (b, OH, NH), 1677 (cyclic C=O), 1633 (NHCO), 1602 (C=N), 1315 (cyclic NCSN), 1225 (C-F), 1156 (C-S), 827, 796, 768 (p-substituted phenyl), 635 (C-F). ¹H NMR (DMSO-d₆) δ (ppm): 13.09 (s,1H, NH), 8.02, 8.05, 8.01, (m, 3H, aromatic protons), 7.97, 7.96, 7.95, 7.94 (m, 4H, aromatic protons), 7.38-7.32 (m, 4H, aromatic protons). ¹³C NMR (DMSO-d₆) δ (ppm): 166.36, 164.04 (2 C=O), 132.11, 132.05, 130.65, 130.38, 130.36, 127.33 (aromatic carbons), 116.15, 116.0 (C-F), 115.69, 115.54 (C-N). CHNSF analysis for $C_{22}H_{12}N_5SF_3O_2$(467). Calcd: C, 56.53; H, 2.56; N, 14.98; S, 6.85; F, 12.20%. Found: C, 56.28; H, 2.50; N, 14.53; S, 6.67; F, 12.00%.

3-(5′-Fluoro-2′-carboxymethylanilino)-7-(4′-fluorophenyl)-1,2,4-triazino[3,4-b] [1,3,4]thiadiazol-4-one (9)

A mixture of 1(0.01 mol) and monochloroacetic acid (0.01 mol) in DMF (20 ml) refluxed for 1h, cooled then poured onto ice. The solid produced filtered off and crystallized from THF to give 9 as yellowish crystals, yield 72%, m.p. 180°C - 182°C. IR (γ) cm⁻¹: 3500 - 3100 (b, OH, NH), 1720, 1633 (2 C=O), 1603 (C=N), 1490, 1413 (deformation CH₂), 1321 (cyclic NCSN), 1226 (C-F), 1156 (C-S), 826, 796 (substituted phenyl), 636 (C-F). ¹H NMR (DMSO-d₆) δ (ppm): 8.73 (s,1H, NH), 7.97, 7.96, 7.95, 7.94 (m, 4H, aromatic), 7.38, 7.37, 7.35 (m, 3H, aromatic), 4.55 (s, 1H, OH), 2.51 (s, 2H, J,8.7, CH₂). ¹³C NMR (DMSO-d₆) δ (ppm): 164.71 (C=O), 163.06 (C=O), 130.70, 130.65, 130.38, 130.36(aromatic carbons), 116.14, 116.0 (C-F), 39.03 (CH₂). CHNSF analysis for $C_{18}H_{11}N_5SFO_3$(415). Calcd: C, 52.53; H, 2.65; N, 16.86; S, 7.71; F, 9.15%. Found: C, 52.33; H, 2.62; N, 16.59; S, 7.62; F, 9.04%.

3-(5′-Fluoro-2′-carboxymethinicanilino)-7-(4′-fluorophenyl)-1,2,4-triazino [3,4-b][1,3,4]thiadiazol-4-one (10)

A mixture of 1 (0.01 mol) and 1,1′-dichloroacetic acid (0.01 mol) in DMF (20 ml) refluxed for 30 min, cooled then poured onto ice. The solid thus obtained filtered off and crystallized from EtOH to give 10 as yellowish crystals, yield

60%, m.p. 154-155°C. IR (γ) cm⁻¹: 3500 - 3300 (OH), 1696, 1632 (C=O), 1602 (C=N), 1321 (cyclic NCSN), 1226 (C-F), 1155 (C-S), 826, 796 (substituted phenyl), 635 (C-F). ^{1}H NMR (DMSO-d$_6$) δ (ppm): 8.8 (s,1H, N=CH), 7.97, 7.96, 7.95, 7.94 (m, 4H, aromatic), 7.38, 7.37, 7.35 (m, 3H, aromatic), 4.41 (s, 1H, OH). ^{13}C NMR (DMSO-d$_6$) δ (ppm): 164.71 (C=O), 163.06 (C=O), 130.71, 130.65, 130.38, 130.36, 128.72, 127.25 (aromatic carbons), 116.14, 116.0 (C-F). CHNSF analysis for C$_{18}$H$_9$N$_5$SFO$_3$ (413). Calcd: C, 52.30; H, 2.17; N, 16.94; S, 7.74; F, 9.20%. Found: C, 52.08; H, 2.11; N, 16.55; S, 7.54; F, 8.89%.

3-(5″-Fluoro-2′-cyanomethylanilino)-7-(4′-fluorophenyl)-1,2,4-triazino[3,4-b] [1,3,4]thiadiazol-4-one (11)

Equimolar amounts of 1 and chloroacetonitrile in DMF (20 ml) refluxed for 1h, cooled then poured onto ice. The solid produced filtered off and crystallized from THF to give 11 as brown crystals, yield 65%, m.p. 177°C - 178°C. IR (γ) cm⁻¹: 3180 (NH), 2220 (C N), 1635 (C=O), 1507, 1413 (deformation CH$_2$), 1320 (Cyclic NCSN), 1225 (C-F), 1155 (C-S), 226, 796 (Substituted phenyl), 635 (C-F).^{1}H NMR (DMSO-d$_6$) δ (ppm): 8.73 (s,1H, NH), 7.97, 7.96, 7.95, 7.94 (4H, aromatic), 7.38, 7.37, 7.35 (m, 3H, aromatic), 3.36 (2H, J, 6.6 p.c.s, CH$_2$).^{13}C NMR (DMSO-d$_6$) δ (ppm): 160.50 (C=O), 130.71, 130.65, 130.38, 130.36, 128.91, 127.88 (aromatic carbons), 116.14, 116.0 (C-F), 39.03 (CH$_2$CN). CHNSF analysis for C$_{18}$H$_{10}$N$_6$SF$_2$O(396). Calcd: C, 54.54; H, 2.52; N, 21.21; S, 8.08; F, 9.59%. Found: C, 54.31; H, 2.50; N, 20.89; S, 7.98; F, 9.41%.

3-(5′-fluoro-2′-Methylanilino)-7-(4′-Fluorophenyl)-1,2,4-Triazino[3,4-b][1,3,4] Thiadi-Azol-4-One (12)

A mixture of 9 (0.20 gm) and K$_2$CO$_3$solution (5%, 50 ml) refluxed for 1h, cooled then acidification use 5% HCl. The solid obtained filtered off and crystallized from EtOH to give 12 as brownish crystals, yield 55%, m.p. 178°C - 180°C. IR (γ) cm⁻¹: 3062 (aromatic CH), 1700, 1632 (C=O), 1600 (C=N), 1507, 1413 (deformation CH$_3$), 1321 (cyclic NCSN), 1225 (C-F), 1155 (C-S), 820, 795 (Substituted phenyl), 635 (C-F). ^{1}H NMR (DMSO-d$_6$) δ (ppm): 8.73 (s,1H, NH), 7.97, 7.96, 7.95, 7.94 (m, 4H, aromatic), 7.38, 7.37, 7.35 (m, 3H, aromatic), 2.22 (s, 3H, CH$_3$N). ^{13}C NMR (DMSO-d$_6$) δ (ppm): 160.11 (C=O), 130.71, 130.65, 130.38, 130.36, 128.72 (aromatic carbons), 116.14, 116.0 (C-F), 22.65(CH$_3$). CHNSF analysis for C$_{17}$H$_{11}$N$_5$SF$_2$O(371). Calcd: C, 54.98; H, 2.96; N, 18.86; S, 8.62; F, 10.24%. Found: C, 54.69; H, 2.22; N, 18.43; S, 8.49; F, 9.98%.

Formation of 9

A mixture of 11 (0.20 gm) and diluted HCl (10%, 50 ml) refluxed for 1h, coled. The solid produced filtered off and crystallized from THF to give 9 as yellowish crystals, yield 60%, m.p. 178°C - 179°C. Mixed melting point no depression.

N'[2'-(4"-Fluorophenyl)-5-oxo-6-(5'-fluorophenyl-2"-yl)-N3-(4'-fluorophenyl)- thiobarbituric acid (13)

Equimolar mixture of 3 and malonic acid in glacial acetic acid (20 ml) refluxed for 4 h, cooled and poured onto ice. Extracted the organic layer by diethyl ether and leaf at room temperature. The solid obtained crystallized from dioxan to give 13 faint yellow crystals, yield 65%, m.p. 150°C - 151°C. IR (γ) cm^{-1}: 3530 (OH), 1660 (C=O), 1488 (deformation CH_2), 1385 (NCSN), 1255 (C-F), 1205 (C=S), 670 (C-F). ^{1}H NMR (DMSO-d$_6$) δ (ppm): 10.04 (s,1H,OH), 8.2 - 8.0 (m, 4H, aromatic protons), 7.60-7.44 (m, 4H, aromatic protons),7.0 - 6.98 (m, 3H, aromatic protons), 3.55, 2.59 - 2.58 (s, 2H, CH_2). ^{13}C NMR (DMSO-d$_6$) δ (ppm): 181.10, 165.71, 160.00, 159.38, 138.2, 133.40, 134.20, 126.91, 126.80, 125.1, 121.69, 115.45, 115.21, 77.79, 77.57, 77.36, 44.36, 40.46-39.8). CHNSF analysis for $C_{36}H_{13}N_6S_2F_3O_3$ (578). Calcd: C, 53.97; H, 2.24; N, 14.53; S, 11.67; F, 9.86%. Found: C, 53.79; H, 2.11; N, 14.33; S, 11.55; F, 9.59%.

PHARMACOLOGICAL EVALUATION

1,2,4-Triazine derivatives showed a wide biocidal spectrum [13] -[15] . Also, 1,3,4-thiadiazoles exhibited a large biocidal agents [11] [12] . In addition, introduction of both fluorine atoms and/or amino groups to heterocyclic systems often improve their medicinal properties [5] -[9] . Thus, in search for new drugs as potential anthelmintic to control on the smoke diseases, the present work, aim to obtain new drugs. All the new synthesized compounds were screened for their anthelmintic activity against H. nana infection in mice, by using the standard method of steward. The oral dose was 200 mg/Kg given for 2 days. Only the compounds 1, 3, 5, 6, 7, 9 and 13 recorded a weak activity (10 <%). On the other hand, evaluation of these compounds against N-brasiliensis infection in rats ta the same oral dose, by using other standard method [26] [27] . The obtained results showed that the activity in range of 25% - 60% (Table 1).

From the obtained results (Table 1) we can be conclude that compounds containing $COCF_3$ are highly effect than aromatic C-F. also, presence of phosphate group bonded to NH enhance that activity. Full fluorinated N,N›-diarylthiourea showed a rise activity towards N-brasifiensis.

Table 1: Anthelmintic activity of the new synthesized compounds.

Compound No.	H. nana	N. brasiliensis
1	10<	55
2	10<	29
3	10<	60
4	10<	35
5	10<	60
6	10<	45
7	10<	58
8	10<	32
9	10<	48
10	10<	29
11	10<	30
12	10<	30
13	10<	49

As well as N-alkyl systems exhibited a moderate activity. Thus, atype of both compounds 5 and 7 would present a fruitful matrix for the future development of a new class of potential anthelmintic agents, that deserves further investigation and derivation. A simple of nucleophilic attack of amino group of fluorinated 1,2,4-triazino[2,3-c]thiadiazolone to various electrophilic agents was deduced to give N-substituted analogues. The anthelmintic activity of these systems was evaluated. Among these tested analogs, compounds 5 and 7 showed 50% - 60% activity, while all the tested compounds exhibited below 10% activity towards H. nana.

CITE THIS PAPER

Mohammed Saleh TawfekMakki,Reda Mohammady AbdelRahman,Ola Ahmad AbuAli, (2015) Synthesis of New Fluorinated 1,2,4-Triazino [3,4-b] [1,3,4]thiadiazolones as Antiviral Probes-Part II-Reactivities of Fluorinated 3-Aminophenyl-1,2,4-triazinothiadiazolone. *International Journal of Organic Chemistry*,**05**,153-165. doi: 10.4236/ijoc.2015.53017

REFERENCES

1.   Al-Harbi, A.S., Abdel-Rahman, R.M. and Asiri, A.M. (2014) Synthesis of Some New Fluorine Substituted Thiobarbituric Acid Derivatives as Anti-HIV1 and Cyclin-Dependent Kinase 2(CDK2) for Cell Tumor

Division—Part II. International Journal of Organic Chemistry, 4, 142-153. http://dx.doi.org/10.4236/ijoc.2014.42016

2. Makki, M.S.T., Ab-del-Rahman, R.M. and Khan K.A. (2014) Fluorine Substituted 1,2,4-Triazinones as Potential Anti-HIV-1 and CDK2 Inhicitors. Journal of Chemistry, 2014, Article ID: 430573.

3. Makki, M.S.T., Abdel-Rahman, R.M., Faidallah, H.M. and Khan K.A. (2013) Synthesis of New Fluorine Substituted Heterocyclic Nitrogen Systems Derived from p-Aminosalicyclic Acid as Anti-Mycobacterial Agents. Journal of Chemistry, 2013, Article ID: 819462.

4. Makki, M.S.T., Abdel-Rahman, R.M., Faidallah, H.M. and Khan, K.A. (2013) Synthesis of Substituted Thioureas and Their Sulfur Heterocyclic Systems of p-Aminosalicylic Acid as Anti-Mycobacterial Agents. Journal of Chemistry, 2013, Article ID: 862463.

5. Abdel-Rahman, R.M. and Ali, T.S. (2013) Synthesis and Biological Evaluation of Some New Polyfluorinated 4-Thiazolidinone and α-Aminophosphonic Acid Derivatives. Monatshefte für Chemie, 144, 1243-1252. http://dx.doi.org/10.1007/s00706-013-0934-6

6. Makki, M.S.T., Bakhotmah, D.A., Abdel-Rahman, R.M. and Elshahawy, M.S. (2012) Designing and Synthesis of New Fluorine Substituted Pyrimidine-Thion-5-Carbonitriels and the Related Derivatives as Photochemical Probe Agents for Inhibition of Vitiligo Diseases. International Journal of Organic Chemistry, 2, 311-320.

7. Makki, M.S.T., Bakhotmah, D.A. and Abdel-Rahman, R.M. (2012) Highly Efficient Synthesis of Novel Fluorine Bearing Quinolone-4-carboxylic Acid the Related Compounds as Amylolytic Agents. International Journal of Organic Chemistry, 2, 49-55. http://dx.doi. org/10.4236/ijoc.2012.21009

8. Abdel-Rahman, R.M., Makki, M.S.T. and Bawazir, W.A. (2011) Synthesis of Some More Fluorine Heterocyclic Nitrogen Systems Derived from Sulfa Drugs as Photochemical Probe Agents for Inhibition of Vitiligo Disease—Part I. E-Journal of Chemistry, 8, 405-414. http:// dx.doi.org/10.1155/2011/586063

9. Abdel-Rahman, R.M., Makki, M.S.T. and Bawazir, W.A. (2010) Synthesis of Some More Fluorine Heterocyclic Nitrogen Systems Derived from Sulfa Drugs as Photochemical Probe Agents for Inhibition of Vitiligo Disease—Part II. E-Journal of Chemistry, 7, S93-S102.

10. Abdel-Rahman, R.M. (2001) Chemistry of Uncondensed 1,2,4-Triazines, Part IV—Synthesis and Chemistry of Bioactive 3-Amino-1,2,4-triazines and Related Compounds. Pharmacies, 56, 275-286.

11. Shawali, A.S, (2014) 1,3,4-Thiadiazoles of Pharmalogical Interest: Resent Trends in Their Synthesis via Tandem 1,3- Dipolar-Cycloaddition: Review. Journal of Advanced Research, 5, 1-17. http://dx.doi.org/10.1016/j.jare.2013.01.004

12. Kushwaha, N., Kushwaha, S.K.S. and Rai, A.K. (2011) Biological Activities of Thiadiazole Derivatives. International Journal of ChemTech Research, 4, 517-531.

13. Abdel-Rahman, R.M., Makki, M.S.T., Ali, T.S. and Ibrahim, M.A. (2012) 1,2,4-Triazine Chemistry Part III-Synthesis Strategies to Functionalized Brideghead Nitrogen Hetero Annulated 1,2,4-Triazine Systems and Their Region Specific and Pharmacological Properties. Current Organic Synthesis, 9, 1-25.

14. Abdel-Rahman, R.M., Makki, M.S.T., Ali, T.S. and Ibrahim, M.A. (2010) 1,2,4-Triazine Chemistry Part I: Orientation of Cyclization Reaction of Functionalized 1,2,4-Triazine Derivatives. European Journal of Chemistr, 1, 236-245.http://dx.doi.org/10.5155/eurjchem.1.3.236-245.54

15. Abdel-Rahman, R.M., Makki, M.S.T., Ali, T.S. and Ibrahim, M.A. (2014) 1,2,4-Triazine Chemistry Part IV: Synthesis and Chemical Behavior of 3-Functionalized-5,6-Diphenyl-1,2,4-Triazines towards Some Nucleophilic and Electrophilic Reagents. Journal of Hetero-cyclic Chemistry, ID JHET, 12-0734. http://dx.doi.org/10.1002/jhet.2014

16. El-Gendy, Z., Morsy, J.M., Allimony, H.A., Abdel-Monem, W.R. and Abdel-Rahman, R.M. (2001) Synthesis of Heterobicyclic Nitrogen Systems Bearing the 1,2,4-Triazine Moiety as Anti-HIV and Anticancer Drugs, Part III. Pharmazie, 56, 376-383.

17. Makki, M.S.T., Ab-del-Rahman, R.M. and Abu-Ali, O.A.(2015) Synthesis of Some More New Fluorinated 1,2,4-Triazino[3,4-b][1,3,4] Thiadiazolones and Their Molluscicidal Against Selective Snails—Part I. Journal of Chemistry and Chemical Engineering, 9, 162-175.

18. Abdel-Rahman, R.M. (1991) Synthesis and Anti-Human Immune Virus Activity of Some New Fluorine Con-taining Substituted 3-Thioxo-1,2,4-Triazin-5-Ones. Farmaco, 46, 379-389.

19. Abdel-Rahman, R.M. (1992) Synthesis of New Fluorine Bearing Trisubstituted 3-Thioxo-1,2,4-Triazin-5-Ons as Potential Anticancer Agents. Farmaco, 47, 319-326.

20. Al-Romazian, A.N., Makki, M.S.T. and Abdel-Rahman, R.M. (2014) Synthesis of New Fluorine/Phosphorus Substituted 6-(2'-Aminophenyl)-3-Thioxo-1,2,4-5(2H,4H)One and Their Related Alkylated Systems as Molluscicidal Agent as against the Snails Responsible for Bilharziiasis

Diseases. International Journal of Organic Chemistry, 4, 154-168. http://dx.doi.org/10.4236/ijoc.2014.42017

21. Breuer, E. (1996) The Chemistry of Organophosphorus Compounds. John Wiley and Sons, NewYork.

22. Blakley, B., Broussea, U., Fournier, M. and Voccia, A.I. (1999) Immunotoxicity of Pesticides: A Review. Toxicology and Industral Health, 15, 119-132. http://dx.doi.org/10.1177/074823379901500110

23. Ibrahim, M.A., Abdel-Rahman, R.M., Abdel-Halim, A.M., Ibracim, S.S. and Allimony, H.A., (2008) Synthesis and Antifungal Activity of Novel Polyheterocyclic Compounds Containing 1,2,4-Triazine Moiety. Archive for Organic Chemistry, 2008, 202-215. http://dx.doi.org/10.3998/ark.5550190.0009.g19

24. Abdel-Rahman, R.M. and Islam, E.I. (1993) Synthesis and Reactions of Acetonitrile Dervatives Bearing a 5,6-Diphenyl-1,2,4-Tryazin-3-yl Moiety. Indian Journal of Chemistry, 32, 526-529.

25. Al-Harbi, A.S., Abdel-rahman, R.M. and Asiri, A.M. (2015) Synthesis of Some New Fluorine Substituted Thiobarbituric Acid Derivatives as Anti HIV-1 and Cyclin-Dependent Kinase 2(CDK2) for Cell Tumor Division-Part I. European Journal of Chemistry, 6, 63-70. http://dx.doi.org/10.5155/eurjchem.6.1.63-70.1147

26. Steward, J.S. (1955) Anthelmintic Studies: II. A Double Entero-Nemacidal Anthelmintic Test Covering a Wide Range of Activities. Parasitology, 45, 242-254. http://Dx.Doi.Org/10.1017/S003118200002761x

27. Stadden, O.P. (1963) Experimental Chemotherapy. Academic Press, New York.

Chapter 14

SYNTHESIS, REACTIONS AND ANTIMICROBIAL ACTIVITY OF SOME NEW 3-SUBSTITUTED INDOLE DERIVATIVES

Asmaa S. Salman[1], Naema A. Mahmoud[1], Anhar Abdel-Aziem[1], Mona A. Mohamed[2], Doaa M. Elsisi[1]

[1]Department of Chemistry, Faculty of Science, Al-Azhar University, Girls' Branch, Nasr City, Cairo, Egypt

[2]Biochemistry Division, Faculty of Science, Al-Azhar University, Girls' Branch, Nasr City, Cairo, Egypt

ABSTRACT

Reaction of indole-3-carboxaldehydes 4 with hydrazine derivatives and different substituted acid hydrazides afforded the corresponding hydrazine derivatives 5a-c and acid hydrazide derivatives 7-11 respectively. Condensation of indole-3-carboxaldehydes 4 with phenacyl bromide and thiourea gives 1,3-thiazol-2-amine derivative 18. On the other hand, reaction 4 with 3-acetylchromene-2-one afforded chalcone derivative 19. Compound 4 undergoing Knoevenagel condensation with cyanoacetamide, ethyl cyanoacetate, benzimidazol-2-ylacetonitrile, rhodanine-3-acetic acid, 2,3- dihydropyrimidin-4-one derivative and 2,4-dihydropyrazol-3-one afforded the compounds 20a,b, 22, 23, 27 and 28 respectively. The structure of the newly synthesized compounds has been confirmed by elemental analysis and spectra data. The antimicrobial activities of the some newly synthesized compounds were measured and showed that most of them have high activities.

INTRODUCTION

In the recent past, bacterial infections have increased at an alarming rate causing deadly diseases and wide- spread epidemics in humans. All types of bacterial diseases have taken a high toll on humanity. The resistance of antibiotics to control emerging and pre-emerging bacterial pathogens focused the medicinal chemists to search potential new antimicrobial agents to cure microbial infections effectively [1] .

Heterocyclic compounds containing nitrogen have been described for their biological activity against various micro-organisms. The indole unit is the key building block for a variety of compounds which have crucial roles in the functions of biologically important molecules. Many indole alkaloids are recognized as one of the rapidly growing groups of marine invertebrate metabolites for their broad spectrum of biological properties [2] [3] . For example, five new indole alkaloids, meridianins A-E have been isolated from the tunicate Aplidium meridianum, which showed cytotoxicity toward murine tumor cell lines [4] .

Introduction of different groups to the modified indole structure can produce a series of compounds with multiple activities. Various 3-substituted indoles had been used as starting materials for the synthesis of a number of alkaloids, agrochemicals, pharmaceuticals and perfumes. Also 3-substituted indole derivatives possess various types of broad spectrum's biological activities such as antimicrobial, antitumor, hypoglycemic, anti-inflamma-tory, analgesic and antipyretic activities [5] [6] . Moreover the substitution at the 3-position of the indole ring can take place by connecting an additional heterocyclic ring, such as imidazole (topsentins, nortopsentins) [7] [8] , dihydroimidazole (discodermindole) [9] , oxazole (pimprinols A-C, almazole C) [10] [11] , thiazole (bacillamide A) [12] , quinazoline (tremorgens) [13] , and pyrimidine [14] . Therefore, 3-substituted indoles still represent a significant synthetic challenge. In view of the important biological properties of the indole ring, we planned to synthesize a new series of 3-substituted indole derivatives bearing side chains with different structures; as such derivatives could possess interesting and useful antimicrobial activity.

MATERIALS AND METHODS

Experimental

Melting points were measured on a Gallenkamp apparatus and are uncorrected. IR spectra were recorded on Shimadzu FT-IR 8101 PC infrared spectrophotometer (υ_{max} in cm^{-1}). The 1H NMR and^{13}C NMR spectra were determined in DMSO-d6 at 300 MHz on a Varian Mercury VX 300 NMR spectrometer using TMS as an internal standard. Mass spectra were measured on a GCMS-QP1000 EX spectrometer at 70 Ev. Elemental analyses were carried out at the Microanalytical Center of Cairo University. Spectral data of the synthesized compounds were given in Table 1.

General Procedure for the Synthesis of 2-Substituted-Indole (3a-d)

Synthesis of 2-substituted-1H-indole 3a-d was carried out by the procedure of Fischer indole synthesis. Phenylhydrazone derivatives 2a-d were prepared by warming a mixture of compounds 1a-d (0.04 mol) and phenyl hydrazine (0.072 ml, 0.04 mol) with 60 ml of ethanol and few drops of glacial acetic acid. The resulting reaction mixture was allowed to stirring for about 2 h. The reaction mixture was then poured into ice water (50 ml) where upon the crude compound was precipitated. The residue obtained after filtration was washed with water and used in second step. A mixture of 2a-d (0.01 mol) and polyphosphoric acid (20 ml) was refluxed for 6 h. After the completion of the reaction, it was filtered and filtrate was poured into ice cooled water. The solid obtained was filtered and recrystallized from the ethanol to give 3a-d.

2-(4-Methylphenyl)-1H-indole 3a

Yellow crystals. Yield: (1.24 g, 60%); m.p.: 220-221°C. Anal. calcd. for C_{15} H_{13} N(207.27): C, 86.92; H, 6.32; N, 6.76. Found: C, 86.62; H, 6.12; N, 6.56.

4-(1H-Indol-2-yl)aniline 3b

White crystals. Yield (1.37 g, 66%); m.p.: 250-252°C. Anal. calcd. for C_{14} H_{12} N_2 (208.25):C, 80.74; H, 5.81; N,13.45.Found: C, 80.54; H, 5.61; N, 13.35.

2-(4-Bromophenyl)-1H-indole 3c

Yellow crystals. Yield (1.63, 60%); m.p.: 220-222°C. Anal. calcd. for C_{14} H_{10} Br N (272.14): C, 61.79; H, 3.70; Br, 29.36; N, 5.15. Found: C, 61.59; H, 3.50; Br, 29.16; N, 5.00.

3-(1H-Indol-2-yl)-2H-chromen-2-one 3d

Dark brown crystals. Yield (1. 57, 60%; m.p.: 240-242°C (DMF). Anal. calcd. for C_{17} H_{11} NO_2(261.27): C, 78.15; H, 4.24; N, 5.36. Found: C, 78.00; H, 4.04; N, 5.06.

2-(4-Bromophenyl)-1H-indole-3-carboxaldehyde 4

Phosphorous oxychloride (21.47 ml, 0.14 mol) was added drop wise to N,N′-dimethylformamide (DMF) (10.23 ml, 0.14 mol) under cooling with an ice bath and the reaction mixture was stirred for 2 h. to prepare the Vilsmeier reagent. Then compound 3c (19.59 g, 0.072 mol) in DMF (20 ml) was added drop wise into the Vilsmeier reagent and continuous stirring and kept at room temperature for 2 h.

Table 1: Spectral data of the newly prepared compounds 3-31.

Compd. No.	Spectral Data
3a	FT-IR (KBr v_{max} cm^{-1}): 3436 (NH), 3042, 2911, 2856 (CH), 1610 (C=N). 1H NMR (DMSO-d6) δ ppm: 2.50 (s, 3H, CH$_3$), 6.82 (s, 1H, H-3 indole), 6.97 - 7.07 (m, 4H, Ar-H), 7.25 (d, 1H, indole proton), 7.36 - 7.73 (m, 2H, indole proton), 7.76 (d, 1H, indole proton), 11.45 (s, 1H, NH). MS. m/z (%): 207 (M$^+$, 100), 192 (3.12), 180 (3.18), 116 (1.99), 89 (20.56), 69 (29.66).
3b	FT-IR (KBr v_{max} cm^{-1}): 3366, 3260, 3180 (NH$_2$, NH), 2915, 2855 (CH), 1599 (C=N). 1H NMR (DMSO-d6) δ ppm: 6.54 (s, 2H, NH$_2$), 6.70 - 7.21 (m, 5H, Ar-H and H-3 indole), 7.50 (d, 1H, indole proton), 7.71 - 7.82 (m, 2H, indole proton), 8.19 (d, 1H, indole proton), 9.13 (s, 1H, NH).
3c	FT-IR (KBr v_{max}, cm^{-1}): 3182 (NH), 3057, 2977, 2867 (CH). 1H NMR (DMSO-d6) δ ppm: 6.46 (s, 1H, H-3 indole), 7.22 - 7.32 (m, 4H, Ar-H), 7.47 (d, 1H, indole proton), 8.21 (d, 1H, indole proton), 12.43 (s, 1H, NH).
3d	FT-IR (KBr v_{max} cm^{-1}): 3426 (NH), 3037, 2917, 2851 (CH), 1671 (C=O). 1H NMR (DMSO-d6) δ ppm: 5.46 (s, 1H, H-3 indole proton), 7.14 - 7.95 (m, 9H, Ar-H and indole proton), 8.20 (s, 1H, NH). MS. m/z (%): 261 (M$^+$, 18.48), 193 (2.80), 145 (18.01), 116 (13.51), 69 (100).
4	FT-IR (KBr, v_{max} cm^{-1}): 3181 (NH); 3063, 2978, 2868, 2713 (CH), 1672 (C=O), 1578 (C=N). 1H NMR (DMSO-d6) δ ppm: 7.22 - 7.32 (m, 4H, Ar-H), 7.50 (d, 1H, indole proton), 7.72 -7.82 (m, 2H, indole proton), 8.19 (d, 1H, indole proton), 9.95 (s, 1H, CHO), 12.40 (s, 1H, NH exchanged by D$_2$O). MS. m/z (%): 300 (M$^+$, 100, %), 220 (47), 219 (90.8), 190 (53), 165 (19.1), 143 (15.7).
5a	FT-IR (KBr, v_{max} cm^{-1}): 3421, 3385, 3129 (NH$_2$, NH), 3048, 2965, 2860 (CH), 1607 (C=N), 1578 (C=C). 1H NMR (DMSO-d6) δ ppm: 7.05 - 7.29 (m, 4H, Ar- H), 7.46 (d, 1H, indole proton), 7.66 - 7.68 (m, 2H, indole proton), 8.42 (d, 1H, indole proton), 8.91 (s, 1H, =CH), 12.06 (s, 1H, NH exchanged by D$_2$O), 4.34 (s, 2H, NH$_2$ exchanged by D$_2$O).
5b	FT-IR (KBr, v_{max} cm^{-1}): 3249 (NH), 3056, 2921, 2855 (CH), 1592 (C=N), 1532 (C=C). 1H NMR (DMSO-d6) δ ppm: 7.04 - 8.10 (m, 13H, Ar-H), 6.75 (s, 1H, =CH), 8.22 (s, 1H, NH exchanged by D$_2$O), 4.33 (s, 1H, NH exchanged by D$_2$O). MS. m/z (%): 391(M$^+$ +1, 26.6), 389 (M$^+$ - 1, 30.5), 233 (2.1), 298 (100), 284 (12.3), 271 (46.9), 190 (60.5).
5c	FT-IR (KBr, v_{max} cm^{-1}): 3181 (NH), 3050, 2921 (CH), 1600 (C=N), 1573 (C=C). 1H NMR (DMSO-d6) δ ppm: 7.06 - 7.24 (m, 4H, Ar-H), 7.26 - 7.31 (m, 2H, benzothiazole proton), 7.38 (d, 1H, benzothiazole proton), 7.47 (d, 1H, indole proton),7.61-7.64 (m, 2H, indole proton), 7.73 (d, 1H, benzothiazole proton), 8.36 (d, 1H, indole proton), 8.42 (s, 1H, =CH), 7.95 (s, 1H, NH exchanged by D$_2$O), 11.82 (s, 1H, NH exchanged by D$_2$O). MS. m/z (%): 447 (M$^+$, 11.24), 313 (13.70), 296 (10.33), 204 (15.63), 150 (100).
7	FT-IR (KBr, v_{max} cm^{-1}): 3207(NH), 3054, 3008, 2932, 2862 (CH), 1652 (C=O), 1611 (C=N), 1573 (C=C). 1H NMR (DMSO-d6) δ ppm: 7.14 - 8.80 (m, 14H, -Ar-H), 4.89 (s, 2H, CH$_2$), 8.90 (s, 1H, =CH), 11.92 (s, 1H, NH exchanged by D$_2$O), 12.05 (s, 1H, NH exchanged by D$_2$O).
8	FT-IR (KBr, v_{max} cm^{-1}): 3389, 3165 (NH), 3054, 2962, 2848 (CH), 1654 (C=O), 1604 (C=N). 1H NMR (DMSO-d6) δ ppm: 3.78 (s, 2H,CH$_2$), 7.16 - 7.28 (m, 5H, Ar-H and H-2 indole), 7.45 (d, 2H, indole proton), 7.56 - 7.89 (m, 4H, indole proton), 8.41 (d, 2H, indole proton), 8.90 (s, 1H, =CH), 4.31 (s, 1H, NH exchanged by D$_2$O), 12.03 (s, 2H, NH exchanged by D$_2$O). MS. m/z (%): 471 (M$^+$, 0.94), 211 (4.96), 203 (74.04), 177 (12.11), 159 (13.50), 136 (37.99), 91 (100).
9	FT-IR (KBr, v_{max} cm^{-1}): 3419, 3379, 3273 (NH$_2$, NH), 3051, 2918, 2856 (CH), 1658 (C=O), 1604 (C=N). 1H NMR (DMSO-d6) δ ppm: 1.95 - 2.03 (m, 2H, CH$_2$), 2.20 (t, 2H,CH$_2$), 2.71(t, 2H, CH$_2$), 6.93 - 7.33 (m, 5H, Ar-H and H-2 indole), 7.42 (d, 2H, indole proton), 7.49 - 7.83 (m, 4H, indole proton), 8.12 (d,1H, indole proton), 8.35 (d, 1H, indole proton), 8.27 (s, 1H, =CH), 10.75 (s, 1H, NH exchanged by D$_2$O), 10.99 (s, 1H, NH exchanged by D$_2$O), 11.85 (s, 1H, NH exchanged by D$_2$O).
10	FT-IR (KBr, v_{max} cm^{-1}): 3426, 3168 (NH), 3098, 3048, 2953, 2849 (CH), 1648 (C=O), 1608 (C=N). 1H NMR (DMSO-d6) δ ppm: 7.17 - 7.29 (m, 5H, Ar-H and H-3 benzofuran), 7.46 (d, 2H, indol and benzofuran),7.65 - 7.85 (m, 4H, indole and benzofuran), 8.42 (d, 2H, indole and benzofuran), 8.90 (s, 1H, =CH), 7.95 (s, 1H, NH exchanged by D$_2$O),12.04 (s, 1H, NH exchanged by D$_2$O). MS. m/z (%): 458 (M$^+$, 0.10), 313 (0.11), 296 (5.76), 52 (66.65), 204 (63.53), 133 (43.88), 91 (100).
11	FT-IR (KBr, v_{max} cm^{-1}): 3394, 3171 (NH), 3045, 2961, 2918 (CH), 1650 (C=O), 1604 (C=N), 2205 (CN). 1H NMR (DMSO-d6) δ ppm: 4.33 (s, 2H, CH$_2$), 7.16 - 7.28 (m,4H, Ar-H),7.44 (d, 1H, indole proton), 7.55 - 7.79 (m, 2H, indole proton), 8.11 (d, 1H, indole proton), 8.33 (s, 1H, N=CH), 11.33 (s, 1H, NH exchanged by D$_2$O), 11.93 (s, 1H, NH exchanged by D$_2$O). ^{13}C NMR (DMSO-d6) δ ppm: 24.39 (CH$_2$), 116.19 (CN), 157.73 (C=N), 163.58 (CO), 107.75, 111.44, 111.66, 120.95, 122.14, 122.27, 123.13, 125.09, 130.07, 130.97, 131.16, 131.73, 131.89, 136.46.

12	FT-IR (KBr, v_{max} cm^{-1}): 3391, 3311, 3184 (NH), 3047, 2960, 2918 (CH), 1667 (C=O), 1603 (C=N), 2212 (CN). 1H NMR (DMSO-d6) δ ppm: 2.29 (s, 3H, CH$_3$), 7.18 - 7.44 (m, 8H, Ar-H), 7.46 (d,1H, indole proton), 7.55 - 7.79 (m, 2H, indole proton), 8.19 (d, 1H, indole proton), 8.38 (s, 1H, = CH), 11.24 (s, 1H, NH exchanged by D$_2$O), 11.33 (s, 1H, NH exchanged by D$_2$O), 11.91 (s, 1H, NH exchanged by D$_2$O). ^{13}C NMR (DMSO-d6) δ pm: 10.65 (CH$_3$), 116.19 (CN), 155.22 (C=N), 159 (C=NNH), 111.41, 119.24, 122.18, 129.79, 130.07, 130.85, 131.04, 131.15, 131.24, 131.79, 133.31, 136.46, 136.46, 136.76.
13a	FT-IR (KBr, v_{max} cm^{-1}): 3335, 3126(NH), 3055, 2921 (CH), 1686 (C=O), 2204 (CN). 1H NMR (DMSO-d6) δ ppm: 7.19-8.57 (m, 19H, Ar-H and H-5 pyrazole ring), 8.96 (s, 1H, =CH), 9.21 (s, 1H, N=CH), 11.73 (s, 1H, NH exchanged by D$_2$O), 12.04 (s, 1H, NH exchanged by D$_2$O). MS, m/z (%): 611 (M$^+$, 0.63), 392 (1.05), 379 (0.51), 358 (11.477), 327 (1.42), 217 (33.11), 77.02 (100).
13b	FT-IR (KBr,v_{max} cm^{-1}): 3233, 3166 (NH), 3091, 2971 (CH), 1682 (C=O), 2208 (CN). 1H NMR (DMSO-d6) δ ppm: 7.22 - 8.39 (m, 12H, Ar-H), 8.68 (s, 1H, =CH), 8.88 (s, 1H, N=CH), 11.93 (s, 1H, NH exchanged by D$_2$O), 12.42 (s, 1H, NH exchanged by D$_2$O).
14	FT-IR (KBr, v_{max} cm^{-1}): 3372, 3155 (NH), 3093, 2930, 2853 (CH), 1712 (C=O). 1H NMR (DMSO-d6) δ ppm: 5.17 (s, 1H, H-4 coumarin ring), 6.56 - 8.22 (m, 12 H, Ar-H), 8.72 (s, 1H, N=CH), 11.44 (s, 1H, NH exchanged by D$_2$O), 12.14 (s, 1H, NH exchanged by D$_2$O). MS. m/z (%): 485(M$^+$, 0.30), 470 (0.4), 457 (0.4), 442 (0.1), 417 (3.00), 271 (100), 191(61.6), 165 (77.9).
15	FT-IR (KBr, v_{max} cm^{-1}): 3475, 3396, 3369, 3234, 3180 (2NH$_2$, NH), 3059, 2934, 2838 (CH), 1721, 1670 (2 C=O), 2210 (CN). 1H NMR (DMSO-d6) δ ppm: 1.26 (t, 3H, CH$_2$-CH$_3$), 4.23 (q, 2H, CH$_2$-CH$_3$), 7.22 - 8.225 (m, 12 H, Ar-H and 2 NH$_2$), 8.25 (s, 1H, N= CH), 12.91 (s, 1H, NH exchanged by D$_2$O), 12.45 (s, 1H, NH exchanged by D$_2$O).
16	FT-IR (KBr, v_{max} cm^{-1}): 3395, 3269, 3164 (NH$_2$, NH), 3049, 2974, 2866 (CH), 1665 (C=O), 1237 (C=S). 1H NMR (DMSO-d6) δ ppm: 7.16 - 8.48 (m, 15 H, Ar-H and NH$_2$), 8.90(s, 1H, N=CH), 11.00 (s, 1H, NH exchanged by D$_2$O), 12.43 (s, 1H, NH exchanged by D$_2$O). ^{13}C NMR (DMSO-d6) δ ppm: 147.39 (C=N), 153 (C-NH$_2$), 161.36 (CO), 183.91 (CS), 112.10, 113.61, 121.01, 122.38, 123.44, 125.7, 126.93, 130.7, 131.14, 131.87, 135.88, 136.51, 136.51, 141.72.
17	FT-IR (KBr, v_{max} cm^{-1}): 3129 (NH), 3091, 3052, 2947, 2860 (CH), 1681 (C=O), 1239 (C=S). 1H NMR (DMSO-d6) δ ppm: 2.25 (s, 3H, CH$_3$), 7.05-7.81(m, 13 H, Ar-H), 8.22(s, 1H, =CH), 12.40 (s, 1H, NH exchanged by D$_2$O).
18	FT-IR (KBr, v_{max} cm^{-1}): 3269 (NH), 3090, 2970, 2869 (CH). 1H NMR (DMSO-d6) δ ppm: 6.85 (s, 1H, N=CH), 6.98-7.32 (m, 10 H, Ar-H and H-5 thiazole ring), 7.50 (d, 1H, indole proton), 7.65 - 7.82 (m, 2H, indole proton), 8.20 - 8.22 (d, 1H, indole proton), 12.43 (s, 1H, NH exchanged by D$_2$O).
19	FT-IR (KBr, v_{max} cm^{-1}): 3127 (NH), 3084, 2967, 2862 (CH), 1671, 1718 (2C=O). 1H NMR (DMSO-d6) δ ppm: 7.22 - 7.32 (m, 5 H, Ar-H and H-4 chromene), 7.44 (d, 2H, indole and chromene), 7.70 - 7.92 (m, 4H, indole and chromene), 7.93 - 7.97 (m, 2H, CH= CH), 8.00 - 8.22 (d, 2H, indole and chromene), 12.43 (s, 1H, NH exchanged by D$_2$O).
20a	FT-IR (KBr, v_{max} cm^{-1}): 3466, 3310, 3149 (NH$_2$, NH), 3048, 2970, 2865 (CH), 1688 (C=O) and 2203 (CN). 1H NMR (DMSO-d6) δ ppm: 7.22 - 7.32 (m, 6 H, Ar-H and NH$_2$ proton), 7.47 - 7.83 (m, 3H, indole proton), 7.96 (d, 1H, indole proton), 8.22 (s, 1H, C=CH), 12.43 (s, 1H, NH exchanged by D$_2$O).
20b	FT-IR (KBr, v_{max} cm^{-1}): 3269 (NH), 3047, 2973, 2861 (CH), 2208 (CN), 1708 (C=O). 1H NMR (DMSO-d6) δ ppm: 1.03 (t, 3H, CH$_2$CH$_3$), 4.23 (q, 2H, CH$_2$-CH$_3$), 7.29 - 7.35 (m, 4H, Ar-H), 7.56 - 7.86 (m, 3H, indole proton), 8.15 - 8.17 (d, 1H, indole proton), 8.25 (s, 1H, =CH), 4.34 (s, 1H, NH exchanged by D$_2$O). MS. m/z (%): 395 (M$^+$, 11.8), 321 (44.6), 242 (100), 215 (17.6), 214 (32.8).
21	FT-IR (KBr, v_{max} cm^{-1}): 3303 (NH), 2203(CN), 3052, 2961, 2921, 2859 (CH), 1672 (C=O). 1H NMR (DMSO-d6) δ ppm: 4.20 (d, 1H, pyrazoline), 6.11 (d, 1H, pyrazoline), 6.97 - 7.00 (m, 4H, Ar-H), 7.02 - 7.87 (m, 3H, indole proton), 7.97 (d, 1H, indole proton), 10.18 (s, 1H, NH exchanged by D$_2$O), 11.41 (s, 1H, NH exchanged by D$_2$O), 12.22 (s, 1H, NH exchanged by D$_2$O).
22	FT-IR (KBr, v_{max} cm^{-1}): 3384 (NH), 2212 (CN), 3054, 2957, 2852 (CH). 1H NMR (DMSO-d6) δ ppm: 7.21 - 7.32 (m, 4H, Ar-H), 7.57 - 7.95 (m, 8H, indole and benzimidazole proton), 8.31 (s, 1H, =CH), 12.57 (s, 1H, NH exchanged by D$_2$O), 12.88 (s, 1H, NH exchanged by D$_2$O).
23	FT-IR (KBr, v_{max} cm^{-1}): 3355 (OH), 3179 (NH), 1623, 1707 (2C=O), 3095, 2974, 2865 (CH), 1242 (C=S). 1H NMR (DMSO-d6) δ ppm: 4.71 (s, 2H, CH$_2$), 7.32 - 7.33 (m, 4H, Ar-H), 7.50 (d, 1H, indole proton), 7.66 - 7.79 (m, 2H, indole proton), 7.99 (s, 1H, CH=C), 8.19 (d, 1 H, indole proton), 12.43 (s, 1H, NH exchanged by D$_2$O), 12.60 (s, 1H, OH exchanged by D$_2$O). MS. m/z (%): 473 (M$^+$, 1.4), 472 (M$^+$-1, 4.1), 456 (0.2), 382 (0.7), 327 (15.6), 271 (100), 202 (4.9).

26	FT-IR (KBr, v_{max} cm^{-1}): 3249, 3154 (2 NH), 3026, 2814 (CH), 1689 (C=O), 1283 (C=S). 1H NMR (DMSO-d6) δ ppm: 3.21 (s, 2H, CH$_2$), 6.79 - 7.79 (m, 6H, Ar-H and H-5 thiazole), 8.31 (s, 1H, NH exchanged by D$_2$O), 9.34 (s, 1H, NH exchanged by D$_2$O).
27	FT-IR (KBr, v_{max} cm^{-1}): 3232, 3260, 3176 (NH), 1668 (C=O), 3053, 2990, 2873 (CH), 1242(C=S). 1H NMR (DMSO-d6) δ ppm: 6.95 - 7.31 (m, 10 H, Ar-H and H-5 thiazole), 7.50 - 7.90 (m, 4H, indole proton), 9.96 (s, 1H, =CH), 12.18 (s, 1H, NH exchanged by D$_2$O), 12.22 (s, 1H, NH exchanged by D$_2$O), 12.40 (s, 1H, NH exchanged by D$_2$O). ^{13}C NMR (DMSO-d6) δ ppm: 185.37 (CS), 168 (CO), 155.46 (C=C), 166 (C-NH), 108.01, 112.01, 112.01, 113.74, 121.01, 123.44, 125.70, 127.62, 128.93, 129.01, 130.21, 130.8, 134.46, 135.88, 144.51, 146.01, 148.01.
28	FT-IR (KBr, v_{max} cm^{-1}): 3275 (NH), 1706 (C=O), 3037, 2988, 2817(CH). 1H NMR (DMSO-d6) δ ppm: 2.38 (s, 3H, CH$_3$), 7.31 - 7.35(m, 9H, Ar-H), 7.36 - 7.53 (m, 4H, indole proton and =CH), 7.82 (d, 1H, indole proton), 12.40 (s, 1H, NH exchanged by D$_2$O). ^{13}C NMR (DMSO-d6) δ ppm: 12.58 (CH$_3$), 153(C=C), 160 (C=O), 108.82, 120.62, 120.74, 126.43, 126.63, 128.87, 129.07, 137.15, 140.16, 140.34. MS. m/z (%): 456 (M$^+$, 1.9), 441 (0.3), 413 (0.4), 336 (3.4), 359 (48.1), 358 (100), 341 (62.3), 266 (20.3), 77 (92.7).
29a	FT-IR (KBr, v_{max}, cm^{-1}): 3428, 3198 (NH), 3039, 2865(CH). 1H NMR (DMSO-d6) δ ppm: 2.26 (s, 3H, CH$_3$), 6.93 - 7.18 (m, 9H, Ar-H), 7.29 - 7.82 (m, 4H, indole proton), 8.04 (s, 1H, NH exchanged by D$_2$O), 11.56 (s, 1H, NH exchanged by D$_2$O).
29b	FT-IR (KBr, v_{max} cm^{-1}): 3209 (NH), 3048, 2917, 2863 (CH). 1H NMR (DMSO-d6) δ ppm: 2.21 (s, 3H, CH$_3$), 6.93 - 7.32 (m, 14H, Ar-H), 7.37 - 7.82 (m, 4H, indole proton), 11.56 (s, 1H, NH exchanged by D$_2$O). MS. m/z (%): 544 (M$^+$, 3.3), 529 (2.3), 467 (22.9), 375 (15.41), 295 (6.2), 298 (100), 273 (71.8), 271 (17.2).
30	FT-IR (KBr,v_{max} cm^{-1}): 3323, 3174 (NH), 3055, 2969, 2864 (CH), 1242 (C=S). 1H NMR (DMSO-d6) δ ppm: 2.24 (s, 3H, CH$_3$), 2.28 (s, 3H, CH$_3$), 7.11 - 7.35 (m, 13H, Ar-H), 7.37 - 7.86 (m, 4H, indole proton), 11.48 (s, 1H, NH exchanged by D$_2$O), 12.00 (s, 1H, NH exchanged by D$_2$O). MS. m/z (%): 617 (M$^+$, 0.46), 587 (0.54), 522 (0.46), 467 (0.63), 414 (1.35), 330 (2.73), 252 (57.01), 125 (100).
31	FT-IR (KBr, v_{max}, cm^{-1}): 3436 (NH), 3050, 2972, 2865 (CH), 1277 (C=S). 1H NMR (DMSO-d6) δ ppm: 2.28 (s, 3H, CH$_3$), 7.93 - 8.31 (m, 17H, Ar-H), 11.54 (s, 1H, NH exchanged by D$_2$O).

The reaction mixture was allowed to stand overnight and was then refluxed for 2 h. under vigorous stirring. The mixture was then poured onto ice cold water and neutralized with dilute ammonia solution till the precipitation occurs. The formed precipitate was collected by filtration and recrystallized from ethanol to give 4 as yellow crystals. Yield (15.13 g, 70%, m.p.: 270-272°C. Anal. calcd. for C$_{15}$ H$_{10}$ Br NO (300.15): C, 60.02; H, 3.36; Br, 26.62; N, 4.67. Found: C, 59.89; H, 3.16; Br, 26.42; N, 4.37.

General Procedure for the Synthesis of 5a-c

An equimolecular mixture of 4 (3 g, 0.01 mol) and the hydrazine derivatives (0.5 ml, 0.01 mol) were refluxed in absolute ethanol (20 ml) in the presence of 2 - 3 drops of glacial acetic acid for the appropriate time. The reaction mixture was cooled to room temperature and poured into ice-cold water. The separated product was filtered, washed with cold water, dried and recrystallized from the appropriate solvent to give 5a-c.

1-[2-(4-Bromophenyl)-1H-indol-3-ylmethylene]hydrazine 5a

Compound 5a was prepared from hydrazine hydrate for 1 h. Orange crystals. Yield (2.48 g, 79%); m.p.: 338-340°C (xylene). Anal. calcd. for C$_{15}$H$_{12}$Br N$_3$ (314.18): C, 57.34; H, 3.85; Br, 25.43; N, 13.37. Found: C, 57.24; H, 3.65; Br, 25.33; N, 13.17.

1-[2-(4-Bromophenyl)-1H-indol-3-ylmethylene]-2-phenyl-hydrazine 5b

Compound 5b was prepared from phenyl hydrazine for 4 h. Pale brown powder. Yield (2.5 g, 64%); m.p.: 115-117°C (hexane). Anal.calcd. for $C_{21}H_{16}BrN_3$ (390.28): C, 64.63; H, 4.13; Br, 20.47; N, 10.77. Found: C, 64.43; H, 4.00; Br, 20.27; N, 10.57.

2-{2-[2-(4-Bromophenyl)-1H-indol-3-ylmethylene]hydrazine}-1,3-benzothiazole 5c

Compound 5c was prepared from 2-hydrazinyl-1,3-benzothiazole for 4 h. Pale yellow crystal. Yield (2.46 g, 55%); m.p.: 280-282°C (ethanol/DMF). Anal. calcd. for $C_{22}H_{15}BrN_4S$ (447.35): C, 59.07; H, 3.38; Br, 17.86; N, 12.52; S, 7.17. Found: C, 58.98; H, 3.18; Br, 17.66; N, 12.40; S, 7.00.

General Procedure for the Synthesis of 7-10

An equimolecular mixture of 4 (3 g,0.01 mol) and the acid hydrazide derivatives 6a-d (0.01 mol) was refluxed for 2 h. in absolute ethanol (20 ml) in the presence of 2 - 3 drops of glacial acetic acid. The reaction mixture was cooled to room temperature and poured into ice-cold water. The separated product was filtered, washed with cold water, dried and recrystallized from the appropriate solvent.

N'-[2-(4-Bromophenyl)-1H-indol-3-ylmethylene]-2-(quinolin-8-yloxy) acetohydrazide 7

Yellow crystals. Yield (3 g, 60%); m.p.: 290-292°C (ethanol/DMF). Anal. calcd for $C_{26}H_{19}BrN_4 O_2$(499.35): C, 62.54; H, 3.84; Br, 16.00; N, 11.22. Found: C, 62.24; H, 3.62; Br, 15.88; N, 11.02.

N'-(2-(4-Bromophenyl)-1H-indol-3-ylmethylene)-2-(1H-indol-3-yl) acetohydrazide 8

Red crystals. Yield (2.83 g, 60%); m.p.: 330-332°C (ethanol). Anal. calcd for $C_{25}H_{19}BrN_4O$ (471.35): C, 63.70; H, 4.06; Br, 16.95; N, 11.89. Found: C, 63.40; H, 4.00; Br, 16.65; N, 11.59.

N'-[2-(4-Bromophenyl)-1H-indol-3-ylmethylene]-4-(1H-indol-3-yl)-butanehydrazide 9

Dark yellow powder. Yield (3.75 g, 75%); m.p.: 170-172°C (xylene). Anal.calcd for $C_{27}H_{23} BrN_4O$ (499.40): C, 64.94; H, 4.64; Br, 16.00; N, 11.22. Found: C, 64.74; H, 4.34; Br, 15.88; N, 11.00.

N'-[2-(4-Bromophenyl)-1H-indol-3-ylmethylene]-benzofuran-2-carbohydrazide 10

Yellow crystals. Yield (2.75 g, 60%); m.p.: 320-322°C (ethanol/DMF). Anal. calcd for $C_{24}H_{16}BrN_3O_2$ (458.30): C, 62.90; H, 3.52; Br, 17.43; N, 9.17. Found: C, 62.70; H, 3.32; Br, 17.23; N, 9.00.

N'-[2-(4-Bromophenyl)-1H-indol-3-ylmethylene]-2-cyanoaceto-hydrazide 11

An equimolecular mixture of 4 (3 g, 0.01 mol) and cyanoacetohydrazide (1.98 g, 0.02 mol) in absolute ethanol (30 ml) was heated under reflux for 2 h. The precipitate formed after cooling was filtered off, washed with cold ethanol, dried and recrystallized from DMF to give 11 as pale brown powder. Yield (3.24 g, 85%); m.p.: 290-292°C. Anal. calcd for $C_{18}H_{13}BrN_4O$ (381.22): C, 56.71; H, 3.44; Br, 20.96; N, 14.70. Found: C, 56.41; H, 3.24; Br, 20.86; N, 14.50.

N'-[2-(4-Bromophenyl)-1H-indol-3-ylmethylene]-2-cyano-2-[(4-methylphenyl)hydrazono]- acetohydrazide 12

To a cold solution of 11 (3.81 g, 0.01 mol) in ethanol (20 ml) containing sodium acetate (3.0 g) was added with continuous stirring 4-methylbenzene diazonium salt (0.01 mol) [prepared by adding sodium nitrite (1.38 g, 0.02 mol) in water (8 ml) to a cold solution of p-toluidine (1.07 g, 0.01 mol) in the appropriate amount of hydrochloric acid]. The reaction mixture was stirred for 2 h. and the formed solid was collected by filtration and recrystallized from ethanol to give 12 as orange crystals. Yield (3.5 g, 70%); m.p.: 240-242°C. Anal. calcd for $C_{25}H_{19}$ Br N_6O (499.36): C, 60.13; H, 3.84; Br, 16.00; N, 16.83. Found: C, 60.00; H, 3.54; Br, 15.98; N, 16.75.

General Procedure for the Synthesis of 13a,b and 14

Equimolecular mixture of 11 (3.81 g, 0.01 mol) and the selected aldehydes such as 1,3-diphenyl-1H-pyrazole-4- carboxaldehyde, p-nitrobenzaldehyde and salicyaldehyde (0.01 mol) in 1,4-dioxane (20 ml) containing piperidine (0.5 ml) was heated under reflux for 3 h. The reaction mixture was left to cool then poured onto ice/water containing few drops of hydrochloric acid and the formed solid product was collected by filtration and recrystallized from the appropriate solvent.

N'-[2-(4-Bromophenyl-1H-indol-3-ylmethylene]-2-cyano-3-(1,3-diphenyl-1H-pyrazol-4-yl) acrylohydrazide 13a

Yellow crystal. Yield (3.6 g, 55%); m.p.: 255-257°C (ethanol/DMF). Anal. calcd for $C_{34}H_{23}BrN_6O$ (611.49): C, 66.78; H, 3.79; Br, 13.07; N, 13.74. Found: C, 66.48; H, 3.59; Br, 13.00; N, 13.55.

N'-[2-(4-Bromophenyl)-1H-indol-3-ylmethylene]-3-(4-nitrophenyl)-2-cyanoacrylohydrazide 13b

Yellow crystals. Yield (2.31 g, 45%); m.p.: 230-232°C (ethanol). Anal. calcd for $C_{25}H_{16}BrN_5O_3$(514.33): C, 58.38; H, 3.14; Br, 15.54; N, 13.62. Found: C, 58.18; H, 3.00; Br, 15.34; N, 13.32.

N'-[2-(4-Bromophenyl)-1H-indol-3-ylmethylene]-2-imino-2H-chromene-3-carbohydrazide 14

Brown crystals. Yield (2.9 g, 60%); m.p.: 130-132°C (hexane). Anal. calcd for $C_{25}H_{17}BrN_4O_2$(485.33): C, 61.87; H, 3.53; Br, 16.46; N, 11.54. Found: C, 61.57; H, 3.33; Br, 16.26; N, 11.34.

General Procedure for the Synthesis of 15 and 16

To a solution of compound 11 (3.81 g, 0.01 mol) in absolute ethanol (50 ml) containing triethylamine (1 ml) either ethyl cyanoacetate (1.13 g, 0.01 mol) or phenylisothiocyanate (1.39 g, 0.01 mol) together with elemental sulfur (0.32 g, 0.01 mol) were added. Reaction mixture was heated under reflux for 8 h. then poured onto ice/water mixture and the formed solid product, in each case, was collected by filtration recrystallized from ethanol.

Ethyl 2,4-diamino-5-{[2-(2-(4-bromophenyl)-1H-indol-3-ylmethylene)hydrazino]-carbonyl}thiophene-3- carboxylate 15

Dark brown crystals. Yield (2.63 g, 50%); m.p.: 190-192°C. Anal. calcd for $C_{23}H_{20}BrN_5O_3S$ (526.40): C, 52.48; H, 3.83; Br,15.18; N, 13.30; S, 6.09. Found: C, 52.28; H, 3.53; Br, 15.00; N, 13.00; S, 6.00.

4-Amino-N'-[2-(4-bromophenyl)-1H-indol-3-ylmethylene]-3-phenyl-2-thioxo-2,3-dihydro-1,3-thiazole-5- carbohydrazide 16

Brown crystals. Yield (3.35 g, 61%); m.p.: 245-247°C. Anal. calcd for $C_{25}H_{18}Br N_5OS_2$ (548.47): C, 54.75; H, 3.31; Br, 14.57; N,12.77; S, 11.69. Found C, 54.55; H, 3.11; Br, 14.37; N, 12.57; S, 11.49.

6-[(2-(4-Bromophenyl)-1H-indol-3-ylmethylene)amino]-5-methyl-2-thioxo-3-phenyl-2,3- dihydro-1,3-thiazolo[4,5-d]pyrimidin-7(6H)-one 17

A solution of compound 16 (5.48 g, 0.01 mol) in a mixture of acetic acid (5 ml) and acetic anhydride (10 ml) was heated under reflux for 8 h. and then allowed to cool. The precipitate that formed was collected by filtration, dried and recrystallized from acetic acid to give compound 17 as yellow crystals; Yield (3.4 g, 60%); m.p.: 316-318°C. Anal. Calcd. $C_{27}H_{18}Br N_5 O S_2$ (572.50):

C, 56.64; H, 3.17; Br, 13.96; N, 12.23; S, 11.2. Found: C, 56.44; H, 3.07; Br, 13.66; N, 12.03; S, 11.00.

N-[2-(4-Bromophenyl)-1H-indol-3-ylmethylene]-4-phenyl-1,3-thiazol-2-amine 18

A mixture of compound 4 (3 g, 0.01 mol), phenacyl bromide (0.01 mol) and thiourea (0.78 g, 0.01 mol) in absolute ethanol (30 ml) containing acetic acid (1 ml) were heated under reflux for 8 h. Reaction mixture poured in an ice cold water, the solid obtained was filtered, dried and recrystallized from ethanol to give 18 as brown powder. Yield (3.21 g, 70%); m.p.: 230-232°C. Anal. calcd. for $C_{24}H_{16}Br N_3S$ (458.37): C, 62.89; H, 3.52; Br, 17.43; N, 9.17; S, 7.00. Found: C, 62.59; H, 3.22; Br, 17.23; N, 9.00; S, 6.81.

3-[3-(2-(4-Bromophenyl)-1H-indol-3-yl)prop-2-enoyl]-2H-chromen-2-on 19

A mixture of 4 (3 g, 0.01 mol), 3-acetyl-2H-chromen-2-one (1.88 g, 0.01 mol) in 20 ml absolute ethanol and 0.5 ml piperdine was refluxed for 30 min. The reaction mixture was left overnight at room temperature, the obtained solid was filtered off and recrystallized from ethanol to give 19 as yellow crystals. Yield (4.01 g, 64%); m.p.: 280-282°C. Anal. calcd. for $C_{26}H_{16}BrNO_3$ (470.31): C, 66.40; H, 3.43; Br, 16.99; N, 2.98. Found: C, 66.30; H, 3.23; Br, 16.79; N, 2.78.

General Procedure for the Synthesis of 20a,b and 22

To a solution of compound 4 (3 g, 0.01 mol) in 20 ml ethanol, the appropriate active methylene compounds such as cyanoacetamide, ethyl cyanoacetate and 1H-benzimidazol-2-ylacetonitrile (0.01 mol) and few drops of triethylamine was added. The reaction mixture was refluxed for 5 h. and then allowed to cool. The formed solid product was collected by filtration, washed with ethanol and recrystallized from the appropriate solvent.

3-[2-(4-Bromophenyl)-1H-indol-3-yl]-2-cyanoprop-2-enamide 20a

Yellow powder. Yield (2.38 g, 65%); m.p.: 210-212°C (ethanol).Anal. calcd for $C_{18}H_{12}BrN_3O$ (366.21): C, 59.03; H, 3.30; Br, 21.82; N, 11.47. Found: C, 58.89; H, 3.00; Br, 21.52; N, 11.17.

Ethyl 3-[2-(4-bromophenyl)-1H-indol-3-yl]-2-cyanoprop-2-enoate 20b

Yellow powder. Yield (2.37 g, 60%); m.p.: 245-247°C (ethanol\DMF). Anal. calcd. for $C_{20}H_{15}BrN_2O_2$ (395.24): C, 60.78; H, 3.83; Br, 20.22; N, 7.09. Found: C, 60.48; H, 3.53; Br, 20.02; N, 7.00.

2-(1H-Benzimidazol-2-yl)-3-[2-(4-bromophenyl)-1H-indol-3-yl] acrylonitrile 22

Brown powder. Yield (2.64 g, 60%); m.p. 250-253°C (ethanol\DMF). Anal. calcd. for $C_{24}H_{15}BrN_4$(439.30): C, 65.62; H, 3.44; Br, 18.19; N, 12.75. Found: C, 65.42; H, 3.24; Br, 18.09; N, 12.65.

3-[2-(4-Bromophenyl)-1H-indol-3-yl]-5-oxopyrazolidine-4-carbonitrile 21

A mixture of compound 20b (3.95 g, 0.01 mole) and hydrazine hydrate (0.75 ml, 0.015 mole) in ethanol (20 ml) was refluxed for 3 h, then poured into water. The resulting solid was collected and recrystallized from ethanol to give 21 as yellowish white crystals. Yield (1.91 g, 50%); m.p.: 198-200°C. Anal. calcd for $C_{18}H_{13}BrN_4O$ (381.23): C, 56.71; H, 3.44; Br, 20.96; N, 14.70. Found: C, 56.51; H, 3.24; Br, 20.76; N, 14.50.

{5-[2-(4-Bromophenyl)-1H-indol-3-ylmethylene]-4-oxo-2-thioxo-1,3-thiazolidin-3-yl}acetic acid 23

To a solution of rhodanine-3-acetic acid (1.91 g, 0.01 mol) and anhydrous sodium acetate (0.5 g) in glacial acetic acid was added the 1H-indole-3-carboxaldehyde 4 (3 g, 0.01 mol). The mixture was stirred under reflux for 6 h and then poured into ice-cold water. The precipitate was filtered, washed with water, dried and recrystallized from xylene to gives 23 as orange powder. Yield (3.08 g, 65%); m.p.: 223-225°C. Anal. calcd. For $C_{20}H_{13}BrN_2O_3S_2$ (473.36): C, 50.75; H, 2.77; Br, 16.88; N, 5.92; S, 13.55. Found: C, 50.45; H, 2.57; Br, 16.58; N, 5.62; S, 13.25.

Ethyl 3-Oxo-3-[(4-phenyl-1,3-thiazol-2-yl)amino]propanoate 25

A mixture of an equimolar amount of 4-phenyl-2-aminothiazole 24 (1.76 g, 0.01 mol) and diethylmalonate 1.6 g, 0.01 mol) was heated in an oil bath at 180°C for 2 hours then left to cool. The product was collected and used in second step.

6-[(4-Phenyl-1,3-thiazol-2-yl)amino]-2-thioxo-2,3-dihydro-pyrimidin-4(5H)-one 26

A mixture of ester 25 (2.9 g, 0.01 mol) and thiourea (0.76 g, 0.01 mol) in ethanol (30 ml) containing sodium ethoxide was heated under reflux for 6 h. The reaction mixture was poured into cold water and the formed solid product was collected by filtration, washed, dried and recrystallized from ethanol to gives 26 as yellow crystals. Yield (3.8 g, 60%); m.p.:200-202°C. Anal. calcd. for $C_{13}H_{10}N_4OS_2$ (302.37): C, 51.64; H, 3.33; N, 18.53; S, 21.21. Found: C, 51.44; H, 3.13; N, 18.33; S, 21.01.

5-[2-(4-Bromophenyl)-1H-indol-3-ylmethylene]-6-[(4-phenyl-1,3-thiazol-2-yl)amino]- 2-thioxo-2,5-dihydropyrimidin-4(3H)-one 27

A mixture of pyrimidine derivative 26 (3 g, 0.01 mol) and 1H- indole-3-carboxaldehyde 4 (3 g, 0.01 mol) in ethanol (30 ml) was heated under reflux for 6 h., then left to cool. The solid product was collected by filtration and recrystallized from xylene to give 27 as green powder. Yield (2.34 g, 40%. m.p.: 208-210°C. Anal. calcd. for $C_{28}H_{18}BrN_5OS_2$ (584.50): C, 57.54; H, 3.10; Br, 13.67; N,11.98; S,10.97. Found: C, 57.34; H, 3.00; Br, 13.47; N, 11.68; S, 10.67.

4-[2-(4-Bromophenyl)-1H-indol-3-ylmethylene]-3-methyl-1-phe-nyl-1H-pyrazol-5(4H)-one 28

A mixture of 1H-pyrazol-5(4H)-one (1.74 g, 0.01 mol) and 1H- indole-3-carboxaldehyde 4 (3 g, 0.01 mol) in acetic acid in the presence of anhydrous sodium acetate was refluxed for 5 h. The reaction mixture was cooled to room temperature and poured into ice cold water. The solid separated out was filtered washed with water and recrystallized from (ethanol/DMF) to give 28 as orange crystals. Yield (3.01g, 66%); m.p.: 190-192°C. Anal. calcd. for $C_{25}H_{18}BrN_3O$ (456.33): C, 65.80; H, 3.98; Br,17.51; N, 9.21. Found: C, 65.56; H, 3.68; Br, 17.31; N, 9.00.

2-(4-Bromophenyl)-3-(4-methyl-6-phenyl-2,6-dihydropyrazolo[3,4-c]pyrazol-3-yl)- 1H-indole 29a

A mixture of compound 28 (4.56 g, 0.01 ml) and hydrazine hydrate (0.5 ml, 0.01 ml) in ethanol in present of few drops of acetic acid was refluxed for 7 h. Reaction mixture was cooled at room temperature and poured in ice cold water. The solid separated was filtered, washed with water and recrystallized from ethanol to give 29a as yellow crystals. Yield (2.81 g, 60%); m.p.: 180-182°C. Anal.calcd. for $C_{25}H_{18}BrN_5$ (468.34): C, 64.11; H, 3.87; Br, 17.06; N, 14.95. Found: C, 64.00; H, 3.57; Br, 16.89; N, 14.65.

2-(4-Bromophenyl)-3-(4-methyl-2,6-diphenyl-2,6-dihydropyrazolo[3,4-c]pyrazol-3-yl)- 1H-indole 29b

A mixture of compound 28 (4.56 g, 0.01 ml), phenyl hydrazine (1.08 ml, 0.01 mol), anhydrous sodium acetate (0.5 g) and acetic acid (20 ml) was refluxed for 7 h. Reaction mixture was cooled to room temperature and poured in ice cold water. The solid separated out was filtered, washed with water and recrystallized from ethanol to give 29b as yellow crystals. Yield (3.54 g, 65%); m.p.: 105-106°C. Anal. calcd. for $C_{31}H_{22}BrN_5$ (544.44): C, 68.39; H, 4.07; Br, 14.68; N, 12.86. Found: C, 68.09; H, 4.00; Br, 14.48; N, 12.66.

3-(2-(4-Bromophenyl)-1H-indol-3-yl)-4-methyl-N-(4-methylphenyl)-6-phenyl-pyrazolo [3,4-c]-pyrazole-2(6H)-carbothioamide 30

A mixture of compound 28 (4.56 g, 0.01 mol) and N-(4-methylphenyl) thiosemicarbazide (1.81 g, 0.01 mol) was refluxed in ethanol in the presence of $NaOH/H_2O$ (10%, 5 ml) for 8 h. Reaction mixture was cooled to room temperature and poured in ice-cold water. The solid separated out was filtered, washed with water and recrystallized from ethanol to give 30 as yellow crystals, Yield (3.89 g, 63%); m.p.: 240-242°C (ethanol). Anal. Calcd. for $C_{33}H_{25}BrN_6S$ (617.56): C, 64.18; H,4.08; Br,12.94; N, 13.61; S, 5.19. Found: C, 64.00; H, 4.00; Br, 12.70; N, 13.36; S, 5.09.

4-(2-(4-Bromophenyl)-1H-indol-3-yl)-3-methyl-5-(4-nitrophenyl)-1-phenyl-1,5-dihydro- 6H-pyrazolo[3,4-d]pyrimidine-6-thione 31

A mixture of compound 28 (4.56 g, 0.01 mol), N-(4-nitrophenyl)thiourea (1.97 g, 0.01 mol) and potassium hydroxide (0.5 g) in ethanol (20 ml) was refluxed with stirring for 4 h. The reaction mixture was left overnight and then concentrated under reduced pressure. The solid residue was collected, washed with water and recrystallized from ethanol to gives 31 as orange powder, Yield (3.48 g, 55%); m.p.: 150-152°C. Anal. calcd. for $C_{32}H_{21}BrN_6O_2S$ (633.52): C, 60.67; H, 3.34; Br, 12.61; N, 13.27; S, 5.06. Found: 60.47; H, 3.14; Br, 12.41; N, 13.07; S, 4.89.

Antimicrobial Assays

Synthesized compounds 5c, 7, 9, 11, 13a, 27, 30 and 31 were screened for their antimicrobial activitiesin vitro against two species of Gram-positive bacteria, namely Staphylococcus saureus RCMB 0100010 (SA), Bacillus subtilis RCMB 010067 (BS) and two negative bacteria, namely Pseudomonas aeuroginosa RCMB 010043 (PA), Escherichia coli CMB 010052 (EC). Two fungal strains Aspergillus fumigatus RCMB 02568 (AF) and Candida albicans RCMB 05036 (CA) are used for antifungal activity. The antibacterial and antifungal activities were determined by means of inhibition% ± standard deviation at a concentration of 100 µg/ml of tested samples [15] [16] . Optical densities of antimicrobial were measured after 24 hours at 37°C to bacteria and measured after 48 hours at 28°C to fungal using a multidetection microplate reader at the Regional Center for Mycology and Biotechnology (Sun Rise-Tecan, USA at 600 nm) Al-Azhar University. Ampicillin, gentamicin were used as bacterial standards and amphotericin B was used as fungal standards for references to evaluate the efficacy of the tested compounds under the same conditions. The MICs of the compounds assays were determined by using microbroth kinetic system [17] .

RESULTS AND DISCUSSION

Chemistry

The synthesis of the new compounds is outlined in Schemes 1-6. 2-Substituted-indole reported to be obtained via Fischer indole synthesis using phenyl hydrazine and acetophenone derivatives 1a-d in present of polyphosphoric acid as catalysis [18] . Synthesis of 1H-indole-3-carboxaldehyde derivative 4 from the 2-(4-bromo- phenyl)-1H-indole 3c via Vilsmeir Haack's formylation using phosphorus oxychloride ($POCl_3$) and N,N'-dime- thylformamide (DMF) [19] (Scheme 1). The IR spectrum of 4 revealed C=O stretching band of formyl group at 1672 cm^{-1}. 1H NMR spectrum showed an D_2O-exchangeable signal at 12.40 ppm assigned to the NH proton and a non exchangeable signal at δ 9.95 ppm corresponding to the formyl proton. The mass spectrum showed the molecular ion peak at m/z 300 corresponding to the molecular formula $C_{15}H_{10}BrNO$.

The hydrazine derivatives 5a-c were obtained by the reaction of 1H-indole-3-carboxaldehyde derivative 4 with different substituted hydrazines [20] namely, hydrazine hydrate, phenyl hydrazine and 2-hydrazinyl-1,3-benzothiazole (Scheme 1). The molecular structure of the synthesis compounds were established based on analytical and spectral data. For example, 1H NMR spectrum of compound 5c showed an D_2O -exchangeable signal at 7.95 and 11.82 assigned to the two NH protons and a non exchangeable signal at 8.42 ppm corresponding =CH proton. On other hand mass spectrum of 5c showed a molecular ion peak m/z at 447 corresponding to the molecular formula $C_{22}H_{15}BrN_4S$.

Reaction of 1H-indole-3-carboxaldehyde derivative 4 with different substituted acid hydrazides [21] such as 2-(quinolin-8-yloxy)acetohydrazide 6a, 2-(1H-indol-3-yl)acetohydrazide 6b, 4-(1H-indol-3-yl)butanehydrazide 6c and 1-benzofuran-2-carbohydrazide 6d in presence of catalytic amount of acetic acid in absolute ethanol afforded the corresponding the acid hydrazide derivatives 7-10 respectively (Scheme 2).

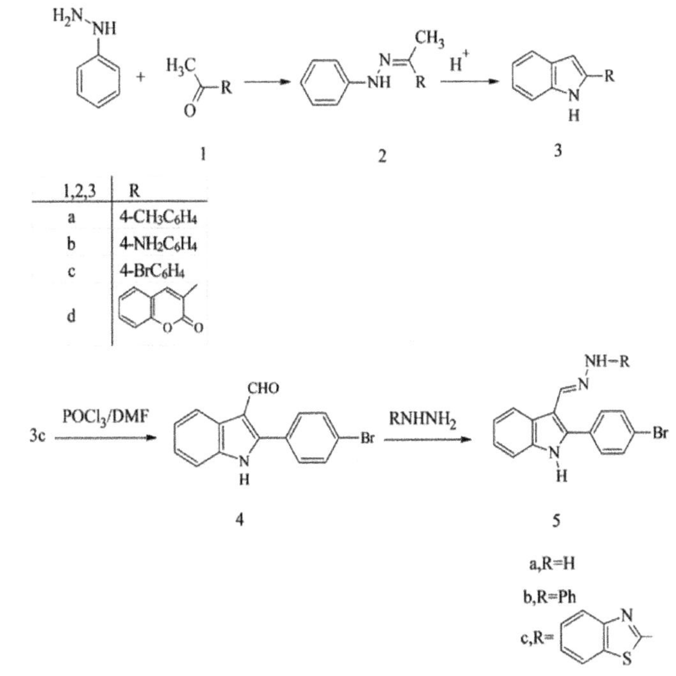

1,2,3 R
a 4-CH3C6H4
b 4-NH2C6H4
c 4-BrC6H4
d
POCl3/DMF
RNHNH2
a,R=H
b,R=Ph
c,R=

Scheme 1: Synthesis of compounds 3-5.

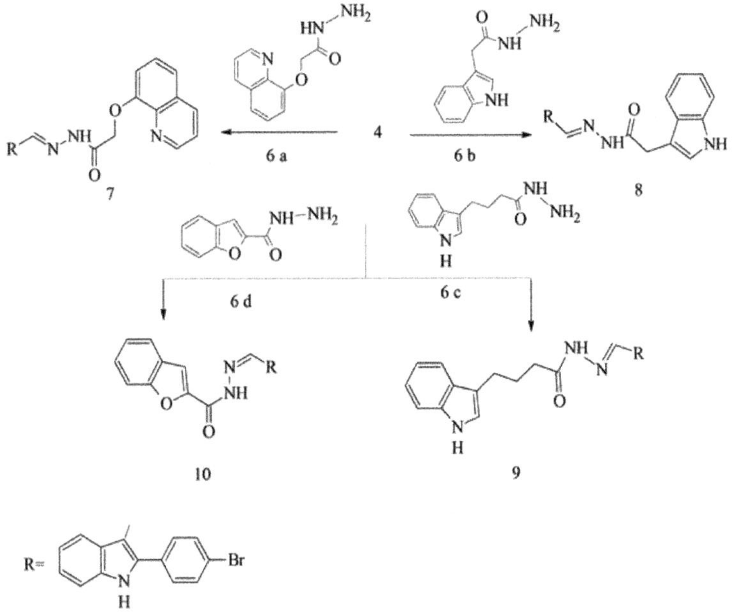

6 a
6 b
6 c
6 d
R=

Scheme 2: Synthesis of compounds 7-10.

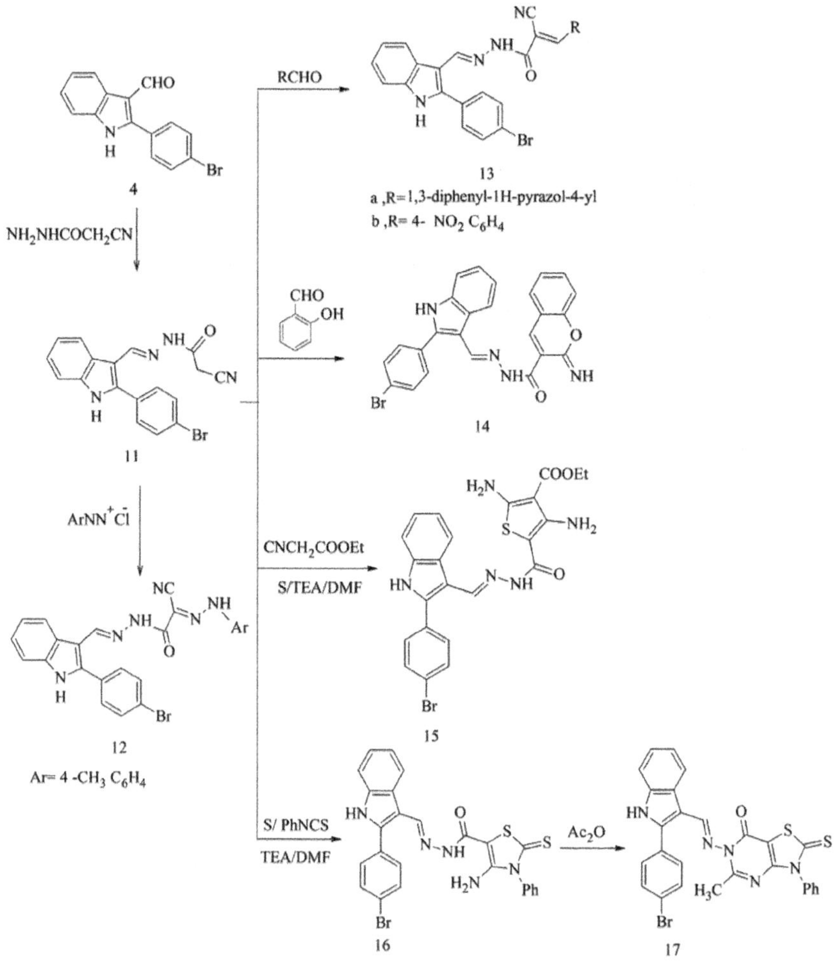

Scheme 3: Synthesis of compounds 11-17.

The assignment of the structure of the synthesis compounds were based on analytical and spectroscopic data. For example IR spectrum of 8 exhibit absorption band at 1604 cm^{-1} and 1654 cm^{-1} due to -C=N and CO groups. 1H NMR of 8 exhibits signal at δ 8.90 ppm for =CH proton and D$_2$O-exchangeable signal at δ 4.31 and 12.03 ppm assigned to the 2NH protons. The mass spectrum of compound 8 showed the molecular ion peak at m/z 471 corresponding to the molecular formula C$_{25}$H$_{19}$BrN$_4$O.

Reaction 1H-indole-3-carboxaldehyde 4 with cyanoacetohydrazide in absolute ethanol [22] to form the N'- [1H-indol-3-ylmethylene]-2-cyanoacetohydrazide derivative 11. The assignment of the structure of compound

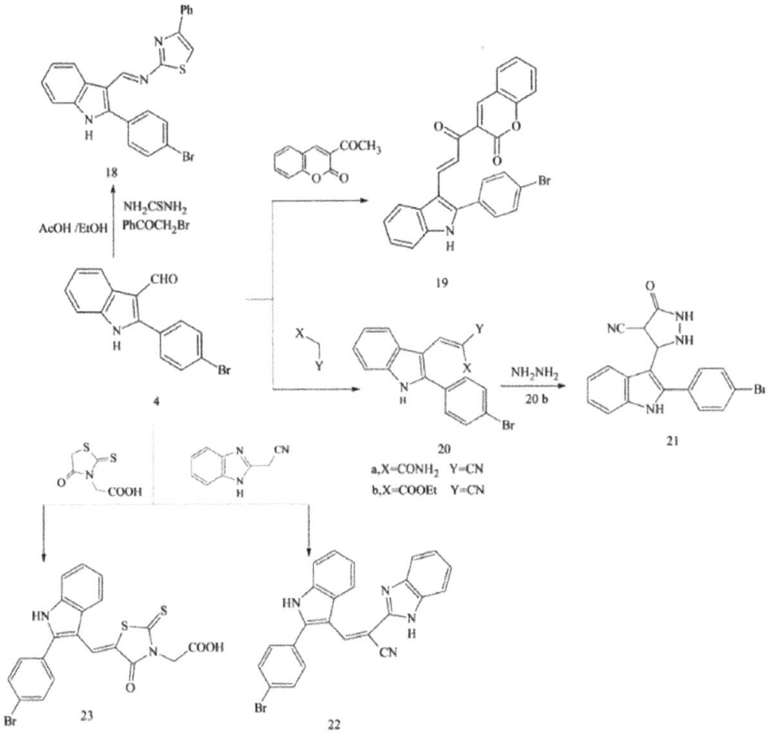

Scheme 4: Synthesis of compounds 18-23.

11 was based on analytical and spectroscopic data. Thus, the 1H NMR showed a singlet at δ 4.33 for the CH$_2$ group, a singlet at δ 8.33ppm for the =CH proton and D$_2$O-exchangeable single at δ 11.33, 11.93 ppm for the two NH protons. ^{13}C NMR spectrum of 11 displayed signals at δ 24.39, 116.19, 157.73 and 163.58 ppm for CH$_2$, CN, C=N and CO respectively. Further structure elucidation of compound 11 was obtained through the study of its reactivity towards chemical reagents. Thus, the reaction of 11 with 4-methylbenzene diazonium chloride [23] gave the hydrazone derivatives 12 (Scheme 3). The structures of the compound 12 were determined from spectroscopic and elemental analytical data (see Experimental section).

Knoevengel condensation of the 2-cyanoacetohydrazide derivatives 11 with aromatic aldehydes namely 1,3-diphenyl-1H-pyrazole-4-carboxaldehyde and p-nitrobenzaldehyde [24] afforded benzylidene derivatives 13a, b (Scheme 3). The IR spectrum of compound 13a, taken as a typical example of the series prepared, revealed absorption bands at 1686 cm^{-1}, 2204 cm^{-1}, 3335 cm^{-1}and 3126 cm^{-1} corresponding to carbonyl, nitrile and 2 NH groups, respectively. Where the 1H NMR spectra showed the absence of the active methylene proton and showed signals at δ 9.21 ppm for N=CH proton and D$_2$O-exchangeable signal at δ 11.73 and 12.04 assigned to the 2NH protons. Its mass spectrum showed a molecular ion peak at m/z 611 corresponding to the molecular formula C$_{34}$H$_{23}$BrN$_6$O.

Cyclocondensation of 2-cyanoacetohydrazide derivatives 11 with salicyaldehyde in dioxane in the presence of a catalytic amount of piperidine afforded 2-imino-2H-chromene-3-carbohydrazide 14.

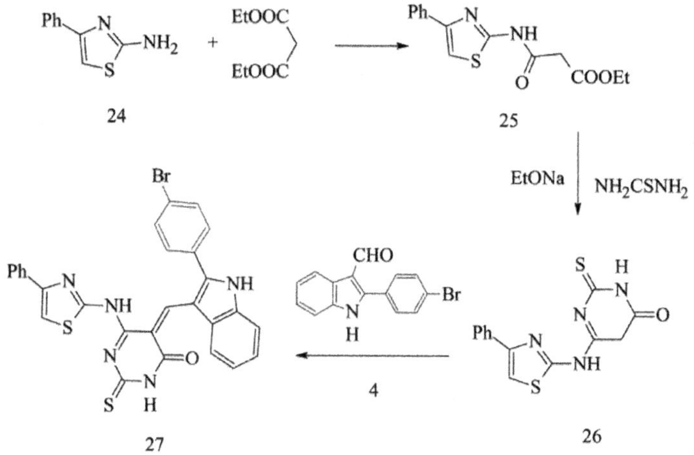

Scheme 5: Synthesis of compounds 25-27.

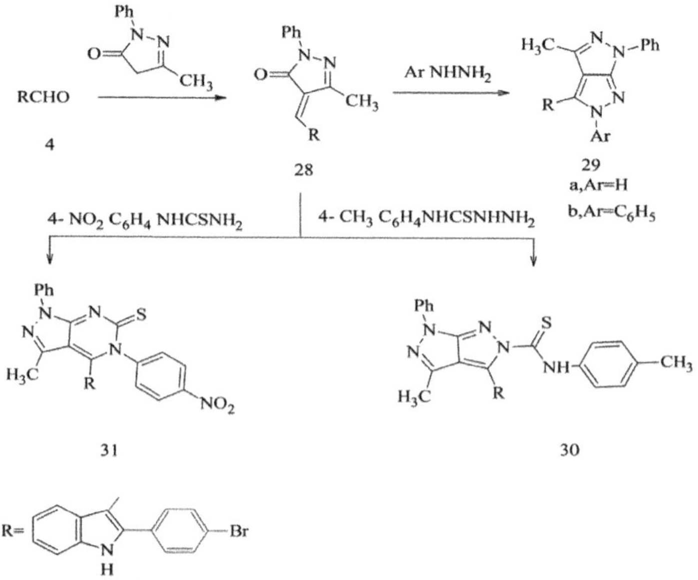

Scheme 6: Synthesis of compounds 28-31.

The plausible mechanism for the formation of compound 14 may be attributed to the initial Knoevenagel condensation of the active methylene nitrile of 11 with carbonyl group of salicyaldehyde followed by an intramolecular 1,6-dipolar cyclization via the addition of the phenolic OH group to the cyano function to afford the target compounds [25] (Figure 1). 1H NMR spectrum of 14 showed three D_2O-exchangeable signal at δ 11.44, δ 11.56 and δ 12.14 ppm due to three NH protons. Its mass spectrum showed a molecular ion peak at m/z 485 corresponding to the molecular formula $C_{25}H_{17}BrN_4O_2$.

The reaction of 11 with ethyl cyanoacetate and elemental sulfur in the presence of triethylamine gave the thiophene derivatives [26] 15. The structure of compound 15 was confirmed by its infrared spectrum which indicated the absence of CN absorption band and contain the characteristic absorption bands for NH and CO functional groups. On the other hand the reaction of 11 with elemental sulfur and phenylisothiocyanate [27] gave the thiazole derivative 16. Compounds 15 and 16 were obtained according to the proposed following mechanism (Figure 2). The structure of compounds 15 and 16 were elucidated on the basis of elemental analysis and spec- tral data. The IR spectrum of thiazoline 16 revealed the absence of CN absorption band and the presence of new absorption bands at 3395, 3269 cm^{-1} assignable to NH_2 group and band at 1237 cm^{-1} due to C=S group. The ^{13}C NMR data showed signals at δ 183.91, δ 161.36, δ 153 and δ 147.39 ppm to CS, CO, C-NH_2 and C=N. Cyclization of thiazoline 16 with acetic anhydride afforded 1,3-thiazolo[4,5-d]pyrimidin-7(6H)-one derivative 17 (Scheme 3).

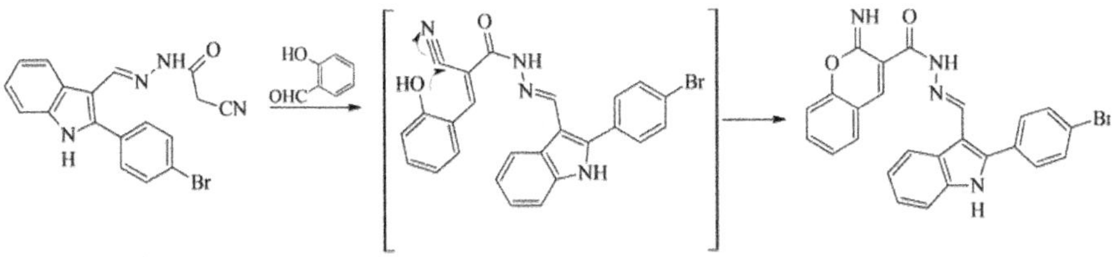

Figure 1: Proposed mechanism of formation of compound 14.

N-[1H-indol-3-ylmethylene]-1,3-thiazol-2-amine 18 was synthesized by the one-pot three compounds. Thus, condensation of phenacyl bromide, 1H-indole-3-carboxaldehyde 4 and thiourea [28] under conventional heating in absolute ethanol using catalytic amount of acetic acid. The 1H NMR spectra of compound 18 showed the absence of the aldehyde proton, moreover D_2O-exchangeable signal at δ 12.43 ppm due to the NH proton and signal at δ 6.85 ppm for =CH proton. Condensation of 1H-indole-3-carboxaldehyde 4 with 3-acetyl-2H-chromen- 2-one [29] afforded 3-[3-(1H-indol-3-yl)prop-2-enoyl]-2H-chromen-2-one 19 (Scheme 4).

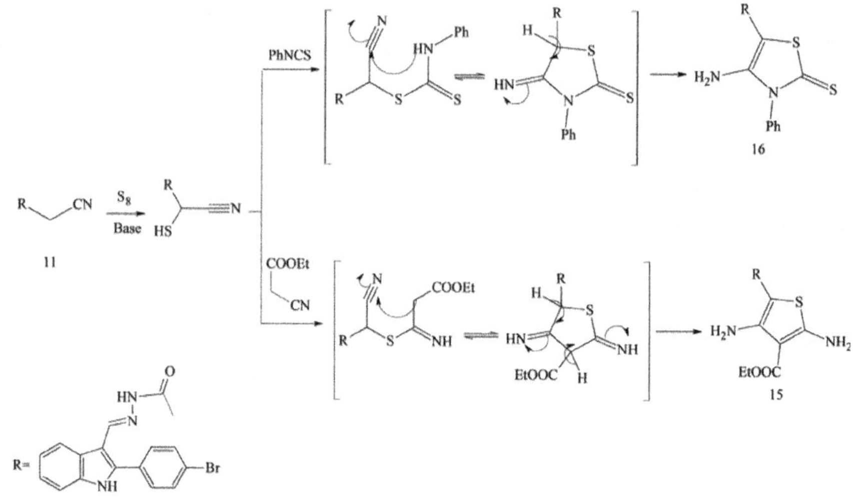

Figure 2: Proposed mechanism of formation of compounds 15 and 16.

Condensation of 4 with cyanoacetamide, ethyl cyanoacetate and 1H-benzimidazol-2-yl-acetonitrile [30] afforded 3-(1H-indol-3-yl)-2-cyanoprop-2-enamide, ethyl 3-(1H-indol-3-yl)-2-cyanoprop-2-enoate 20a,b and 2- (1H-benzimidazol-2-yl)-3-(1H-indol-3-yl)crylonitrile 22 respectively (Scheme 4). The structure of the reaction product 20a,b and 22 were ascertained on the basis of its elemental analysis and spectral data. The IR spectrum of compound 20a exhibited characteristic absorption bands at 3466 cm^{-1}, 3310 cm^{-1}, 2203 cm^{-1} and 1688 cm^{-1} corresponding to NH$_2$, CN and CO groups respectively. The 1H NMR spectrum of 20a indicated the presence of one singlet peak at δ 8.22 ppm of the =CH proton and the disappearance of a singlet at δ 9.95 ppm of CHO proton. Cyclization of 20b by hydrazine hydrate to 5-oxopyrazolidine-4-carbonitrile derivative 21 was achieved by refluxing in ethanol. The 1H NMR spectrum of compound 21 indicated the presence of D$_2$O-ex- changeable singlet at δ 10.18, δ 11.41 and δ 12.22 ppm which correspond to three NH groups.

On the other hand, the reaction of 1H-indole-3-carboxaldehyde 4 with rhodanine-3-acetic acid [31] afforded [5-(1H-indol-3-ylmethylene)-1,3-thiazolidin-3-yl]acetic acid derivative 23. The 1H-NMR spectrum of compound 23 indicated the presence of singlet at δ 4.71 ppm of the CH$_2$ and D$_2$O-exchangeable singlet at δ 12.60 ppm of the OH proton. Mass spectrum of 23 showed a molecular ion peak m/z at 473 corresponding to the molecular formula C$_{20}$H$_{13}$BrN$_2$O$_3$S$_2$ (Scheme 4).

The key substrate ester 25 was synthesized from the reaction of 4-phenyl-2-amino-thiazole 24 and diethyl malonate. Reaction of ester 25 with thiourea in ethanolic sodium ethoxide solution afforded 6-[1,3-thiazol- 2-ylamino]-2-thioxo-2,3-dihydropyrimidin-4(5H)-one derivative [32] 26. Treatment of pyrimidinone derivative 26 with 1H-indole-3-carboxaldehyde 4 afforded 5-[1H-indol-3-ylmethylene]-2-thioxo-2,5-dihydropyrimidin- 4(3H)-one 27 (Scheme 5). The structure of 27 was identified as the reaction product on the basis of its elemental analysis and spectroscopic data. The 1H NMR spectrum of compound 27 indicated the presence of singlet signal at δ 9.96 ppm of the =CH proton and D_2O-exchangeable singlet at δ 12.18, δ 12.22 and δ 12.40 ppm corresponding to the three NH protons.^{13}C NMR spectrum showed signal at δ 185.37 (C=S) and 168.01 (C=O).

Condensation of 1H-indole-3-carboxaldehyde 4 with 5-methyl-2-phenyl-2,4-dihydro-3H-pyrazol-3-one afforded 4-[1H-indol-3-ylmethylene]-1H-pyrazol-5(4H)one 28 (Scheme 6). The IR spectrum of 28 exhibited characteristic absorption bands at 3275 cm^{-1} and 1706 cm^{-1} corresponding to NH and CO groups, respectively.^{13}C NMR spectrum showed signal at δ 12.58 (CH_3), δ 153 (C=C) and δ 160 (C=O). Mass spectrum of 28 showed a molecular ion peak m/z at 456 corresponding to the molecular formula $C_{25}H_{18}BrN_3O$.

Compound 28 was used as key intermediates in the synthesis of novel pyrazolo[3,4-c]pyrazolone and pyrazole[3.4-d] pyrimidine derivatives via their interaction with different reagents. Thus, the reaction of 28 with hydrazine hydrate, phenyl hydrazine and N-(4-methylphenyl)thiosemicarbazide by cyclocondensation reaction [33] afforded 3-(4-methyl-pyrazolo[3,4-c] pyrazol-3-yl)-1H-indole derivatives 29a,b and pyrazolo[3,4-c]-pyrazole-2(6H)-carbothioamide 30 respectively (Scheme 6). The structure of the newly synthesis compounds were based on their correct elemental analysis and spectral data. 1H NMR spectrum of 30 exhibited a singlet signal at δ 2.28 ppm due to CH_3 protons of tolyl moiety. Its mass spectrum, the compound displayed the molecular ion peak at m/z 617 corresponding to the molecular formula $C_{33}H_{25}BrN_6S$. Alternatively, treatment of the compound 28 with N-(4-nitrophenyl)thiourea [34] afforded pyrazolo[3,4-d]pyrimidine-6-thione derivatives 31. The structures of the compound 31 were determined from spectroscopic and elemental analytical data (see Experimental section).

Antimicrobial Activity

The newly synthesized compounds 5c, 7, 9, 11, 13a, 27, 30 and 31 were evaluated for their in vitro antibacterial activity against Gram-positive namely Staphylococcus aureus RCMB 0100010 (SA) and Bacillus subtilis RCMB 010067 (BS) and Gram-negative Pseudomonas aeuroginosa RCMB 010043 (PA) and Escherichia coli RCMB 010052 (EC). They were also evaluated for their in vitro antifungal activity against Aspergillus fumigatus RCMB 02568

(AF) and Candida albicans RCMB 05036 (CA). Ampicillin was the standard used for the evaluation of antibacterial activity against gram positive bacteria and Gentamicin was used as a standard in assessing the activity of the tested compounds against gram negative bacteria, while Amphotericin B was taken as a reference for the antifungal effect. The inhibitory effects of the synthetic compounds against these organisms are given in Table 2, Figure 3 and Figure 4.

In general, most of the tested compounds revealed better activity against the Gram-positive rather than the Gram-negative bacteria. All test compounds were found to be inactive against Pseudomonas aeuroginosa RCMB 010043 (PA). It was shown (Figure 3) that the majority of the compounds studied possessed significant antibacterial activity towards Staphylococcus aureus RCMB 0100010 (SA), Bacillus subtilis RCMB 010067 (BS) and Escherichia coli RCMB 010052 (EC). The highest activities were observed for compounds 9 and 30, followed by 11, 13a and 31.Compounds 5c, 7 and 27 showed the least antibacterial activity.

It was shown (Figure 4) that the compounds 5c, 11, 31 strong antifunger activity against Aspergillus fumigates RCMB 02568 (AF) and Candida albicans RCMB 05036 (CA) comparable to Amphotericin B. The compounds 9, 13a and 30 showed moderate activities against Aspergillus fumigates RCMB 02568 (AF) and Candida albicans RCMB 05036 (CA) comparable to Amphotericin B.

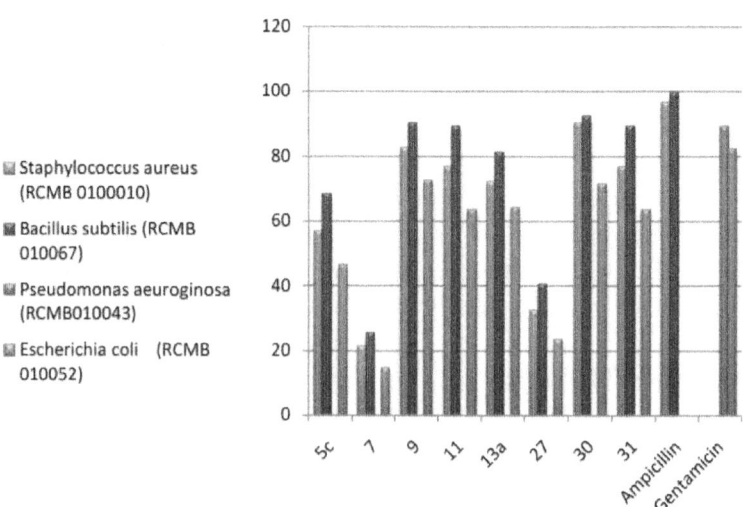

Figure 3: Graphical representation of the antibacterial activity of tested compounds compared to Ampicillin and Gentamicin.

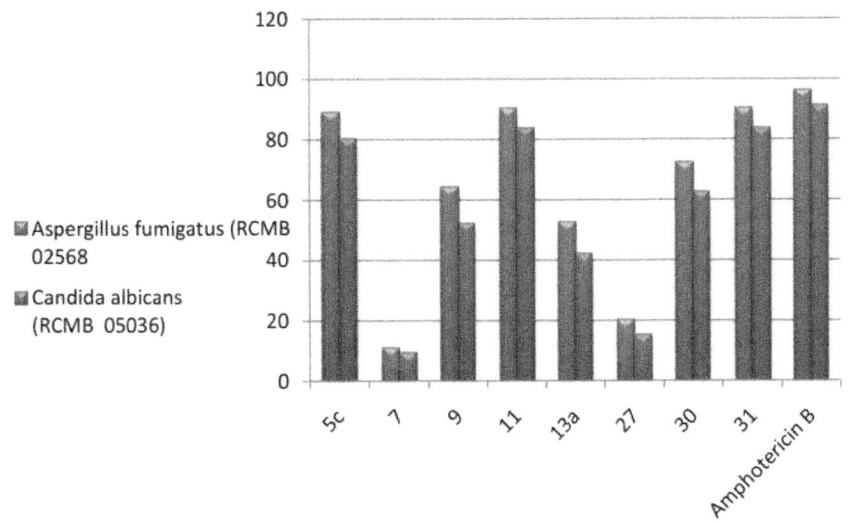

Figure 4: Graphical representation of the antifungal activity of tested compounds compared to Amphotericin B.

While the compounds 7 and 27 weak antifungal activity against Aspergillus fumigates RCMB 02568 (AF) and Candida albicans RCMB 05036 (CA) comparable to Amphotericin B.

The minimum inhibitory concentration (MIC) was considered to be the lowest concentration of the tested compound which inhibits growth of the microorganisms. The initial screening of the tested compounds showed promising activity of the compounds 5c, 9, 30 and 31 which encouraged the determination of their minimum inhibitory concentration (MIC) (Table 3).

The best results were demonstrated by compounds 9, 30 and 31 as antibacteria, it possessed double the activity of the standard, Ampicillin against Bacillus subtilis RCMB 010067 (BS) 1.95 and 3.9 µg/ml respectively. Moderate activity against Staphylococcus aureus RCMB 0100010 (SA) and Escherichia coli RCMB 010052 (EC) were also demonstrated by compounds 9 and 30. On other hand moderate activity against Aspergillus fumigatus RCMB 02568 (AF) and Candida albicans RCMB 05036 (CA) were also demonstrated by compounds 5c and 31.

Table 2: Antimicrobial evaluation of the some synthesized compounds.

Comp. No.	Inhibition % ± standard deviation					
	Gram positive bacteria		Ggram negative e bacteria		Fungal	
	SA	BS	PA	ES	AF	CA
5c	56.85 ± 0.58	68.32 ± 1.2	NA	46.32 ± 0.58	89.25 ± 0.72	80.23 ± 1.2
7	21.25 ± 0.58	25.63 ± 0.28	NA	14.63 ± 0.2	11.22 ± 0.28	9.32 ± 0.72
9	82.63 ± 0.75	90.42 ± 0.28	NA	72.46 ± 0.28	64.35 ± 0.58	52.14 ± 0.63
11	76.8 ± 0.58	89.4 ± 0.63	NA	63.5 ± 0.72	90.3 ± 0.58	83.6 ± 0.28
13a	72.14 ± 0.58	81.32 ± 0.63	NA	64.21 ± 0.63	52.63 ± 0.58	42.18 ± 1.2
27	32.6 ± 0.63	40.4 ± 0.85	NA	23.6 ± 1.2	20.3 ± 1.2	15.4 ± 0.58
30	90.3 ± 1.2	92.4 ± 0.72	NA	71.6 ± 0.93	72.4 ± 0.58	62.5 ± 0.28
31	76.8 ± 0.58	89.4 ± 0.63	NA	63.5 ± 0.72	90.3 ± 0.58	83.6 ± 0.28
Ampicillin	96.52 ± 0.2	99.65 ± 0.3				
Gentamicin			89.23 ± 0.1	82.14 ± 0.3		
Amphotericin B					96.25 ± 0.1	91.29 ± 0.1

(SA): *Staphylococcus aureus* RCMB 0100010, (BS): *Bacillus subtilis* RCMB 010 067 (BS), (PA): *Pseudomonas aeuroginosa* RCMB 010043, (EC): *Escherichia coli* RCMB 010052, (AF): *Aspergillus fumigatus* RCMB02568 and (CA): *Candida albicans* RCMB 05036.

Table 3: Minimum inhibitory concentration of compounds 5c, 9, 29 and 31.

Comp. No.	Minimum inhibitory concentration (µg/ml)					
	Gram positive bacteria		Gram negative e bacteria		Fungi	
	SA	BS	PA	EC	AF	CA
5c	31.25	15.63	NA	62.5	3.9	3.9
9	3.9	1.95	NA	7.81	15.63	31.25
30	1.95	1.95	NA	7.81	7.81	15.63
31	7.81	1.95	NA	15.63	1.95	3.9
Ampicillin	0.98	3.9				
Gentamicin			1.95	3.9		
Amphotericin B					0.98	1.95

CONCLUSION

In the present work, we synthesized novel series of 3-substituted indole by reaction of indole-3-carboxaldehyde derivative with different reagents. Screening for some selected compounds was carried for their potential antibacterial, antifungal activity. Most of the tested compounds revealed better activity against the Gram-positive rather than the Gram negative bacteria. All test compounds were found to be inactive against Pseudomonas aeuroginosa. Compounds 9, 30 and 31 exhibited excellent activity against Staphylococcus aureus, Bacillus subtilis and Escherichia coli compared with the standards drugs, while compounds 5c, 11 and 31 have strong antifunger activity against Aspergillus fumigatus and Candida albicans comparable to Amphotericin B.

REFERENCES

1. Nagai, K., Davies, T.A., Jacobs, M.R. and Appelbaum, P.C. (2002) Effects of Amino Acid Alterations in Penicillin- Binding Proteins (Pbps) 1a, 2b, and 2× on PBP Affinities of Penicillin, Ampicillin, Amoxicillin, Cefditoren, Cefuroxime, Cefprozil, and Cefaclor in 18 Clinical Isolates of Penicillin-Susceptible, Intermediate, and Resistant Pneumococci. Antimicrobial Agents and Chemotherapy, 46, 1273-1280.http://dx.doi. org/10.1128/AAC.46.5.1273-1280.2002

2. Kaniwa, K., Arai, M.A., Li, X. and Ishibashi, M. (2007) Synthesis, Determination of Stereochemistry, and Evaluation of New Bisindole Alkaloids from the Myxomycete Arcyriaferruginea: An Approach for Wnt Signal Inhibitor. Bioorganic & Medicinal Chemistry Letters, 17, 4254-4257. http://dx.doi.org/10.1016/j.bmcl.2007.05.033

3. Mingxing Teng, M. (2014) Total Synthesis of the Monoterpenoid Indole Alkaloid (±)-Aspidophylline A. Angewandte Chemie International Edition, 53, 1814-1817.

4. Pilar, M.F., Pedro, M. and Juan, A.B. (2001) Synthesis of the Indole Alkaloids Meridianins from the Tunicate Aplidium Meridianum. Tetrahedron, 57, 2355-2363.http://dx.doi.org/10.1016/S0040-4020(01)00102-8

5. Sharma, A. and Pathak, D.M. (2013) The Synthesis and Antimicrobial Activity of Indole Thiacarbamide Derivatives. International Journal of Scientific & Engineering Research, 4, 203-206.

6. Chavan Rajashree, S. and More Harinath, N. (2011) Synthesis, Characterization and Evaluation of Analgesic and Anti- Inflammatory Activities of Some Novel 2-(4,5-Dihydro-1H-pyrazol-3-yl)-3-phenyl-1H-indole. Journal of Pharmacy Research, 4, 1575-1578.

7. Kawasaki, I., Katsuma, H., Nakayama, Y., Yamashita, M.Y. and Ohta, S. (1998) Total Synthesis of Topsentin, Antiviral and Antitumor Bis(indolyl) imidazole. Heterocycles, 48, 1887-1901. http://dx.doi.org/10.3987/ COM-98-8225

8. Kawasaki, I., Yamashita, M.Y. and Ohta, S. (1996) Total Synthesis of Nortopsentins A-D, Marine Alkaloids. Chemical and Pharmaceutical Bulletin, 44, 1831-1839.http://dx.doi.org/10.1248/cpb.44.1831

9. Sun, H.H. and Sakemi, J. (1991) A Brominated (Aminoimidazolinyl) indole from the Sponge Discodermia Polydiscus. The Journal of Organic Chemistry, 56, 4307-4308.http://dx.doi.org/10.1021/jo00013a045

10. Raju, R., Gromyko, O., Fedorenko, V., Luzhetskyy, A. and Müler, R. (2012) Pimprinols A-C, from the Terrestrial Actinomycete, Streptomyces

sp. Tetrahedron Letters, 53, 3009-3011. http://dx.doi.org/10.1016/j. tetlet.2012.03.134

11. Fresneda, P.M., Castaneda, M., Blug, M. and Molina, P. (2007) Iminophosphorane-Based Preparation of 2,5-Disub- stituted Oxazole Derivatives: Synthesis of the Marine Aalkaloid Almazole C. Synlett, 2, 324-326. http://dx.doi.org/10.1055/s-2007-967995

12. Aaron, M.S., Richard, A.L. and David, C.R. (2007) Bacillamides from a Hypersaline Microbial Mat Bacterium. Journal of Natural Products, 70, 1793-1795.http://dx.doi.org/10.1021/np070126a

13. Gao, X., Chooi, Y.-H., Ames, B.D., Wang, P., Walsh, C.T. and Tang,Y. (2011) Fungal Indole Alkaloid Biosynthesis: Genetic and Biochemical Investigation of the Tryptoquialanine Pathway in Penicillium aethiopicum. Journal of the American Chemical Society, 133, 2729-2741. http://dx.doi. org/10.1021/ja1101085

14. Mohamed, A.R. and Mahmound, E.S. (2007) Synthesis and Antitumor Activity of Indolylpyrimidines: Marine Natural Product Meridianin D Analogues. Bioorganic & Medicinal Chemistry Letters, 15, 1206-1217. http://dx.doi.org/10.1016/j.bmc.2006.11.023

15. Sofy, A.R., Hmed, A.A., Sharaf, A.M. and El-Dougdoug, K.A. (2014) Structural Changes of Pathogenic Multiple Drug Resistance Bacteria Treated With T. Vulgaris Aqueous Extract. Nature and Science, 12, 83-88.

16. Caldeira, E.M., Osório, A., Oberosler, E.L., Vaitsman, D.S., Alviano, D.S. and Nojima, M.D.G. (2013) Antimicrobial and Fluoride Release Capacity of Orthodontic Bonding Materials. Journal of Applied Oral Science, 21, 327-334. http://dx.doi.org/10.1590/1678-775720130010

17. Kaya, E.G., Özbilge, H. and Albayrak, S. (2009) Determination of the Effect of Gentamicin against Staphlococcus au- reus by Using Microbroth Kinetic System. ANKEM Dergisi, 23,110-114.

18. March, S.J. (2000) Advanced Organic Chemistry: Reactions, Mechanisms and Structure. 5th Edition, John Wiley and Sons, New York, 1824.

19. Attaryan, O., Antanosyan, S., Panosyan, G., Asratyan, G. and Matsoyan, S. (2006) Vilsmeier Formyatiom of 3,5-Di- methylpyrazoles. Russian Journal of General Chemistry, 76, 1817-1819. http://dx.doi.org/10.1134/ S1070363206110260

20. El-Nakkady, S.S., Hanna, M.M., Roaiah, H.M. and Ghannam, I.A. (2012) Synthesis, Molecular Docking Study and Antitumor Activity of Novel 2-Phenylindole Derivatives. European Journal of Medicinal Chemistry, 47, 387-398.http://dx.doi.org/10.1016/j.ejmech.2011.11.007

21. Saundane, A.R., Katkar, V.T. and Vaijinath, V. (2013) Synthesis, Antimicrobial, and Antioxidant Activities of N-[(5)- Substituted-2)-phenyl-1H-indol-3-yl)methylene]-5H-dibenzo[b,f]azepine-5-carbohydrazide Derivatives. Journal of Chemistry, 2013, 1- 9.http://dx.doi.org/10.1155/2013/530135

22. Mahmouda, M.R., El-Ziatya, A.K., Abu El-Azma, F.S.M., Ismail, M.F. and Shibab, S.A. (2013) Utility of Cyano-N- (2-oxo-1,2-dihydroindol-3-ylidene)acetohydrazide in the Synthesis of Novel Heterocycles. Journal of Chemical Research, 37, 80-85.http://dx.doi.org/10.3184/17475191 2X13567100793191

23. Mohareb, R.M., El-Arab, E.E. and El-Sharkawy, K.A. (2009) The Reaction of Cyanoacetic Acid Hydrazide with 2- Acetylfuran: Synthesis of Coumarin, Pyridine, Thiophene and Tthiazole Derivatives with Potential Antimicrobial Activities. Scientia Pharmaceutica, 77; 355-366. http://dx.doi.org/10.3797/scipharm.0901-20

24. Darwish, E.S., Abdel Fattah, A.M., Attaby, F.A. and Al-Shayea, O.N. (2014) Synthesis and Antimicrobial Evaluation of Some Novel Thiazole, Pyridone, Pyrazole, Chromene, Hydrazone Derivatives Bearing a Biologically Active Sulfonamide Moiety. International Journal of Molecular Sciences, 15, 1237-1254. http://dx.doi.org/10.3390/ijms15011237

25. Volmajer, J., Toplak, R., Leban, I. and Le Marechal, A.M. (2005) Synthesis of New Iminocoumarins and Their Transformations into N-Chloro and Hydrazono Compounds. Tetrahedron, 61, 7012-7021. http://dx.doi.org/10.1016/j.tet.2005.05.020

26. Alafeefy, A.M., Alqasoumi, S.I., Ashour, A.E., Masand, V., Al-Jaber, N.A., Hadda, T.B. and Mohamed, M.A. (2012) Quinazoline-Tyrphostin as a New Class of Antitumor Agents, Molecular Properties Prediction, Synthesis and Biological Testing. European Journal of Medicinal Chemistry, 53, 133-140. http://dx.doi.org/10.1016/j.ejmech.2012.03.044

27. El-Gaml, K.M. (2015) Synthesis and Antimicrobial Evaluation of New Polyfunctionally Substituted Heterocyclic Compounds Derived from 2-Cyano-N-(3-cyanoquinolin-2-yl)acetamide. American Journal of Organic Chemistry, 5, 1- 9.

28. Ravibabu, V., Janardhan, B., Rajitha, G. and Rajitha, B. (2013) Efficient One-Pot Three-Component Synthesis of Thia- zolylpyrazole Derivatives under Conventional Method. Der Chemica Sinica, 4, 79-83.

29. Dilli Varaprasad, E., Mastan, M. and Sobha Rani, T. (2012) Synthesis and Evaluation of Analgesic Activity of Novel Series of Indole Derivatives Linked to Isoxazole Moiety. Der Pharmacia Lettre, 4, 1431-1437.

30. El-Nakkady, S., Abbas, S., Roaiah, H. and Ali, I. (2012) Synthesis, Antitumor and Anti-Inflammatory Activities of 2- Thienyl-3-Substitued Indole Derivatives. Global Journal of Pharmacology, 6, 166-177.

31. Chen, Z.-H., Zheng, C.-J., Sun, L.-P. and Piao, H.-R. (2010) Synthesis of New Chalcone Derivatives Containing Rhodanine-3-Acetic Acid Moiety with Potential Anti-Bacterial Activity. European Journal of Medicinal Chemistry, 45, 5739-5743.http://dx.doi.org/10.1016/j.ejmech.2010.09.031

32. Chaudhari, S.A., Sandeep, R., Patil, S.R., Patil, V.M., Patil, S.V., Jachak, M.N. and Desai, A. (2015) Synthesis of Pyrano[2,3-d]pyridine, Pyrazolo[3,4-b]pyridine Derivatives by Microwave Irradiation and Study of Their Insecticidal Activity. Journal of Chemical and Pharmaceutical Research, 9,476-482.

33. Shabaan, M., Taher, A.T. and Osma, E.O. (2011) Synthesis of Novel 3,4-Dihydroquinoxalin-2(1H)-one Derivatives. European Journal of Chemistry, 2, 365-371.http://dx.doi.org/10.5155/eurjchem.2.3.365-371.289

34. Abdelhafez, O.M., Amin, K.M., Batran, R.Z., Maher, T.J., Somaia, A., Nada, S.A. and Sethumadhavan, S. (2010) Syn- thesis, Anticoagulant and PIVKA-II Induced by New 4-Hydroxycoumarin Derivatives. Bioorganic & Medicinal Chemistry, 18, 3371-3378.http://dx.doi.org/10.1016/j.bmc.2010.04.009

CITATION

CHAPTER 1

Kofie, W. and Caddick, S. (2015) Palladium/Imadazolium Salt Mediated Cyclisations for the Synthesis of Heterocyclic Compounds. *International Journal of Organic Chemistry*, **5**, 223-231. doi: 10.4236/ijoc.2015.54022.

CHAPTER 2

Vuram, P. , Kabilan, C. and Chadha, A. (2015) Catalyst and Solvent-Free Microwave Assisted Expeditious Synthesis of 3-Indolyl-3-hydroxy Oxindoles and Unsymmetrical 3,3-Di(indolyl)indolin-2-ones. *International Journal of Organic Chemistry*, **5**, 108-118. doi: 10.4236/ijoc.2015.52012.

CHAPTER 3

Razzak, M. , Karim, M. , Hoq, M. and Mirza, A. (2015) New Schiff Bases from 6,6'-Diformyl-2,2'-Bipyridyl with Amines Containing O, S, N and F: Synthesis and Characterization. *International Journal of Organic Chemistry*, **5**, 264-270. doi: 10.4236/ijoc.2015.54026.

CHAPTER 4

Hassaneen, H. and Gomaa, Z. (2015) Utility of Styrylpyrazoloformimidate in the Synthesis of Fused Heterocyclic Compounds. International Journal of Organic Chemistry, 5, 213-222. doi: 10.4236/ijoc.2015.54021.

CHAPTER 5

Makki, M. , Abdel-Rahman, R. and Aqlan, F. (2015) Synthesis of Fluorinated Heterobicyclic Nitrogen Systems Containing 1,2,4-Triazine Moiety as CDK2 Inhibition Agents. *International Journal of Organic Chemistry*, **5**, 200-211. doi: 10.4236/ijoc.2015.53020.

CHAPTER 6

Said, S. , Mashaly, M. , Sheta, A. and Elmorsy, S. (2015) New Method for Preparation of 1-Amidoalkyl-2-Naphthols via Multicomponent Condensation Reaction Utilizing Tetrachlorosilane under Solvent Free Conditions. International Journal of Organic Chemistry, 5, 191-199. doi: 10.4236/ijoc.2015.53019.

CHAPTER 7

Al-Romaizan, A. , Makki, M. and Abdel-Rahman, R. (2014) Synthesis of New Fluorine/Phosphorus Substituted 6-(2'-Amino Phenyl)-3-Thioxo-1,2,4-Triazin-5(2H, 4H)One and Their Related Alkylated Systems as Molluscicidal Agent as against the Snails Responsible for Bilharziasis Diseases. *International Journal of Organic Chemistry*, **4**, 154-168. doi: 10.4236/ijoc.2014.42017.

CHAPTER 8

Abdel-Rahman, R. , Makki, M. and Al-Romaizan, A. (2014) Synthesis of Novel Fluorine Substituted Isolated and Fused Heterobicyclic Nitrogen Systems Bearing 6-(2'-Phosphorylanilido)-1,2,4-Triazin-5-One Moiety as Potential Inhibitor towards HIV-1 Activity. *International Journal of Organic Chemistry*, **4**, 247-268. doi:10.4236/ijoc.2014.44028.

CHAPTER 9

Segura-Campos, M. , López-Sánchez, S. , Castellanos-Ruelas, A. , Betancur-Ancona, D. and Chel-Guerrero, L. (2015) Physicochemical and Functional Characterization of *Mucuna pruries* Depigmented Starch for Potential Industrial Applications. *International Journal of Organic Chemistry*, **5**, 1-10. doi: 10.4236/ijoc.2015.51001.

CHAPTER 10

Darvish, F. and Khazraee, S. (2015) FeCl3 Catalyzed One Pot Synthesis of 1-Substituted 1H-1,2,3,4-Tetrazoles under Solvent-Free Conditions.

International Journal of Organic Chemistry, 5, 75-80. doi: 10.4236/ijoc.2015.52009.

CHAPTER 11

Senapati, B. and Mal, D. (2015) Synthetic Studies of Naphtho[2,3-b]furan Moiety Present in Diverse Bioactive Natural Products. *International Journal of Organic Chemistry*, 5, 63-74. doi: 10.4236/ijoc.2015.52008.

CHAPTER 12

Bashandy, M. (2015) 1-(4-(Pyrrolidin-1-ylsulfonyl)phenyl) ethanone in Heterocyclic Synthesis: Synthesis, Molecular Docking and Anti-Human Liver Cancer Evaluation of Novel Sulfonamides Incorporating Thiazole, Imidazo[1,2-a]pyridine, Imidazo[2,1-c] [1,2,4]triazole, Imidazo[2,1-b]thiazole, 1,3,4-Thiadiazine and 1,4-Thiazine Moieties. International Journal of Organic Chemistry, 5, 166-190. doi: 10.4236/ijoc.2015.53018.

CHAPTER 13

Makki, M. , Rahman, R. and Ali, O. (2015) Synthesis of New Fluorinated 1,2,4-Triazino [3,4-b][1,3,4]thiadiazolones as Antiviral Probes-Part II-Reactivities of Fluorinated 3-Aminophenyl-1,2,4-triazinothiadiazolone. *International Journal of Organic Chemistry*, **5**, 153-165. doi: 10.4236/ijoc.2015.53017.

CHAPTER 14

Salman, A. , Mahmoud, N. , Abdel-Aziem, A. , Mohamed, M. and Elsisi, D. (2015) Synthesis, Reactions and Antimicrobial Activity of Some New 3-Substituted Indole Derivatives. International Journal of Organic Chemistry, 5, 81-99. doi: 10.4236/ijoc.2015.52010.

INDEX